Methods in Molecular Biology

Series Editor
John M. Walker
School of Life and Medical Sciences
University of Hertfordshire
Hatfield, Hertfordshire, AL10 9AB, UK

For further volumes:
http://www.springer.com/series/7651

Barley

Methods and Protocols

Edited by

Wendy A. Harwood

Crop Transformation Group, Department of Crop Genetics, John Innes Centre, Norwich, UK

Humana Press

Editor
Wendy A. Harwood
Crop Transformation Group
Department of Crop Genetics
John Innes Centre
Norwich, UK

ISSN 1064-3745 ISSN 1940-6029 (electronic)
Methods in Molecular Biology
ISBN 978-1-4939-8942-3 ISBN 978-1-4939-8944-7 (eBook)
https://doi.org/10.1007/978-1-4939-8944-7

Library of Congress Control Number: 2018962159

This Humana Press imprint is published by the registered company Springer Science+Business Media, LLC, part of Springer Nature.
The registered company address is: 233 Spring Street, New York, NY 10013, U.S.A.

Preface

Barley is the world's fourth most important cereal, both in terms of area and production, as well as one of the world's first domesticated crops. In addition to being an important crop for both food and feed, it is also a valuable diploid model cereal. Barley has extensive genetic resources, continually advancing genomic resources, and a range of efficient biotechnological tools.

In this volume, we start with review chapters that introduce barley as both crop and model, and consider the history and importance of mutation breeding in the development of the crop. Protocol chapters then cover practical techniques such as crossing barley, a range of tissue culture methods, the preparation of barley tissues for different forms of microscopy, and the assessment of sensitivity to abiotic stresses. Efficient protocols are provided for transformation, TILLING, virus-induced gene silencing, and genome editing. There is also particular emphasis on a range of protocols for genotyping and for the analysis of gene expression.

This book should be a valuable reference volume for cereal researchers and breeders by providing detailed protocols covering important traditional skills such as crossing and tissue culture through to the latest technologies for genotyping, expression analysis, and genome editing.

Norwich, UK *Wendy A. Harwood*

Contents

Contributors

AMINA ABED • *Department de Phytologie and Institut de Biologie Intégrative et des Systèmes (IBIS), Université Laval, Québec, QC, Canada*
SHAHIN S. ALI • *SPCL, USDA/ARS Beltsville Agricultural Research Center, Beltsville, MD, USA*
MIKEL ARRIETA • *Cell and Molecular Sciences, The James Hutton Institute, Dundee, Scotland, UK*
MENTEWAB AYALEW • *Biology Department, Spelman College, Atlanta, GA, USA*
MICHA BAYER • *The James Hutton Institute, Dundee, Scotland, UK*
FRANÇOIS J. BELZILE • *Département de Phytologie and Institut de Biologie Intégrative et des Systèmes (IBIS), Université Laval, Québec, QC, Canada*
ALLAN BOOTH • *The James Hutton Institute, Dundee, Scotland, UK*
CLARE BOOTH • *Cell and Molecular Sciences, The James Hutton Institute, Dundee, Scotland, UK*
BRIAN BOYLE • *Plateforme d'Analyses Génomiques, Institut de Biologie Intégrative et des Systèmes (IBIS), Université Laval, Québec, QC, Canada*
JOHN W. S. BROWN • *Cell and Molecular Sciences, The James Hutton Institute, Dundee, Scotland, UK; Division of Plant Sciences, University of Dundee at The James Hutton Institute, Dundee, Scotland, UK*
HAZEL BULL • *James Hutton Limited, Dundee, Scotland, UK; Syngenta UK Ltd., Market Stainton, UK*
CRISTIANE P. G. CALIXTO • *Division of Plant Sciences, University of Dundee at The James Hutton Institute, Dundee, Scotland, UK*
KAROLINA CHWIALKOWSKA • *Centre for Bioinformatics and Data Analysis, Medical University of Bialystok, Bialystok, Poland; Department of Genetics, University of Silesia in Katowice, Katowice, Poland*
ISABELLE COLAS • *Cell and Molecular Sciences, The James Hutton Institute, Dundee, Scotland, UK*
AGATA DASZKOWSKA-GOLEC • *Department of Genetics, Faculty of Biology and Environmental Protection, University of Silesia in Katowice, Katowice, Poland*
FIONA M. DOOHAN • *School of Biology and Environmental Science and Earth Institute, College of Science, University College Dublin, Dublin, Ireland*
PATRICIO ESTEVES • *Département de Phytologie, Université Laval, Québec, QC, Canada*
BRIAN P. FORSTER • *BioHybrids International Ltd., Reading, UK*
JEROME D. FRANCKOWIAK • *Agronomy/Plant Genetics, University of Minnesota Twin Cities, St. Paul, MN, USA*
JOHN FULLER • *Cell and Molecular Sciences, The James Hutton Institute, Dundee, Scotland, UK*
KATARZYNA GAJEK • *Faculty of Biology and Environmental Protection, Department of Genetics, University of Silesia, Katowice, Poland*
LOKANADHA R. GUNUPURU • *Department of Plant, Food, and Environmental Sciences, Dalhousie University, Truro, NS, Canada*
RUTH HAMILTON • *University of Dundee at the James Hutton Institute, Dundee, Scotland, UK*

WENDY A. HARWOOD • *Crop Transformation Group, Department of Crop Genetics, John Innes Centre, Norwich, UK*
PETE E. HEDLEY • *Genome Technology, Cell and Molecular Sciences, James Hutton Institute, Dundee, UK*
GOETZ HENSEL • *Plant Reproductive Biology, Leibniz Institute of Plant Genetics and Crop Plant Research (IPK) Gatersleben, Seeland, Germany*
ALISON HINCHLIFFE • *Crop Transformation Group, Department of Crop Genetics, John Innes Centre, Norwich, UK*
MATTHIAS JOST • *Leibniz Institute of Plant Genetics and Crop Plant Research (IPK), Seeland, Germany*
URSZULA KOROTKO • *Department of Genetics, University of Silesia in Katowice, Katowice, Poland*
JOCHEN KUMLEHN • *Plant Reproductive Biology, Leibniz Institute of Plant Genetics and Crop Plant Research (IPK) Gatersleben, Seeland, Germany*
MARZENA KUROWSKA • *Department of Genetics, Faculty of Biology and Environmental Protection, University of Silesia in Katowice, Katowice, Poland*
MIROSLAW KWASNIEWSKI • *Centre for Bioinformatics and Data Analysis, Medical University of Bialystok, Bialystok, Poland; Department of Genetics, University of Silesia in Katowice, Katowice, Poland*
TOM LAWRENSON • *John Innes Centre, Norwich, UK*
GAÉTAN LÉGARÉ • *Plateforme d'Analyses Génomiques, Institut de Biologie Intégrative et des Systèmes (IBIS), Université Laval, Québec, QC, Canada*
MAREK MARZEC • *Faculty of Biology and Environmental Protection, Department of Genetics, University of Silesia, Katowice, Poland; Department of Physiology and Cell Biology, Leibniz Institute of Plant Genetics and Crop Plant Research (IPK), Seeland, Germany*
CLAUS-DIETER MAYER • *Biomathematics and Statistics Scotland, Dundee, UK*
MICHAEL MELZER • *Department of Physiology and Cell Biology, Leibniz Institute of Plant Genetics and Crop Plant Research (IPK), Seeland, Germany*
LINDA MILNE • *Information and Computational Sciences, The James Hutton Institute, Dundee, Scotland, UK*
SYBILLE MITTMANN • *Cell and Molecular Sciences, The James Hutton Institute, Dundee, Scotland, UK; Division of Plant Sciences, University of Dundee at The James Hutton Institute, Dundee, Scotland, UK*
JENNY A. MORRIS • *Genome Technology, Cell and Molecular Sciences, James Hutton Institute, Dundee, UK*
LUDMILA OHNOUTKOVA • *Faculty of Science, Centre of the Region Hana for Biotechnological and Agricultural Research, Department of Chemical Biology and Genetics, Palacky University, Olomouc, Czech Republic*
ALEXANDRE PEROCHON • *School of Biology and Environmental Science and Earth Institute, College of Science, University College Dublin, Dublin, Ireland*
SONIA POMERLEAU • *Plateforme d'Analyses Génomiques, Institut de Biologie Intégrative et des Systèmes (IBIS), Université Laval, Québec, QC, Canada*
LUKE RAMSAY • *Cell and Molecular Sciences, The James Hutton Institute, Dundee, Scotland, UK*
PAULO RAPAZOTE-FLORES • *Information and Computational Sciences, The James Hutton Institute, Dundee, Scotland, UK*
JOANNE R. RUSSEL • *The James Hutton Institute, Dundee, Scotland, UK*

Steven R. Scofield • *Crop Production and Pest Control Research Unit, USDA-ARS, West Lafayette, IN, USA; Department of Agronomy, Purdue University, West Lafayette, IN, USA*
Craig G. Simpson • *Cell and Molecular Sciences, The James Hutton Institute, Dundee, Scotland, UK*
Anna Skubacz • *Department of Genetics, Faculty of Biology and Environmental Protection, University of Silesia in Katowice, Katowice, Poland*
Michał Słota • *Department of Genetics, Faculty of Biology and Environmental Protection, University of Silesia in Katowice, Katowice, Poland*
Jérôme St-Cyr • *Plateforme d'Analyses Génomiques, Institut de Biologie Intégrative et des Systèmes (IBIS), Université Laval, Québec, QC, Canada*
Nils Stein • *Leibniz Institute of Plant Genetics and Crop Plant Research (IPK), Seeland, Germany*
Dorota Swiergolik • *Department of Genetics, Faculty of Biology and Environmental Protection, University of Silesia in Katowice, Katowice, Poland*
Iwona Szarejko • *Faculty of Biology and Environmental Protection, Department of Genetics, University of Silesia, Katowice, Poland*
Miriam Szurman-Zubrzycka • *Faculty of Biology and Environmental Protection, Department of Genetics, University of Silesia, Katowice, Poland*
William T. B. Thomas • *The James Hutton Institute, Dundee, Scotland, UK*
Niki Uzrek • *The James Hutton Institute, Dundee, Scotland, UK*
Tomas Vlcko • *Faculty of Science, Centre of the Region Hana for Biotechnological and Agricultural Research, Department of Chemical Biology and Genetics, Palacky University, Olomouc, Czech Republic*
Robbie Waugh • *Cell and Molecular Sciences, The James Hutton Institute, Dundee, Scotland, UK; Division of Plant Sciences, University of Dundee at The James Hutton Institute, Dundee, Scotland, UK*

Chapter 1

An Introduction to Barley: The Crop and the Model

Wendy A. Harwood

Abstract

Barley has a number of unique features among our crop plants. It was one of the earliest plants to be domesticated and continues to play an important role in modern agriculture today. It is a versatile crop, used both for human nutrition and for animal feed, and plays an important role as an experimental model plant allowing advances in plant genetics, plant physiology, plant pathology, plant biochemistry, and more recently in plant biotechnology. In this introductory chapter, the key features of barley as both crop and model plant are considered.

Key words Barley, Mutation breeding, Model cereal, Biotechnology tools, Genetic resources, Genomic resources

1 Barley the Crop

Cultivated barley, *Hordeum vulgare*, is a member of the Poaceae grass family. It is thought that barley was first domesticated around 10,000 years ago from its wild relative *Hordeum spontaneum*. There is evidence that barley was first cultivated in the Israel-Jordan area of the Fertile Crescent [1] (with *H. spontaneum* still to be found in this region), and evidence also for the additional domestication of barley in Tibet [2].

Barley is an annual grass with both spring and winter types, with winter annuals being planted in autumn and requiring a period of cold before they will flower. Ancient barley was originally two-rowed but modern barley cultivars vary in spike morphology, with both two-row and six-row types. Both hulled and hulless types are also present among modern cultivars. Barley has a diploid chromosome number of 14 and is self-pollinating.

In 2016/2017 barley production worldwide was 145 million metric tons. This puts barley in fourth place in terms of production behind corn, wheat, and rice. The main barley producing countries are Russia, Germany, France, Ukraine, Australia, and Canada (Fig. 1). Barley is a very adaptable and robust crop that is now

Wendy A. Harwood (ed.), *Barley: Methods and Protocols*, Methods in Molecular Biology, vol. 1900,
https://doi.org/10.1007/978-1-4939-8944-7_1, © Springer Science+Business Media, LLC, part of Springer Nature 2019

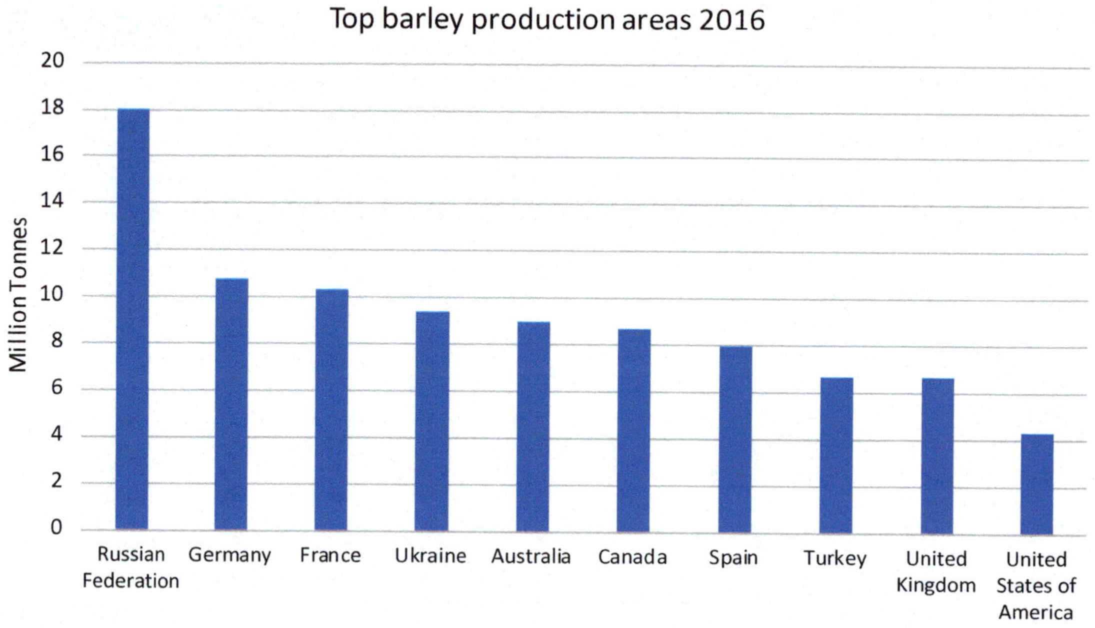

Fig. 1 Barley production during 2016 in million tons from the top 10 producers. Data source: http:/www.fao.org/statistics/en/

grown in over 100 countries worldwide. Although mainly grown in temperate countries it is also grown in some tropical regions. At least 60% of the barley crop is used for animal feed with the rest being used for human consumption, mainly in the brewing industry. Outside of use in the beer making industry, barley now only forms a minor component of human food except in specific regions of the world such as Tibet. However, barley is a good source of beta-glucan. This soluble fiber is thought to offer a range of health benefits, including lowering levels of LDL cholesterol making the crop increasingly popular as a whole food.

Mutation breeding has played a very important role in the development of improved barley genotypes from its first use in the 1930s (discussed further in Chapter 2). The last four decades have seen increases in barley yields of 60% in Europe. This has been largely due to success in breeding more productive cultivars with more efficient disease and insect resistance. Improved agricultural practices have also played an important role [3]. Average barley yield was 6.1 tonnes per hectare in the UK in 2017 [4].

2 Barley the Model

Barley has emerged as a model cereal for a range of applications including genetic studies [5] and biotechnology applications [6]. Barley has several attributes that contribute to its value as a model. Firstly, it has a diploid genome with a low chromosome number ($2n = 14$) making it an obvious choice as a model for the Triticeae tribe, which includes the much more complex hexaploid wheat. Cultivation of barley is easy under a wide range of environmental conditions, it has extensive genetic resources and cross-breeding is also straightforward (*see* Chapter 3, which provides a detailed protocol for crossing barley). Although barley has a very large genome size, at approximately 5.3 Gbp for the haploid genome, there are extensive genomic resources available including a high-quality barley reference genome sequence [7]. Barley also has a range of biotechnology tools including efficient transformation, TILLING, and more recently efficient genome editing.

2.1 Genetic Resources

There are extensive genetic resources available for barley. For example, the public barley collection in the Germplasm Resource Unit of the John Innes Centre is a sub-collection of the BBSRC Small Grain Cereals Collections which forms part of the UK contribution to the Multilateral System (MLS) of the International Treaty for Plant Genetic Resources for Food and Agriculture (ITPGRFA). The collection originated from the Plant Breeding Institute in the mid-1970s and moved to the JIC in 1990. It comprises over 10,000 accessions of *Hordeum vulgare* and about 200 accessions of *Hordeum spontaneum*, commonly known as wild barley in addition to four accessions of *Hordeum chilense*. Of the cultivated barley, 710 are named UK varieties and 924 are UK breeder's lines. A total of 1954 are named varieties originating in other countries and 2038 are breeder's lines of non-UK origin. About 4500 accessions are landraces or were derived from landraces. The collection can be browsed and seed can be ordered via SeedStor: A Germplasm Information Management System and Public Database (https://www.seedstor.ac.uk) [8].

2.2 Genomic Resources

The barley genome at 5.3 Gbp contains over 39,000 genes. The genome structure is complex with large regions of repetitive DNA. Annotated transposable elements were found to make up 80.8% of the genome sequence and it was discovered that almost 30,000 genes belonged to gene families with multiple members [7]. The high-quality barley reference sequence is providing important insights into the evolution of the barley genome and greatly facilitating gene isolation.

Additional genomic resources include a variety of methods for analysis of the gene-rich region of the barley genome including a custom microarray (*see* Chapter 12), exome capture methods (*see* Chapter 18), and methods for the analysis of alternative barley transcripts (*see* Chapter 17). Polymorphism in barley can be analyzed using genotyping by sequencing (GBS) (*see* Chapter 15). There are also established protocols for the analysis of DNA methylation (*see* Chapter 16). The availability of improved genomic resources and analysis methodologies in barley has been vital in allowing more rapid identification of genes involved in important agronomic traits such as yield, malting quality, and resistance to pests and disease.

2.3 Biotechnology Tools

One of the most important biotechnology tools for functional genomics studies in barley is transformation. Fortunately, barley has available very efficient Agrobacterium-mediated transformation protocols [9] (*see* Chapter 8). Although transformation of the spring barley Golden Promise is routine, the method is still genotype dependent with many newer cultivars being recalcitrant to transformation success. This may be due to difficulties regenerating plants from the explants used for transformation. Tissue culture regeneration protocols are generally well developed in barley with efficient methods for the generation of doubled haploid plants from both anther (Chapter 4) and microspore culture (Chapter 5). Virus-induced gene silencing (VIGS) is an alternative tool for the validation of gene function that can provide results more rapidly than the generation of stable transgenic plants (*see* Chapter 7).

Barley is an amenable model for the study of developmental processes and responses to various biotic and abiotic stresses. As barley is a relatively robust crop, it has frequently been used in studies looking at the response to abiotic stresses such as drought. Chapter 9 details methods for the assessment of barley exposed to abiotic stress during early development. Microscopy techniques are also important members of the toolbox of methods available in barley research. Chapter 10 looks specifically at the preparation of barley roots for study under both light and electron microscopes and Chapter 11 details the preparation of pollen mother cells for microscopic examination.

Methods for the generation of barley mutants are particularly important in a range of research areas as well as in the development of improved varieties. TILLING (Targeting Induced Local Lesions IN Genomes) has been a popular reverse genetics approach in barley (*see* Chapter 6). There are now new opportunities for the creation of targeted mutants using genome editing technologies. Two such technologies are detailed in this volume, the use of TALENs (transcription activator-like effector nucleases) in Chapter 13 and the use of CRISPR/Cas9 in Chapter 14 [10]. These gene editing technologies depend on efficient transformation protocols to introduce the components required for genome editing. Genome editing technologies are being developed at a rapid pace and there will certainly be opportunities and protocols for many more precise genome modification methods in barley in the near future.

References

1. Badr A, Müller K, Schäfer-Pregl R, El Rabey H, Effgen S, Ibrahim HH, Pozzi C, Rohde W, Salamini F (2000) On the origin and domestication history of barley (*Hordeum vulgare*). Mol Biol Evol 17(4):499–510
2. Wang Y, Ren X, Sun D, Sun G (2015) Origin of worldwide cultivated barley revealed by *NAM-1* gene and grain protein content. Front Plant Sci 6:803
3. Friedt W (2011) Barley breeding history, progress, objectives, and technology. In: Ullrich SE (ed) Barley: production, improvement and uses. Blackwell Publishing Ltd, Ames, IA
4. National Statistics (2017) Farming Statistics Final crop areas, yields, livestock populations and agricultural workforce at June 2017–United Kingdom. Department for Environment Food and Rural Affairs (DEFRA) National statistics document
5. Saisho D, Takeda K (2011) Barley: emergence as a new research material of crop science. Plant Cell Physiol 52(5):724–727
6. Harwood W (2016) Barley as a cereal model for biotechnology applications. In: Jones HD (ed) Biotechnology of major cereals. CABI, Wallingford, pp 80–87
7. Mascher M, Gundlach H et al (2017) A chromosome conformation capture ordered sequence of the barley genome. Nature 544:427
8. Horler RSP, Turner AS, Fretter P, Ambrose M (2018) SeedStor: a germplasm information management system and public database. Plant Cell Physiol 59(1):e5
9. Bartlett JG, Alves SC, Smedley M, Snape JW, Harwood WA (2008) High-throughput *Agrobacterium*-mediated barley transformation. Plant Methods 4:22
10. Lawrenson T, Shorinola O, Stacey N, Liu C, Østergaard L, Patron N, Uauy C, Harwood W (2015) Induction of targeted, heritable mutations in barley and *Brassica oleracea* using RNA-guided Cas9 nuclease. Genome Biol 16:258

Chapter 2

Mutation Breeding in Barley: Historical Overview

Ludmila Ohnoutkova

Abstract

The discovery of radioactivity at the end of the nineteenth century played a key role in a series of historical landmarks that would lead to contemporary mutation breeding in agricultural crops. The aim of the earliest experiments was to test the effects of radiation on living organisms beginning with fruit flies. Exposure of plants to X-rays provided the first incontrovertible proof that phenotypic changes could be induced. Chemicals were a second type of mutagen tested from the 1940s and both forms are used today. This chapter is an overview of some of the historical developments that led to the use of mutagenesis in plants, with a focus on barley, a model species for mutation genetics and breeding as well as a major cereal crop. Perhaps the most well-known examples of mutant barley cultivars are Diamant, Golden Promise, and their hybrids.

Key words Barley (*Hordeum vulgare*), Mutation history, Radiation mutagenesis, Chemical mutagenesis

1 Introduction

Mutation breeding plays an important role in improving the yield and quality of barley genotypes. Mutation breeding is based on exploitation of local varieties with the aim of enhancing resistance to disease and biotic and abiotic stress, increasing grain size, and improving the quality of malt. Induced mutation in crop improvement programs was initiated around the 1930s, after the discovery of the mutagenic effects of X-rays. Today both physical and chemical mutagens are used to create cultivars with new traits showing better protein, starch quality, and mineral composition. Phenotypic changes such as greater number of fertile tillers, grain number and 1000 grain weight, stem length reduction and alterations in root architecture have considerable economic value. The FAO/IAEA Mutant Varieties Database lists 312 barley cultivars that have been produced through induced mutation (https://mvd.iaea.org; accessed February 1, 2017). Barley is also a model species for mutation research and study of genome changes.

Wendy A. Harwood (ed.), *Barley: Methods and Protocols*, Methods in Molecular Biology, vol. 1900,
https://doi.org/10.1007/978-1-4939-8944-7_2, © Springer Science+Business Media, LLC, part of Springer Nature 2019

Mutation techniques in combination with tissue culture and molecular methods are proven effective tools in breeding programs.

2 History of Mutation Breeding Research

The following is a look at some of the key figures, concepts, events, and experiments that shaped the field. Understandably, the coverage is sketchy. The first half of the twentieth century was a period of turmoil, expansion, wars, and ideologies. The lives of those involved in genetics were likewise complex, scientifically and politically.

Hugo de Vries (1848–1935) was a Dutch botanist whose observation of Evening primrose plants he found growing on a potato field formed the basis of his mutation theory of evolution. Taking seeds from the primrose plants, he found they produced new varieties and he used the term mutations for them. In mutation theory sudden leaps and sudden changes in genetic traits are the basis for the emergence of new species. He was an influential scientist and his ideas were credible as they resolved some of the problems intellectuals of the day were having with Darwin's theory of gradual evolution. He was one of the "rediscoverers" of Mendel's laws of heredity even if there is some controversy about his prior acquaintance with Mendel's research. He also suggested that X-rays among other forms of radiation might prove useful in plant mutation.

X-rays were discovered by Wilhelm Conrad Röntgen (1845–1923), a mechanical engineer and physicist, who was awarded the first Nobel Prize in Physics in 1901. In 1895, Röntgen was exploring the electrical rays passing from an induction coil through a partially evacuated glass tube. The tube was covered in black paper and the room was dark but the rays were able to penetrate objects. His own hand showed the contrast between the bones and flesh. One of the first of Röntgen's images was the left hand of his wife Bertha. Röntgen's discovery of X-rays was to revolutionize medicine. It was quickly followed by the finding of gamma rays (1896, Henri Becquerel), and radium (Pierre and Marie Curie 1896).

X-rays were used in clinical medicine and experimental physics for some years but their value to genetics research and mutation only became clear when Hermann Joseph Muller (1890–1967) used radioactivity to produce point mutations in the fruit fly *Drosophila*. Muller was interested in genes, specifically, the physical-chemical nature of genes. He was part of the research team at Columbia University that developed the chromosomal theory of heredity under T H Morgan and worked in Morgan's famous fly lab which was described as "cramped, dusty, smelly and cockroach ridden." In 1919, Muller made the important discovery of a mutant (later found to be a chromosomal inversion) that appeared to suppress crossing-over. This had impact on mutation rate studies but the breakthrough came in 1926 with two experiments using varied doses of X-rays, the second of which involved the crossing-over suppressor stock ("ClB") he had found in 1919. A clear, quantitative relation between radiation and lethal mutations quickly emerged. Muller's discovery created a media stir after his public delivery of a paper "The Problem of Genetic Modification" at the Fifth International Congress of Genetics in Berlin. By 1928, others had replicated his results, expanding them to other organisms like wasps and maize. His Nobel Lecture was "The Production of Mutation." In this lecture he emphasized the importance of mutation for plant breeding and evolution.

3 Barley Mutation Breeding

3.1 Radiation Mutagenesis

Lewis John Stadler (1896–1954) was an American geneticist who began and ended his scientific career at Missouri. His initial research was on the field plot technique and related agronomic issues in maize and wheat. Around 1920 he became seriously interested in genetics, and investigated the genetics of maize. From 1925, he researched the mutagenic effects of different forms of radiation on maize and barley. The four plants from three individual families of heterozygous maize were treated in the open field by X-raying the young tassels in four treatment series with different radiation exposures and intervals. The results showed a pronounced increase in the percentage of mosaic endosperms following X-ray treatment though this effect may have partly been due to increase in gene mutation frequency as well as chromosome aberrations.

The X-ray and radium treatment that Stadler [1] used was also applied to barley. Mutations were detected by growing the progeny of the individual treated spike of a plant. The mutations appeared to be chlorophyll related with virescent seedlings and yellowing described. All mutations he tested were found to be recessive, making him believe that recessive mutations were the result of chromosome aberration or the destruction of a gene. Later on, he tried ultraviolet radiation as a new mutagen and was optimistic about the quality of mutants so produced. The three major discoveries emerging from Stadler's mutation work in 1928 with barley were: (1) Seeds soaked in water to initiate germination gave almost eight times more mutations than dormant dry seeds. (2) Mutant rate was independent of temperature at the time of irradiation. (3) The relation of mutation rate to total dosage was linear.

The exhaustive results of the pioneering work of Muller and Stadler proved without doubt that mutations could be artificially induced but Stadler remained skeptical about the benefits to practical breeding, concluding that it is better to make use of the great wealth of existing genetic variations in nature from which valuable traits could be found without accompanying negative effects [2].

3.1.1 Scandinavian Mutation Research in Barley

Undeterred by Sadler's caveats, Swedish research on induced mutations in barley began in 1928 on a small scale at Svalöf, through the initiation of the distinguished Swedish geneticists Herman Nilsson-Ehle and Åke Gustafsson. The first treatments with X-rays and UV irradiation were started, using the local cultivar "Gull." Different pretreatments were also tested since mutation frequency would increase if seeds were soaked in water beforehand. The first chromosome aberrations, phenotypic changes in the seedlings and chlorophyll mutations were observed. Gustafsson introduced mutation frequency that was calculated according to the "spike progeny method." This was later used as the standard method for measuring the induced mutagenic effect [3]. Distinct varieties were established: albina, viridis, xantha seedlings, rare two colored, striped, and zoned. In the mid-1930s, the first viable result emerged and it was possible to distinguish two subtypes "Morphological" and "Physiological" mutations. The most common group at that time were *Erectoides* mutants characterized by compact or dense spikes. Morphologically they resembled the erectum barley, in comparison to the normal spikes in most barley cultivars. In the following years, several mutants were produced that were considered very valuable in terms of yield, straw-stiffness, straw-length, early maturity, tillering capacity, changes in spikes, kernels and awns, and pigmentation among others [4, 5].

Comparison of the various forms of irradiation showed that, seeds were 20–30 times more sensitive to neutrons than to X-rays, and germinating seeds were two to three times more sensitive to neutrons than to X-rays; neutrons were around ten times more effective than X-rays in producing chromosome disturbances and 50–100 times more effective in increasing the mutation rate in the second generation; neutrons produced relatively more chlorophyll mutations than X-rays [6, 7].

In 1953, the "Group for Theoretical and Applied Mutation Research" was established under the Swedish Government with the aim of basic research to improve methods for breeding cultivated plants. The Agricultural Research Council provided funding for most of the Mutation Group's scientific activities approved by the Swedish Parliament. Its peak activities were during the 1950s, 1960s, and 1970s. Promising results induced the Swedish Seed Association to support this new field of research, and to date, a collection of around 12,000 different mutant alleles have been collected [8].

3.1.2 The Effects of Atomic Bomb Radiation and X-Rays on Cereal Seeds

Barley seeds were one of the objectives for determining of the effect of irradiation on mutation after the Second World War. Samples of two varieties of barley cultivars (1-year-old Trebi and 19-year-old Moister) were exposed to radiation from the first aerial atomic bomb explosion at Bikini atoll, on July 1, 1946, and seeds of the variety Trebi were exposed to three doses of X-rays. Plants of the first generation were evaluated by macroscopic and microscopic observation. Spikes from each plant were harvested and sent to two different institutions in the USA for tests of mutation frequency. Comparisons of the germination of control, bomb-irradiated, and X-rayed seeds showed little effect of the radiations on germination even of the most heavily exposed seeds. From the results, there were no remarkable differences in mutation frequency between barley cultivars Trebi and Carleton durum wheat. More effective mutagenicity was from the radiation of atomic bombs in the most heavily X-rayed (16,000r) material [9, 10].

3.2 Chemical Mutagenesis

In the mid-1940s, a number of chemical compounds began to be tested on plants. Institutions and scientists became involved in experiments on irradiation and they tested different types of chemicals for producing mutations. Compared to radiation, the application of chemicals was very simple, without any special equipment. In 1940 Ch. Auerbach, A.J. Clark, and J.M. Robson discovered that mustard gas (Yperite) could cause mutation in fruit flies. Research on chemical mutagenesis in barley was started by the Swedish group. Gustafson and MacKey [11] applied mustard gas in barley, followed by many different compounds. Ehrenberg et al. [12] published the preliminary results on testing chemical components, alkylating agents (Ethylene oxide), and oxidizing agents (Chlorine) for inducing mutation in barley. They evaluated mutation rate and also the frequency and type of chlorophyll mutation types (albina, viridis xanta, and others). Figures 1, 2, 3, 4, 5, 6, and 7 illustrate different types of chlorophyll mutants in the M1 generation and field trials after chemical mutagenic treatment in M2 and M3 generations. Classification of leaf color mutants in spring barley used the system developed by Lamprecht [13].

Ehrenberg et al. [13] compared the effect of ionizing radiations and chemical mutagens from a cytological point of view. Barley and peas where treated by chemicals ethyleneimine (EI), ethyl methanesulfonate (EMS), and propyl methanesulfonate (PMS) and with radiation neutrons and X-rays. Large comparative experiments analyzed mutations in the first, second, and third generation in the greenhouse and in the field. The results showed that the spectrum of chlorophyll mutations varied with the kind of mutagen employed.

Ekberg [14] compared the genetic effects of ionizing radiation and chemical mutagens on barley. Neutrons and X-ray irradiation, three chemical compounds with alkylating effect and oxygen and combined treatments with X-rays and 8-ethoxycaffrene were used for inducing mutagenesis and for the study of frequency of different hereditary types of sterility. Ionizing radiations caused higher percentages of translocation and inversions than chemical mutagens. The first chemically induced mutant barley variety, Luther

Fig. 1 Field-grown mutants of spring barley showing induced chlorophyll chimeras in M1 generations. (**a–c**) Plants with albino and green parts. (**d**) Plants with yellow leaves with a green strip

released in 1966 in the USA, was obtained by treatment of seeds with the chemical mutagen 0.0038 M dES (diethyl sulfate) for a period of 3.5 h. In the IAEA database, 30 varieties or their hybrids are registered that were obtained with the use of chemical mutagenesis, treated by NEU (*N*-nitroso-*N*-ethyl urea), EI (ethylene imine), dES (Diethyl Sulfate), and EMS (ethyl methane sulfonate). The three alkylating agents EMS, ENU (*N*-nitroso-*N*-etyl urea), and MNU (*N*-nitroso-*N*-emethylurea) are currently used for barley mutagenesis due to their effectiveness and efficiency. Each mutagen has a mutational specificity producing a different percentage of

Fig. 2 Chlorophyll mutant of spring barley, *type albina-virescens*

Fig. 3 Chlorophyll mutant of spring barley *type xantha*

nucleotide transitions and transversion [15]. Kurowska et al. [16] presented molecular analyses of point mutation in a barley genome exposed to MNU and gamma rays. The MNU was a much more efficient mutagen than gamma rays in including points mutant in barley, 63% of the MNU-induced nucleotide changes were transitions, with similar numbers of G → A and C → T substitutions. For barley concentrations of 10, 20, 40 mM fresh solutions of EMS, 0.2, 0.4, 0.8 mM fresh solutions of ENU and MNU are commonly used. In a typical protocol, the barley seeds are presoaked in distilled water for 16–24 h in the dark, at 24 °C. The volume of the

Fig. 4 Chlorophyll mutant of spring barley, *chlorina xanthescens*

Fig. 5 Chlorophyll mutant of spring barley, *albino vario maculata*

solution of mutagen is tenfold larger than that of the treated seeds. The exposure times vary from 2 to 12 h.

Finally in the mid-1970s, the chemical mutagen sodium azide (NaN_3) was introduced for practical purposes. Sodium azide is a very efficient mutagen in barley and in some other crop species. Nilan et al. [17] first time tested different concentration of sodium azide (1–4 mM at pH 3) and time of treatment on different varieties of spring barley. Higher frequencies of chlorophyll mutations were observed at a concentration of 4 mM. In order to understand its mutagenic mechanism many studies in barley were performed from 1974. For example, Prina and Favret [18] studied the parabolic effect of sodium azide mutagenesis in barley, the effect of sodium

Fig. 6 M2 generation of spring barley in research plots, after NaN_3 and EMS treatment. Plant Breeding Station Hrubčice, Czech Republic

Fig. 7 M3 generation from selected progeny of spring barley in field trails, after NaN_3 treatment. Plant Breeding Station Hrubčice, Czech Republic

azide concentration in producing mutagenic effects in the two-row barley variety Malteria Heda and its double-mutant MC 90. Crispi et al. [19] analyzed pollen sterility and endosperm development in advanced generation, sodium azide induced lines of spring barley. The results indicated that the sterility was inherited in the mutant lines and was not caused by translocations, inversion, endosperm lethals, embryo-endosperm lethals, or major gene mutation. The sterility was essentially eliminated in the F1 and F2 of crosses between partially sterile lines and fertile parents. It was inherited and not caused by translocation. Also studied was the effect of a metabolite of sodium azide (azidoalanine) and this was compared with sodium azide in terms of the efficiency of DNA damage in barley germinating embryos cultivated under in vitro conditions [20]. Olsen et al. [21] studied the molecular bases for the absence of anthocyanins and proanthocyanidins in four independent sodium azide ant18 barley mutants. It was demonstrated that there were 86% transitions, and 14% transversions, A.T → G.C were about three times more frequent than G.C → A.T transitions. In 2009, the Bulgarian mutant variety IZ Bori, developed by treatment with sodium azide, was officially approved. The metabolism and activity of sodium azide and its cytotoxic and mutagenic effect was published by Gruszka et al. [15].

4 Application of Mutation Breeding

During the 1950–1960s, the type of mutagen, optimal dose of mutagen, and optimal application to seeds and plants were tested in the different counties of the world including the USA, China, India, Japan, the Netherlands, Germany, Scotland, and the Ukraine. A specialized department and breeding stations for the mutation breeding of cereals were established (e.g., from 1960 Plant Breeding Stupice in Czechoslovakia). Meetings, conferences, and symposia on these topics were organized. A large share of the development and application of mutagen techniques are due to programs for the application of atomic energy in food and agriculture of the IAEA, established in 1964, https://www.iaea.org/.

5 Important Barley Mutant Varieties

In the FAO/IAEA Mutant Variety Database plant mutant varieties released officially or commercially worldwide are listed. Three hundred and twelve barley cultivars have been registered as produced through induced mutation. The database includes direct mutants and their hybrids. The information includes the parent name, country, description, character improvement, and development type, https://mvd.iaea.org.

The first mutant variety "Pallas" was officially approved in 1960. It was developed by irradiation of presoaked seeds with X-rays (7.35 Gy). "Pallas" barley is a high-yielding mutant induced by X-ray mutagenesis of parent variety "Bonus" and was among the first cereal mutants released into practice. It is characterized by a high lodging resistance and surpasses its parent variety "Bonus," under good soil conditions and in yielding capacity. The main improved attribute of the mutant variety was stiff straw. It has been widely cultivated in several European countries. The variety "Pallas" was also used in hybridization programs [22].

The two most important barley mutant cultivars are "Diamant" and "Golden Promise" released in the second half of twentieth century.

The mutant variety "Diamant" was officially registered in 1965 in Czechoslovakia (breeder Josef Bouma). It was developed by irradiation of seeds of the local Czech variety "Valtický" with X-rays (100 Gy). The mutant variety was 15 cm shorter in height, with an increased yield of around 12% over than of the parent cultivar. It is a strongly tillering variety, with very good grain and malting quality, lodging resistance, and carries the *denso* allele. "Diamant" was used in crosses leading to more than 150 leading barley cultivars all over the world, including important German varieties such as Trumph. In 1972, "Diamant" and hybrid cultivars derived from it covered almost half of the planted spring barley in Czechoslovakia. In the same year, cultivars derived from "Diamant" accounted for 2.86 million hectares in Europe [23, 24].

The mutant variety "Golden Promise" was officially registered in 1966 in the United Kingdom, originating after γ-irradiation (dose unknown) of salt-sensitive variety "Maythorpe" in 1956. Golden Promise is characterized by short stiff straw, a good yield and good malting quality. In 1967, it was recommended for planting in England and Scotland, and significantly contributed to the increased production of barley for the UK market. "Golden Promise" is a semidwarf type of barley. It is salt-tolerant, but susceptible to powdery mildew. It is a very important cultivar and has been used in hybridization programs and in many genetic studies. "Golden Promise" is the most responsive barley genotype in tissue culture and for this reason it is one of the few useful cultivars for barley transformation.

One later mutant, officially approved in 2009, is a very good Bulgarian winter feed barley variety "IZ Bori." It was developed by treatment with sodium azide. It is tolerant to low temperatures with very good resistance to powdery mildew, as well as to brown, black, and stem rust, high grain yield (15–17%), and high grain protein and lysine content. This variety is well adapted to different agronomic conditions and can be cultivated throughout the country.

6 Summary of Economic Impact

Mutant barley varieties and lines have laid the foundation for an increase in barley yield and production worldwide. The Golden Promise cultivar remains popular up to the present day as a model variety for transformation studies as well as still being grown to a limited extent in Scotland. Ahloowalia et al. [25] estimated that in Scotland itself, Golden Promise contributed US $ 417 million to grain production. The enormous economic contribution of the Diamant cultivar, now crossed into many world barley varieties cultivars is incalculable.

Acknowledgements

This work was supported by the Ministry of Education, Youth and Sports, Czech Republic (Grant LO1204 from the National Program of Sustainability I.). I would like to thank Alexander Oulton, Jiri Vagera, and Tomas Vlcko for their assistance.

References

1. Stadler LJ (1928) Genetic effects of X-rays in maize. Proc Natl Acad Sci U S A 14:69–75
2. Stadler LJ (1930) Some genitic effects of x-rays in plants. J Hered 21:3–20
3. Gustafsson A (1938) Studies on the genetic basis of chlorophyll formation and the mechanism of induced mutating. Hereditas 24:33–93
4. Gustafsson A (1941) Mutation experiments in barley. Hereditas 27:225–242
5. Gustafsson A (1947) Mutations in agricultural plants. Hereditas 33:1–100
6. Ehrenberg L, Gustafsson A (1954) The effects of ionizing radiations in barley. Acta Radiol 41:101–104
7. von Wettstein D, Gustafsson A, Ehrenberg L (1959) Mutationsforschung und Züchtung. Arb gemeinsch f Forsch des Landes Nordrhein-Westfalen 73:7–60
8. Lundquist U (2005) The Swedish collection of barley mutants held at the Nordic Genebank. Barley Gene Newslett 35:150–154
9. Smith L (1950) Effects of atomic bomb radiations and X-rays on seeds of cereals; a comparison of the effects of ionizing radiations from the 'Test Able' atomic bomb and from X-rays on seeds of barley, wheat and oats. J Hered 41:125–130
10. Moh CC, Smith L (1951) An analysis of seedling mutants (spontaneous, atomic bomb-radiation-, and X ray-induced) in barley and durum wheat. Genetics 36:629–640
11. Gustafsson A, MacKey J (1948) The genetical effects of mustard gas substances and neutrons. Hereditas 34:371–286
12. Ehrenberg L, Gustafsson A, Lundqvist U (1956) Chemically induced mutation and sterility in barley. Acta Chem Scand 10:492–494
13. Ehrenberg L, Gustafsson A, Lundqvist U (1961) Viable mutants induced in barley by ionizing radiations and chemical mutagens. Hereditas 47:243–282
14. Ekberg I (1969) Different types of sterility induced in barley by ionizing radiations and chemical mutagens. Hereditas 63:257–278
15. Gruszka D, Szarejko I, Maluszynski M (2012) Mutation categories. Plant Mut Breed Biotechnol 53:47–55
16. Kurowska M, Labocha-Pawłowska A, Gnizda D et al (2012) Molecular analysis of point mutations in a barley genome exposed to MNU and Gamma rays. Mutation Res 738–739:52–70
17. Nilan RA, Sideris EG, Kleinhofs A et al (1973) Azide–a poten mutagen. Mutation Res 17:142–144
18. Prina AR, Favret EA (1983) Parabolic effect in sodium azide mutagenesis in barley. Hereditas 98:89–94

19. Crispi ML, Ullrich SE, Nilan RA (1987) Investigation of partial sterility in advanced generation, sodium azide-induced lines of spring barley. Theor App Genet 74:402–408
20. Veleminsky J, Rosichan JL, Juříček M et al (1987) Interaction of the mutagenic metabolite of sodium azide, synthetized in vitro, with DNA of barley embryos. Mutation Res 181:73–79
21. Olsen O, Wang X, von Wettstein D (1993) Sodium azide mutagenesis: preferential generation of A.T→G.C transitions in the barley Ant18 gene. Proc Natl Acad Sci U S A 90:8043–8047
22. Gustafsson Å, Ekman G, Dormling I (1977) Effects of the Pallas gene in barley: phene analysis, overdominance, variability. Hereditas 86:251–266
23. Bouma J (1967) New variety of spring barley Diamant in Czechoslovakia. Abhandlungen der Deutschen Akademie der Wissenschaften zu Berlin 2:177–182
24. Bouma J, Ohnoutka Z (1991) Importance and application to the mutant 'Diamant' in spring barley breeding. Proceeding of an international symposium on the contribution of plant mutation breeding for crop improvement, Vienna, p 127–133
25. Ahloowalia BS, Maluszynski M, Nichterlein K (2004) Global impact of mutation-derived varieties. Euphytica 135:187–204

Chapter 3

A Practical Guide to Barley Crossing

William T. B. Thomas, Hazel Bull, Allan Booth, Ruth Hamilton, Brian P. Forster, and Jerome D. Franckowiak

Abstract

Barley is naturally an inbreeding hermaphrodite plant so that each generation resembles its parental generation. New variation can be introduced by crossing parents that complement each other for desirable or target characteristics but requires human intervention to ensure that all the resulting seeds are hybrids of the two parents. That means that plants selected to be female parents have to be emasculated and are then fertilized with pollen from plants selected to be male parents. Here we describe how to emasculate and pollinate barley plants with a method that can be used either in the glasshouse or in the field.

Key words Emasculation, Pollination, Barley, Anthers, Spikelets

1 Introduction

Crossing of selected parents is the first step in any breeding program or genetic study. Crossing is therefore fundamental to the generation and exploitation of new combinations of allelic variants at genetic loci that can be used either to select out superior recombinant progeny or to understand the genetic control of key characters.

Cultivated barley (*Hordeum vulgare L.*) is a diploid hermaphrodite with male and female organs carried in sessile spikelets on an inflorescence or spike. Three spikelets are born at each rachis node of the spike and each spikelet is uni-floreted. If all three spikelets at a rachis internode have a fertile floret, then a six-rowed barley spike results, but only the central spikelets carry a fertile floret in a two-rowed barley spike. There is no self-incompatibility system and the anthers tend to ripen and shed pollen inside the spikelet making barley a natural inbreeder with low rates of outcrossing (<5% [1]). Outcrossing can increase if plants are exposed to stress just prior and/or during the spike (ear) emergence phases (Decimal Growth Stage 39–61 [2]). This means that human intervention is required to generate hybrids between two different parents and a

Wendy A. Harwood (ed.), *Barley: Methods and Protocols*, Methods in Molecular Biology, vol. 1900,
https://doi.org/10.1007/978-1-4939-8944-7_3, © Springer Science+Business Media, LLC, part of Springer Nature 2019

working knowledge of barley's reproductive organs helps achieve this.

Each barley floret, enclosed within the lemma and palea, carries three anthers, which are single organ structures in the model proposed by Forster et al. [3]. Each anther is supported on a filament to constitute a stamen and is associated with the paired structures palea, lodicules, and vestigial lodicules [3]. In the 2 + 1 (paired and single) phytomer model proposed by Forster et al. [3], the styles form the next paired structure and the ovary the next single organ structure with the integuments (paired) and ovule (single) the last 2 + 1 unit, which might explain why barley is protandrous (stamens mature before carpels).

Crossing of crops such as barley can be conducted in the field or in the glasshouse. The advantage of crossing in the field is that one can utilize all the material in the field without incurring any glasshouse costs. The main disadvantages are that one is restricted to a limited crossing window within any 1 year and it is really only feasible in climates with generally sunny weather. Crossing in the glasshouse can be conducted at any time during the year provided one has adequate temperature control and lighting, and it is also protected from unfavorable weather. Another advantage of using the glasshouse is that one can conduct successional sowings of parents to help match their flowering times so that crosses can be made. This is particularly valuable when making crosses between diverse parents and when the relative flowering times of at least some of the parents are unknown. In this respect, knowledge of the vernalization requirement of parents is especially useful as a vernalization period of 6–8 weeks has to be built into the crossing schedule. Sowing times for parents are less of a problem when crossing within a gene pool of parents with similar vernalization requirement but becomes more difficult when crossing outside such pools, e.g., crossing winter and spring growth habit parents. However, this can be achieved, provided one has access to a vernalization chamber to satisfy chilling requirements. But such crosses can lead to challenges in terms of predicting when to sow the spring parents so they flower at the same time as the winter parents after vernalization. One can buffer against mismatches in flowering time by vernalization of an excess of the winter parents and then removing a small subset each week to provide a range of flowering times.

2 Materials

Parental plants should be sown in a glasshouse in pots and a medium sufficient to provide around five to ten fertile spikes per plant in order to provide sufficient spikes for emasculation and pollination (Fig. 1). If a particular parent is being used repeatedly in crossing, then a number of separate pots of it can be sown to

Fig. 1 Glasshouse with pots of parents for use in a crossing program

increase the number of available spikes but taking care to separate the pots so that earlier sown plants do not shade later sown ones. While barley is an inbreeder and can be considered homozygous, it is clear from mapping populations that heterogeneity is present quite frequently in so-called inbred lines so it is best to avoid growing more than one plant in a pot so that any obvious off-types can be easily discarded. In general, such heterogeneity is rarely apparent phenotypically.

The potential for genotypic variation among different plants of the same inbred line becomes a potential problem only when a large number of F1 seed are required and therefore a large number of individual pollinations. In plant breeding, one has to consider how many segregating plants can be handled and at which generation. If, for instance, one conducts single plant selections, then the constraints are: (a) the overall number of single plants that can be grown out and assessed, and (b) the number of crosses to be made. For example, a program handling 500,000 single plants from 200 crosses means that 2500 single plants are required from each cross. Assuming that each successful hybridization generates 10–20 hybrid seeds, that means one hybridization would be sufficient because each F1 plant should produce 250 seeds when grown under good conditions. In most crossing situations one hybridization will generally produce sufficient seed for a cross but larger numbers of F1 seed are required when at least one of the parents is heterozygous as is the case in a three-way cross (A/B//C) or a backcrossing program when each gamete from the heterozygous parent is unique. Depending upon the differences between parents, over 100 hybrid seeds may be necessary to sample a sufficient range of the potential variation from the heterozygous parent. In addition, it may be that a cross may not respond particularly well to doubled haploid production and a larger number of F1 plants need to be grown in order to produce in excess of several hundred doubled haploids. In cases where more than one spike hybridization is required to produce sufficient F1 seed, it would be preferable to always use the same female plant and the same male plant, especially if the population is being generated for genetic research purposes. This, however, is not always possible and so, when more than one plant is being used to make bulk F1 seeds for a cross, then the hybrid seeds from each cross should ideally be kept separate and samples of each parental plant retained so that any potential heterogeneity can be evaluated genotypically.

The basic tools necessary for making barley crosses are:

A pair of sharp fine-pointed scissors similar to ones that are used in dissections.

Fine-pointed forceps.

A supply of glassine, cellophane, or paper packets approximately 4 cm wide by 15 cm long.

A supply of jewelers tags approximately 3.7 cm by 2.4 cm.

A hollow tube approximately 2–3 cm internal width by 10–15 cm long.

A supply of alcohol and a bottle brush or cloth to sterilize the tube, scissors, and forceps.

3 Methods

3.1 Choosing the Right Spike for Emasculation

The first step in making a cross is emasculation of the female parent. Choosing the right stage for emasculation is critical but generally occurs just after the awns begin to emerge from the flag leaf auricles (DGS49, Fig. 2). The leaf sheath should be visibly swollen but the degree of swelling will be greater in a six-rowed parent than in a two-rowed one. At this stage, the spike should be in the range of 2–4 days in advance of anthesis (when anther dehiscence via a pore occurs due to swelling of the pollen grains [4] and pollen is shed on the adjacent stigma), which is ideal for emasculation. At earlier development stages, it is very easy to damage the peduncle, which will cause the spike to dry after emasculation and thus render it unsuitable for pollination. Emasculating at a later stage risks anther dehiscence and hence self-pollination of at least some of the florets.

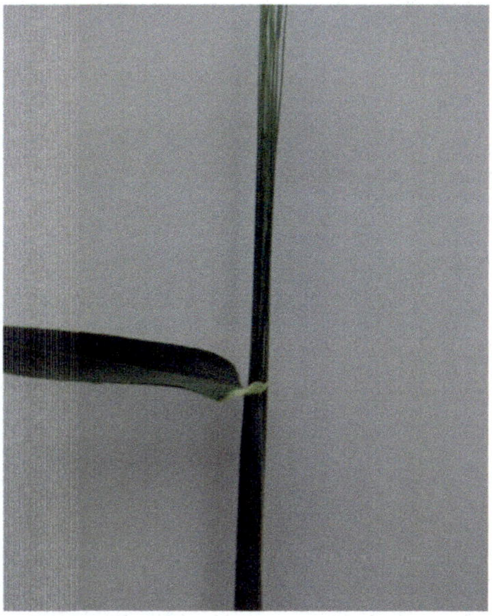

Fig. 2 Barley spike at approximately 2–4 days before anthesis

Fig. 3 Unfurling the flag leaf sheath to expose the barley spike for emasculation

Having chosen a spike at the right stage, the next step is to unwrap it from the flag leaf sheath (Fig. 3) taking care not to damage the spike itself (*see* **Note 1** for an alternative method). Once unwrapped, the flag leaf sheath should be cut away to the collar of the spike (*see* **Note 2** for a variation) and the awns of the spike trimmed off. The spike should then be able to stand erect without any support (Fig. 4); if it flops over, it is generally an

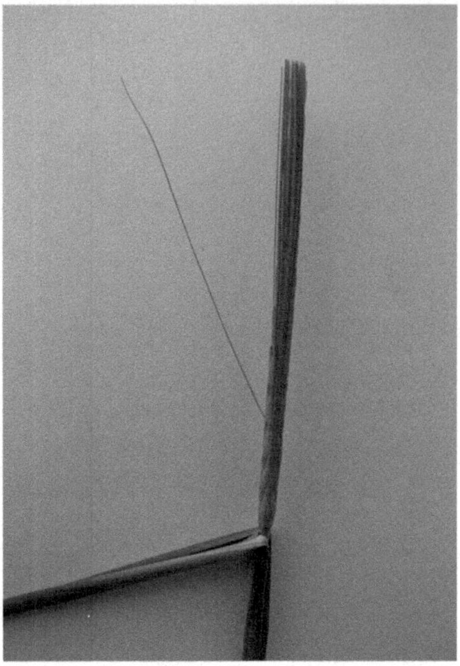

Fig. 4 Barley spike removed from flag leaf sheath with awns trimmed and ready for emasculation

indication that the spike has either been sampled too early or that stress has led to premature development of the spike. The spikelets of a normally developed spike should be light green in color with a slight transparency of the lemma so that the anthers are visible through it. If ready for emasculation, the anther tips should be approximately half way up the lemma. If the anthers are further up this is an indication that they have already shed pollen and it is too late to emasculate the spike.

3.2 Emasculation

The next step is to prepare the spike for emasculation. Here one uses a pair of fine-pointed scissors to cut off the top of the lemma and palea in an "egg-topping" approach ([5]; *see* **Note 3** for the alternative "slit" method). The idea is to cut at the top of the anthers and one can generally see the anthers through the lemma as a dark shadow against it (Fig. 5) so that the cut can be made just above the anthers. If it is not easy to see the anthers, shining a light behind the spikelet can help reveal where they are. It is also advisable to trim the lateral spikelets and this can be done at the same time as trimming the central spikelet (Fig. 6). This is because the lateral spikelets can, under certain glasshouse conditions, produce anthers that shed pollen and thus could potentially self-fertilize the emasculated spikelets.

The next stage in emasculation is to use fine forceps to remove the anthers from the spikelets. The best way to do this is to push the

Fig. 5 Trimming the lemma and palea to expose the anthers for emasculation

Fig. 6 Trimming the lateral spikelets as well as the lemma and palea of the central spikelet

forceps into the cut spikelet with the tips held together. Then relax the grip slightly so that the tips open 2–3 mm and force the lemma and palea apart to make the anthers more clearly visible. Slide the forceps either side of the anthers and near the top to avoid including the stigma and then lift upward with the forceps held tight together (but not too tightly) (Fig. 7). This should generally remove all three anthers (Fig. 8) but if less than three have been removed then one needs to repeat the process to remove any anthers remaining within the spikelet. As one checks that all three anthers have been removed, one should also verify that the anther

Fig. 7 Using fine forceps to remove the anthers from a spikelet. Note that the anthers are a light green color

Fig. 8 All three anthers removed against a human finger

pores have not opened and shed pollen. If this does occur in an isolated case, then one can remove the individual spikelet, but if it occurs for several spikelets, then it is best to discard the entire spike. Anthers, especially the lateral ones, can sometimes be held in a fold of the palea and the same process can be used to find missing anthers (*see* **Note 4** for abnormal anther development).

When all the spikelets on both sides of the spike have been emasculated, the final stage is to label the spike with the date of emasculation and plant identity (Fig. 9). The spike is covered with a bag. Many people use glassine bags and, if necessary, it can be

Fig. 9 Emasculated spike labeled with a jeweler's tag with glassine bag in right hand for placing over the emasculation to prevent pollination

Fig. 10 Emasculated spike about to be bagged with glassine bag and showing all spikelets closed

closed at the bottom and secured with a paper clip. The bag itself can be used as the label or a jeweler's tag can be tied at the base of the spike and used as a label. At this stage, all the emasculated spikelets should be closed as in Fig. 10 and the spike can be left for between 2 and 5 days before it is ready for pollination. Records of the emasculations made each day should be kept and cross-checked against the crossing list in order to avoid spending time making unnecessary emasculations.

3.3 Pollination

The spikelets gape markedly when the spike is ready for pollination as in Fig. 11. Emasculations should be inspected daily and those that are ready set aside for pollination. It is also advisable to check that all the spikelets are gaping; any in the middle three quarters of

Fig. 11 Spike ready for pollination with gaping spikelets

the spike that are closed may well be accidental self-pollinations due to emasculating too late. If one has sufficient plants to make further emasculations, then it is best to discard the spike, but if there is a shortage of that female parent, then one can use forceps to pull the closed spikelet off the spike and still pollinate the floret in each of the remaining spikelets on the spike. The next step is to identify the male parent(s) that was chosen to cross with the female and find a spike that is ready to shed pollen. Spikes that are slightly more advanced than the emasculation stage should be identified.

Depending upon the environment, the top of the spike will generally be emerging from the flag leaf auricles just before pollen shed. Anthers will dehisce early in the day, so it is best to look for pollinating spikes as early in the day as possible, although low light conditions will delay dehiscence (*see* **Note 5**). As with emasculations, the flag leaf sheath is unwrapped to expose the spike and the spikelets should then be inspected to check that the anthers have not extended to the top of the spikelet, a sign that they have already shed pollen and the spike is too old to be used. If the tops of the anthers are still no further than the middle of the spikelet, then a middle spikelet can be clipped and the spike left for 5–10 min to check that they are ready to shed pollen.

Extrusion of the anthers from the cut spikelet after a period of 5–10 min is indicative that they are ready to shed pollen. The whole spike plus at least 20 cm of the peduncle (stem) can then be cut off and all the spikelets trimmed in the same manner as for

Fig. 12 Anthers protruding from cut spikelets of a spike from the male parent

Fig. 13 Pollen grain falling from burst anthers of a barley spike

emasculation. The spike is then set aside for another 5–10 min, by which time, many of the anthers will be protruding from the cut spikelets as in Fig. 12. One can check that pollen is shedding by gently tapping the spike at this stage and a cloud of pollen grains should fall from the anthers (Fig. 13).

Fig. 14 Pollination by the twirl method where a tube is placed over the emasculation and the spike used to pollinate is introduced at the top of the tube. The pollinating spike is twirled between the finger and thumb to produce a cloud of pollen

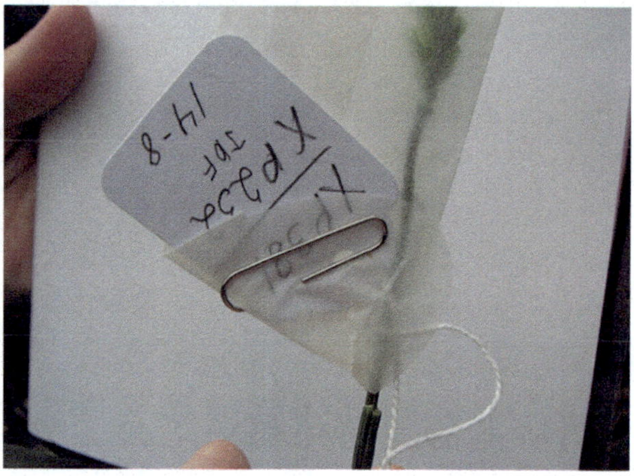

Fig. 15 Pollinated spike labeled with parents and dates

Spikes at this stage can be used for pollination. The simplest method is the "twirl" technique where the glassine bag is removed from the emasculated spike and a 10–15 cm long tube that is 2–3 cm wide and open at both ends is placed over the emasculation. The spike with the extruded anthers is then held upside down over the top of the tube and is twirled and inserted into the tube simultaneously (Fig. 14). The pollen falls from the anthers as a cloud inside the tube with much of it landing on the stigmas of the emasculated spikelets. The pollinating spike and tube are then removed and the glassine bag replaced on the emasculated spike. The date of pollination and the pollen parent are then recorded on the tag (Fig. 15). The tube is cleaned with alcohol between

Fig. 16 Pollinated spikes immediately after pollination (LHS) and 7 days after pollination (RHS)

pollinations to sterilize it and avoid inadvertent cross-pollination. Multiple pollinations using the same spike can be done by this method (*see* **Note 6**) or by the anther transfer method (*see* **Note 7**). Records of the pollinations made each day should also be kept and cross-checked against the crossing list to ensure optimum usage of the emasculated spikes.

If spikelets have been successfully pollinated, they will close the day after, depending upon temperature and water availability. The developing caryopses (kernels) will begin to be visible around 5 days after pollination and will have emerged from the cut spikelets by 7 days after pollination (Fig. 16). Successful pollinations should also be recorded and cross-checked against the crossing list to avoid wasting resources. The spikes of successful pollinations can then be left to ripen for a period that will depend upon local conditions. In cooler climates, they will be physiologically mature approximately 28 days after pollination. The grain will dry from the tip of the spike downward and all should be dry 35–42 days after pollination. In warmer and dryer conditions, these times will be shortened. When all the grain have reached physiological maturity, or dried, the spike can be cut off the plant together with the identifying label. The spike can then be left to complete drying and the F1 hybrid seed picked off and packeted for subsequent sowing to produce F1 plants. To reduce the time between crossing and sowing of F1 hybrid seeds developing F1 embryos can be cultured (*see* **Note 8**).

4 Notes

1. The spike can also be removed from the flag leaf sheath by carefully inserting the point of the fine forceps in the sheath at the base of the spike and moving them upward, taking care not

to damage the spike. The halves of the sheath so formed can then be peeled back to expose the spike.

2. Instead of cutting the flag leaf sheath at the base of the spike, it can also be cut at the tip of the spike to provide some support for the spike itself and determine if elongation of the peduncle occurred after emasculation. If applying this method, the slitting method of exposing the spike for emasculation is not appropriate.
3. An alternative emasculating method is the slit method where only the awns are trimmed off the top of the spike. The point of the forceps is then used to make a slit in the lemma and the point inserted to extract the three anthers. This method prevents excessive drying out of the spikelets and can be used in warmer conditions when drying is a problem and can also be used in the *Hordeum bulbosum* method of haploid/doubled haploid production to prolong longevity of floral structures. The downside of the method is that it takes much longer as it is harder to find the anthers and extract them successfully.
4. Not all the anthers develop normally under stress conditions and some may be small, white, and sterile and not immediately visible but it is best to remove these. They will be located in the usual position, so exploring with the forceps should eventually locate and remove them.
5. If pollinating under low light conditions, exposing a pollinating spike to artificial lighting can encourage anther extrusion and dehiscence.
6. If the same pollen parent is being crossed to several female parents, then two emasculated spikes can be pollinated at once by placing both inside the pollinating tube and then applying the twirl method. One can include more than two but experience shows that the number of pollinated spikelets decreases with more than two emasculated spikes.
7. The anther transfer method is an alternative method to pollinate emasculated spikes. Here, one uses forceps to extract anthers that are about to, or have just, burst (Fig. 17) and then use them to dust the stigmas of emasculated spikelets. One anther can generally be used to pollinate two to three emasculated spikelets. The anther transfer procedure generally gives very good success rates but takes much longer then the twirl method, yet the same pollinating spike can be used for at least three different emasculations. Also, pollinations are possible at any time of the day. The forceps should be sterilized in alcohol after the pollinating spike has been utilized.
8. If one has a small number of crosses that one wants to advance as quickly as possible, then applying an embryo rescue technique [6] to culture the embryos of the developing hybrid kernels can save up to 21 days in the time taken between

Fig. 17 Pollinating each spikelet with a ripe anther

Fig. 18 Developing F1 seed at GS77, a suitable stage for embryo rescue

pollination and having an established F1 seedlings. Briefly, the embryo culture method involves removing the developing caryopsis from the spike at DGS 75-77 (milky to doughy ripe; Fig. 18) and then dissecting the developing embryo from the caryopsis. This is then placed scutellum side initially down on a suitable medium such as modified Murashige and

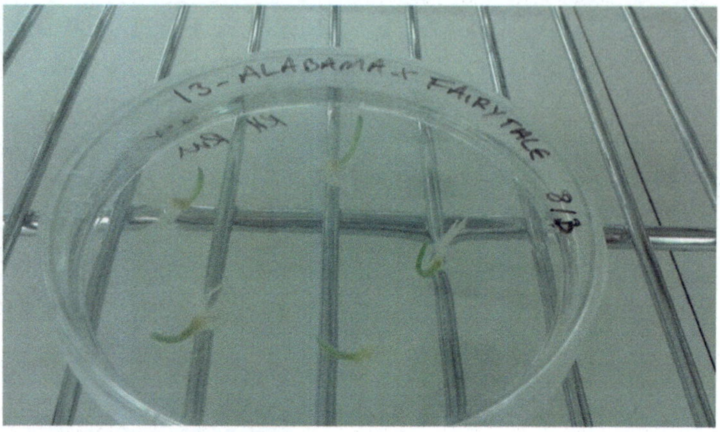

Fig. 19 Shoot and root growth of excised and cultured embryos approximately 5 days after placing on medium

Skoog and the media plates sealed and placed in an incubator with day and night periods. After 1–2 days, root and shoot development occurs and once the developing seedlings have established well (Fig. 19), they can be transferred to a suitable growing medium, e.g., Jiffy 7®, and later transferred to a larger pot if a large number of tillers are required.

9. We have described crossing of two-rowed barley. Exactly the same methods can be applied to six-rowed barley with the lateral spikelets being emasculated at the same time as the centrals or alternatively removed from the spike entirely.
10. A video of the emasculation and pollination process is available at https://www.youtube.com/watch?v=HQ_-0ALP1ig.

Acknowledgements

W.T.B.T., H.B., A.B., and R.H. gratefully acknowledge funding from the Scottish Governments Rural Affairs and the Environment programme.

References

1. Doll H (1987) Outcrossing rates in autumn and spring-sown barley. Plant Breed 98:339–341
2. Tottman DR, Makepeace RJ, Broad H (1979) An explanation of the decimal code for the growth stages of cereals, with illustrations. Ann Appl Biol 93:221–234
3. Forster BP, Franckowiak JD, Lundqvist U, Lyon J, Pitkethly I, Thomas WTB (2007) The barley phytomer. Ann Bot 100:725–733
4. Matsui T, Omasa K, Horie T (2000) Rapid swelling of pollen grains in the dehiscing anther of two-rowed barley (*Hordeum distichum* L. emend. LAM.). Ann Bot 85:345–350
5. Pope MX (1944) Some notes on technique in barley breeding. J Hered 35:99–111
6. Forster BP, Till BJ, Ghanim AMA, Huynh HOA, Burstmayr H, Caligari PDS (2014) Accelerated plant breeding. CAB Rev 9(043):1–16

Chapter 4

Barley Anther Culture

Ludmila Ohnoutkova, Tomas Vlcko, and Mentewab Ayalew

Abstract

The production of doubled haploid (DH) barley plants through anther culture is a very useful yet simple in vitro technique. DH plants derive from divisions of haploid microspores that have undergone a developmental switch under the appropriate conditions. The successive divisions lead to the formation of an embryo or callus rather than the formation of mature pollen grains. Plants that regenerate from these embryos are often either haploid, in which case their chromosome set can be doubled by treatment with colchicine, or spontaneous double haploids. The efficiency of DH plant production is highly variable depending on the genotype of the source material. Despite this limitation, DH plants have been widely used in breeding and research programs. Compared to conventional approaches, breeding strategies that makes use of DH plants achieve a homozygous state, allowing transgene or mutation stabilization in the genome, within a considerably shorter time, thus accelerating workflow or reducing work volume.

Key words Barley (*Hordeum vulgare*), Anther culture, Doubled haploid, Androgenesis

1 Introduction

The ontogenesis of higher plants consists of two developmental phases– a haploid phase (gametophyte) and a diploid stage (sporophyte). In angiosperms, the male haploid phase spans from the inception of microspores to their maturation into pollen grains, and extends to the growth of pollen tubes for fertilization of the egg, the female haploid phase originating from a megaspore that has undergone several divisions leading to the development and differentiation of the embryo sac. In haploid plants however, a single set of chromosomes (gametophytic number) is present in the sporophyte, in contrast to diploid plants that contain two sets of chromosomes. Haploid plants are of major value in the production of homozygous lines (homozygous doubled haploid plants) for breeding programs. Their practical use, however, has been restricted due to the very low frequency of spontaneous occurrence, ranging, according to the specificity of plant material, from 0.001% to 0.01%.

Wendy A. Harwood (ed.), *Barley: Methods and Protocols*, Methods in Molecular Biology, vol. 1900,
https://doi.org/10.1007/978-1-4939-8944-7_4,

Work aimed at increasing the frequency of doubled haploid production was initiated after Guha and Maheshwari (1964) described the development of haploid embryos and plantlets from microspores of *Datura innoxia* using cultures of excised anthers [1]. A few years later, Bourgin and Nitsch (1967) obtained the first haploid plants from *Nicotiana tabacum* [2]. Since then, considerable progress has been made in the anther cultures of maize, wheat, rice, pepper, and a range of other economically important species. In 1973, Claphan was the first to report obtaining haploid plants from anther culture in barley [3].

The process of inducing haploid or doubled haploid (DH) plants from male meiotic spores (microspores) through anther culture or isolated microspores is called androgenesis [4]. Anthers are removed from flowers and either directly cultured under in vitro conditions or microspores can be isolated prior to cultivation. The method of inducing DH plants from microspore cultures may be more efficient [5], but is technically demanding [6].

Microspores or non-differentiated immature pollen grains can develop into mature pollen grains in axenic conditions in vitro in a suitable culture medium. Alternatively, a number of factors such as their removal from the plant, the presence of phytohormones, or the effect of stress can lead to a reprogramming of their development that produces pollen embryos or calli from which plants can be regenerated by organogenesis. Unlike zygotic embryogenesis, embryo development from microspores is more asynchronous. According to Reynolds [7], this development of embryos without fertilization from microspores is a phenomenon regarded as one of most striking examples of cellular totipotency. In the opinion of Bonet et al. [8], totipotence is not enough to explain the phenomenon of androgenesis. They argued that, pollen embryogenesis might be a result of atavism, e.g., morphogenic capacity. Authors hypothesized that microspores, when subjected to a stress, change their pattern of genetic expression and take on morphogenic capacity.

Based on differences of karyokinesis during the early stages of division, embryogenic microspores follow three distinct developmental paths [9]. The first path involves asynchronous mitosis (homeotypic division) which starts with a separation of the vegetative and generative core. The pollen embryo then forms from divisions of the vegetative nucleus, and, in rare cases, with additional splitting of the generative nucleus. The second path involves the symmetrical mitosis (homeotypic division), in which two equivalent nuclei are formed. Each which will give rise to several other rounds of divisions resulting in the embryo. In the third path the symmetric mitosis (homeotypic division) is followed by the merger of the two cores (the first two cells) and the emergence of doubled haploid cells. From this point they can regenerate into a DH plant.

The androgenic response is the result of two distinct hereditary mechanisms involving embryogenesis (induction and development of a microspore embryo) and plant regeneration. The overall frequency of the formation of haploid and dihaploid or double haploid (DH) plants is dependent on many factors. The most significant factors are the genotype of donor plants, the stage of the cultured explants, composition of culture media, and culture conditions. Dihaploidization, subsequent genetic stability, and the appearance of albino plants are other limiting factors.

The genotype of the donor plant is the most critical factor in obtaining microspore embryos. Specific QTLs are known to contribute to the effectiveness at individual stages of androgenesis [10]. In many barley genotypes, the androgenic process is rather inefficient. Lazaridou et al. [11] conducted research on a large set of barley genotypes and showed an average regeneration efficacy of 1.7 green plants (GP) per 100 anthers cultured. However, in the model cultivar "Igri," 1009.8 plantlets per 100 responding anthers were regenerated, with the proportion of green plants reaching 87.6% [12]. The genotype determines the cellular response at each stage of DH plant development, starting from the induction of androgenesis, through embryogenesis, the production of embryos/calli, regeneration of plant, to the undesirable phenomenon of albinism. Until now, it has not been possible to establish the sources of such a large disparity in the efficiency of this process among genotypes [13]. During the induction of androgenesis, cellular homeostasis is disrupted and cells display oxidative stress [14]. The cells ability to cope with this stress and their ability to survive the critical juncture of switching their developmental pathway and the time in vitro culture has a major effect on the efficiency of DH plant regeneration. This determines the ability to produce embryos and the rate of regeneration [13].

The androgenic process induces formation of albino plantlets in varying proportions, depending on the cultivar, which can greatly limit the potential yield of barley dihaploid plants. In some genotypes, the frequency of albino plantlets is insignificant (1%), but in others, almost all regenerated plants are albinos (99.7%), [5, 15]. The mechanism responsible for the inability to produce chloroplasts has not yet been fully clarified [13]. Nonetheless, plastids in *Poaceae* are particularly sensitive to culture conditions, suggesting that there is discrimination between nuclear and plastid behaviors under in vitro conditions. When albino plantlets are formed, the nucleus may conserve totipotency, whereas the plastids may lose the ability to dedifferentiate and follow a new sporophytic program during the culture process [15].

Streamlining DH plant production is particularly valuable for breeding programs of self-pollinating cereal crops such as barley. DH plants allow a reduction in the time required to produce homozygous plants compared to a conventional breeding cycle.

Moreover, DH plants facilitate studies on plant molecular genetics and genomics, such as genome mapping, QTL analysis, molecular marker-assisted selection, gene isolation, and cloning [16]. Recessive mutations, important recombinations, and other genomic changes can be found more easily in DH plants [17]. Reverse genetics utilizing instruments of precise genome editing or transgene over-expression frequently require a homozygous state for the gene under study [18].

2 Materials

Barley (*Hordeum vulgare*, L.), ($2n = 2x = 14$), spring and winter cultivars, variety/breeding material.

2.1 Plant Material and Growth Conditions

1. Barley seeds are germinated on moistened filter paper in closed Petri dishes for 7 days at room temperature in the dark.
2. Germinated seedlings are potted into 15 cm diameter pots containing professional substrate mix for grasses (e.g., Florcom®) and fertilized at the tillering stage with one teaspoon of Osmocote® (14-14-14 N-P-K) from ICL Specialty Fertilizers.
3. Plants are grown in a growth chamber or phytotron at 15 °C day/12 °C night temperatures, approximately 80% relative humidity and under conditions of a 16 h photoperiod with light levels of 500 μmol/m^2/s. The winter barley seedlings are vernalized for 8 weeks at 4 °C, under conditions of a 12 h photoperiod (night and day) at low light intensity (100 μmol/m^2/s).

2.2 Spike Sampling

1. The best time for harvesting barley spikes is when the majority of microspores are at the mid- to late-uninucleate stage (Fig. 1). Anthers with uninucleate microspores are more responsive than the binucleate stage.
2. Donor tillers are cut between the second and third node with scissors, approximately 25 cm in length, at a time when the flag leaf grows apart from the second leaf by a distance of 7–9 cm (Fig. 2a). The first four inflorescences produce a greater percentage of green haploid plants. The developmental stage of microspores is checked by acetocarmine staining (*see* **Note 1**).

2.3 Cold Pretreatment

1. All leaves are cut except the flag leaf. The tillers are placed into Erlenmeyer (conical) flasks which contain sterile distilled water and are covered by a micro-perforated plastic bag to assure humidity.
2. Donor tillers are cold pretreated in the dark for 7–21 days at a temperature of 4 °C, for instance in a refrigerator (Fig. 2b).

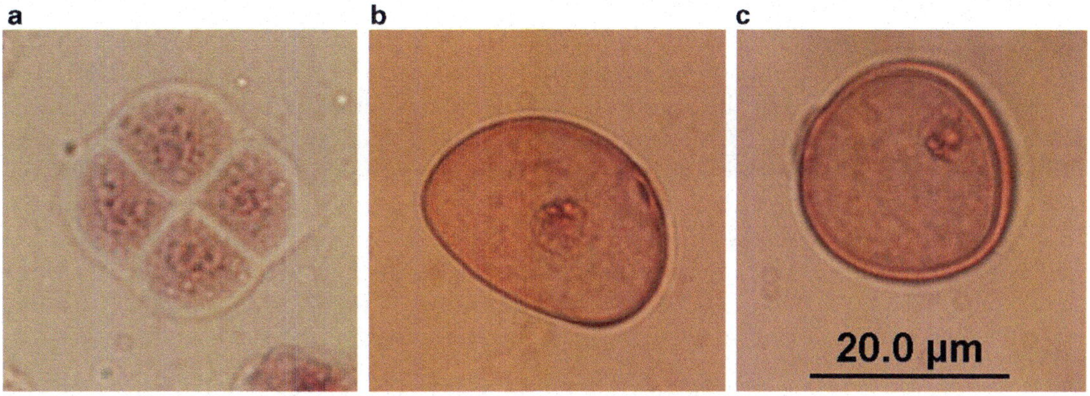

Fig. 1 Barley microspores stained by acetocarmine. (**a**) Microspore tetrade. (**b**) Mid-uninucleate microspore. (**c**) Late-uninucleate microspore

Fig. 2 Spike sampling. (**a**) Donor tillers are cut between the second and third node, arrows indicate distance between the flag leaf from the second leaf, optimum is 7–9 cm. (**b**) All leaves were cut except the flag leaf, the tillers are pretreated at 4 °C

2.4 Plant Tissue Culture Media

Two different plant tissue media are used for the induction and the regeneration steps of barley anther culture. For induction of pollen embryogenesis auxins 2,4-D and NAA, and cytokinin KI are used. For the regeneration process, a combination of low concentrations of NAA and KI are used. The induction medium is supplemented with 10 μM $CuSO_4$ solution, regeneration medium is supplemented with 2 μM $CuSO_4$. Maltose is used at two concentrations depending on the media type. All media components are supplied by Sigma Aldrich or Duchefa which are designed for plant tissue culture. All stock and media are prepared using high-quality water, optimally water from ultrapure water purification systems.

1. Induction medium: **CHU (N_6)** micro and macro elements [19] 4.0 g/L (Duchefa C 0203, Sigma C 1416), 1 mL/L $CuSO_4{\cdot}5H_2O$ stock solution, 80 g/L maltose, 550 mg/L casein hydrolysate, 146 mg/L glutamine, 80 mg/L myo-inositol, 2.0 mg/L nicotinic acid, 2.0 mg/L pyridoxine HCl, 2.0 mg/L thiamine HCl, 3 mg/L glycine, 2.0 mg/L biotin, 2.0 mg/L 2,4-dichlorophenoxy acetic acid (2,4-D), 0.5 mg/L alpha-Naphthaleneacetamide (NAA), 0.5 mg/L kinetin (KI), 3.5 g/L Phytagel. The pH is adjusted to 5.8 with NaOH.
2. Regeneration medium: **190-2** [20], modified (*see* composition Table 1).

   The medium contains: 50 g/L maltose, 0.5 mg/L 1-Naphthaleneacetamide (NAA), 0.5 mg/L kinetin (KI), 3 g/L Phytagel or 6 g/L Phytoagar. The pH is adjusted to 5.8 with NaOH (*see* **Note 2**).

Growth regulator stocks are prepared at a concentration of 1 mg/mL

1. 2,4-Dichlorophenoxy acetic acid (2,4-D), 10 mg can be dissolved in a small volume of 95% ethanol and then brought to volume of 10 mL with sterile water. Heat gently if required.
2. alpha-Naphthaleneacetic acid (NAA), 10 mg can be dissolved in several drops of 1 N NaOH and brought to 10 mL with of sterile water. The stock solution is stored at 2–8 °C.
3. The stock solution of 2,4-D and NAA can be stored in refrigerator for 1 month or at −20 °C for several months.
4. Kinetin (KI), 10 mg dissolved in several drops of 1 N NaOH and diluted in 10 mL sterile water. The stock solution is stored at −20 °C for several months.

Media stock:

$CuSO_4$ stock: 250 mg of $CuSO_4{\cdot}5H_2O$ dissolved in a total volume of 100 mL of sterile water, filter sterilized, and stored at 4 °C.

Table 1
190-2 Regeneration medium

Components	*Mg/l*	*Stock solution*		*Medium 1 l*
Macro-Micro Elements				
KNO_3	1000	10.0 g	*in to 1000 ml*	100 ml
$(NH_4)_2 SO_4$	200	2.0 g		
$Ca(NO_3)_2 . 4H_2O$	100	1.0 g		
$KH_2 PO_4$	300	3.0 g		
$MgSO_4 . 7H_2O$	200	2.0 g		
KCl	40	0.4 g		
$MnSO_4 . H_2O$	6.0	60 mg		
$ZnSO_4 . 7H_2O$	3.0	30 mg		
H_3BO_3	3.0	30 mg		
$CuSO_4 . 5H_2O$	2.5	25 mg		
KI	0.8	8 mg		
$Na_2EDTA . 2H_2O$	37.3	3.7 g	*in to 500 ml*	5 ml
$FeSO_4 . 7H_2O$	27.8	2.8 g		
or FeNaEDTA (Duchefa E0509)	36.7			
Vitamins				
Nicotinic acid	0.5	10 mg	*in to 100 ml*	5 ml
Pyridoxine HCL	0.5	10 mg		
Thiamine HCL	2.0	40 mg		
Glycine	3.0	30 mg	*in to 100 ml*	10 ml
Myo-Inositol	80.0			
Maltose	50 g			
Phytagel	3 g			
Hormones-first step				
NAA	0.5 mg/l			
KI	0.5 mg/l			
Hormones-second step				
NAA	0.5 mg/l			
KI	1.5 mg/l			
pH 5.8				

3 Methods

3.1 Media Preparation

Phytagel at double concentration is autoclaved at 121 °C for 150 min (this can be prepared in advance). Both induction and regeneration media are prepared at twice the required concentration. The pH is adjusted and the media sterilized by filtration (0.22 μm filter). The 2× phytagel and 2× media are both warmed to 55 °C in a water bath, mixed (1:1) under sterile conditions and poured into 6-cm Petri dishes. For regeneration, the media are poured into a sterile Erlenmeyer flask (100 mL) or into a culture box (Duchefa E 1650).

3.2 Isolation of Anthers

1. The stem under the last node is cut, and the spikes in the leaf sheaths are surface sterilized by spraying with 70% ethanol (Fig. 3a), then, under sterile conditions by soaking in a solution of sodium hypochloride (NaOCl 6%) for 5 min and then washed four times in sterile distilled water (Fig. 3b).
2. A scalpel is used to remove the leaf sheath and tweezers are employed to remove the ear. Step by step, under a binocular microscope using tweezers and a scalpel the anthers are isolated (Fig. 4a–c).The anthers without filaments are laid on the

Fig. 3 Sterilization of the samples after pretreatment. (**a**) The spikes are spraying with 70% ethanol. (**b**) Soaking in a solution of sodium hypochloride

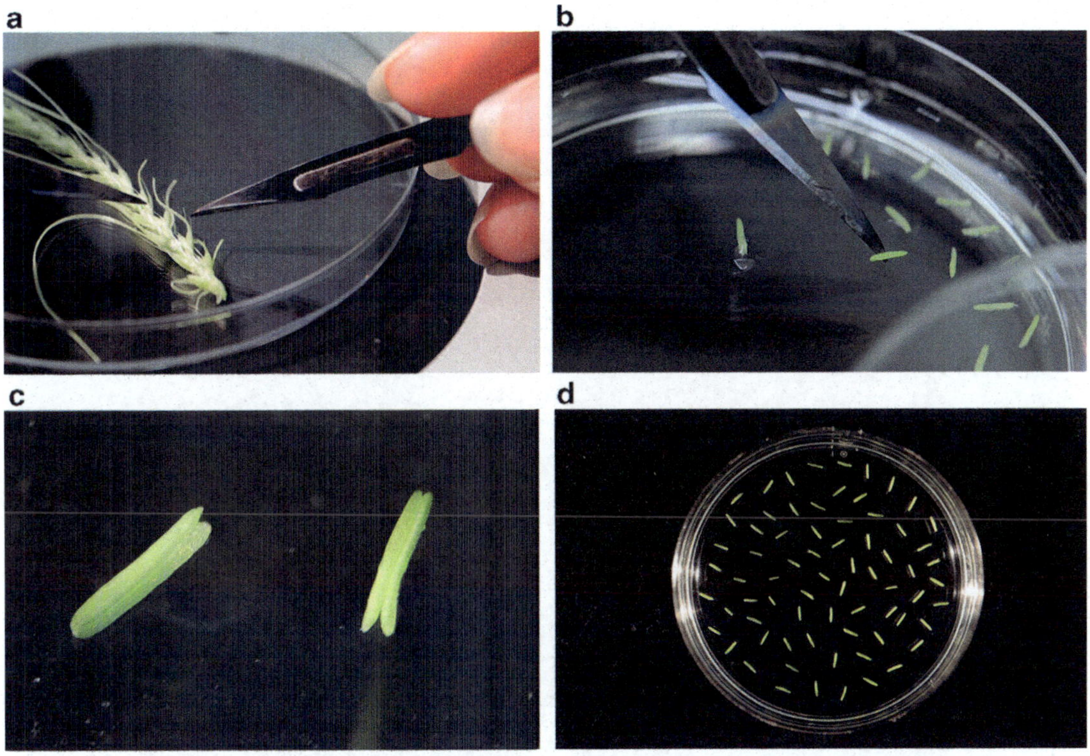

Fig. 4 Isolation of anthers in sterile condition. (**a**) Preparation of anthers under a binocular microscope. (**b**) The anthers without filaments are laid on the induction medium. (**c**, **d**) Cultivation of anthers on 6-cm Petri in the dark at 26 °C

induction medium, preferably with only one sac in direct contact with the medium.

3. Sixty to eighty anthers are placed on each 6-cm Petri dish and cultured in the dark at 26 °C (Fig. 4d), the induction of pollen embryogenesis will start after a few days and can be visualized with acetocarmine staining (Fig. 5a–d).

3.3 Regeneration of Plants

1. After around 4 weeks of anther cultivation, pollen embryos or calli become visible (Fig. 6a–g). The small clumps of embryos and calluses (hard structures) larger than 2 mm are transferred to plant tissue culture container with regeneration media containing 50 g/L maltose, 0.5 mg/L NAA, and 0.5 mg/L KI. If the calli do not produce plants after 3 weeks, they are transferred to regeneration medium containing 30 g/L maltose and 0.5 mg/L NAA and 1.5 mg/L KI (Fig. 7).
2. The plantlets with well-developed roots are transferred to 30 mm Jiffy peat and placed in a plastic bag for several days to maintain humidity until the plants are tested for their ploidy level (Fig. 8a, b).

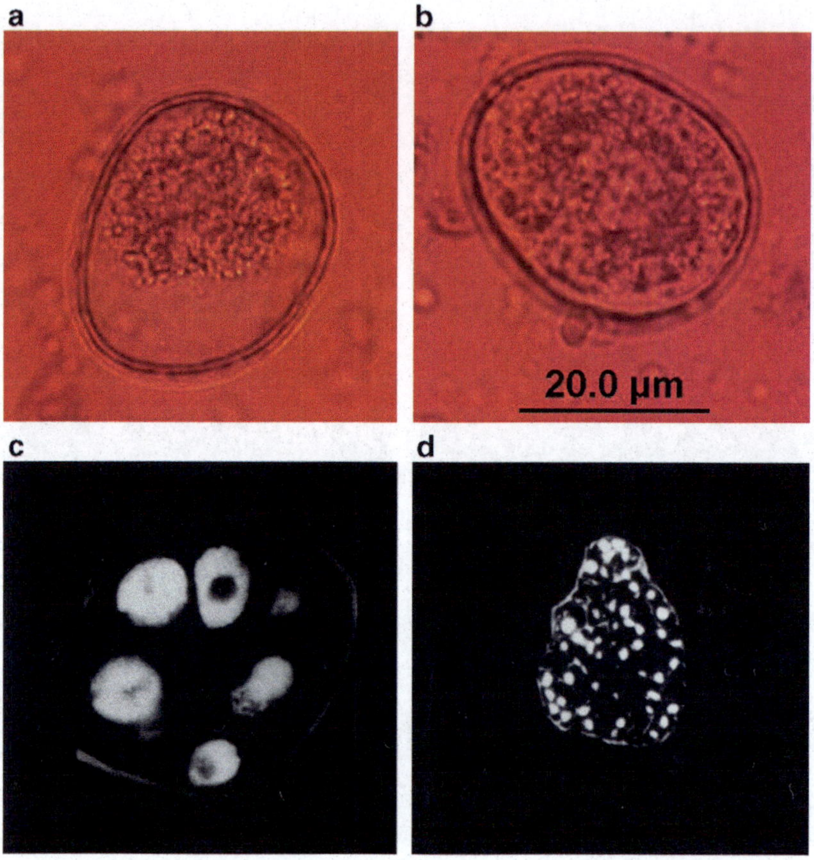

Fig. 5 Formation of multicellular structure can be visualized by acetocarmine staining. (**a**, **b**) After 3–7 days cultivation on induction medium. (**c**, **d**) Development of globular pollen embryos

3. The ploidy of plants regenerated from barley anthers is best determined by flow-cytometric estimation of DNA content. Briefly, suspension of intact cell nuclei is prepared by chopping a small amount of leaf tissue with a sharp razor blade in the appropriate solution [21]. The nuclei in solution are stained by a DNA fluorochrome and their relative fluorescence is analyzed using flow cytometry. Nuclei isolated from a diploid barley variety are used as control. For details *see* [21] (*see* **Note 3**) (Fig. 9a–c). Haploid, diploid, and tetraploid plants have a distinguishable appearance (Fig. 10).
4. Individual dihaploid plants are transferred in the barley growth mix described above in 15 cm diameter pots and transferred to growth room or greenhouse at 15 °C day/12 °C night temperatures, and under a 16 h photoperiod with light levels of 500 µmol/m^2/s (Fig. 8c). The seeds from individual plants are harvested separately. The DH lines are evaluated and tested under greenhouse and field conditions (Fig. 11).

Fig. 6 Induction and regeneration processes in barley anther culture. (**a**–**d**) Formation of pollen embryos or calli on induction medium after 4–5 weeks. (**e**–**g**) Regenerated structures were transferred onto the regeneration medium

4 Notes

1. Acetocarmine staining

   Acetocarmine preparation (1% solution): Heat 100 mL solution of 45% acetic acid, add 0.5 g of carmine (Fisher C579-25) while stirring and continue heating for 30–60 min, add boileezers, and reflux for 24 h. Filter into dark bottles and store at 4 °C. This solution can be stored for a long time. Staining can be intensified by adding ferric chloride

Fig. 7 Regenerating plantles from pollen embryogenic calluses after 3 weeks on regeneration medium

Fig. 8 Regenerated plants are transferred to Jiffy peat (**a**), and well-rooted plants (**b**) are potted and placed in the greenhouse (**c**)

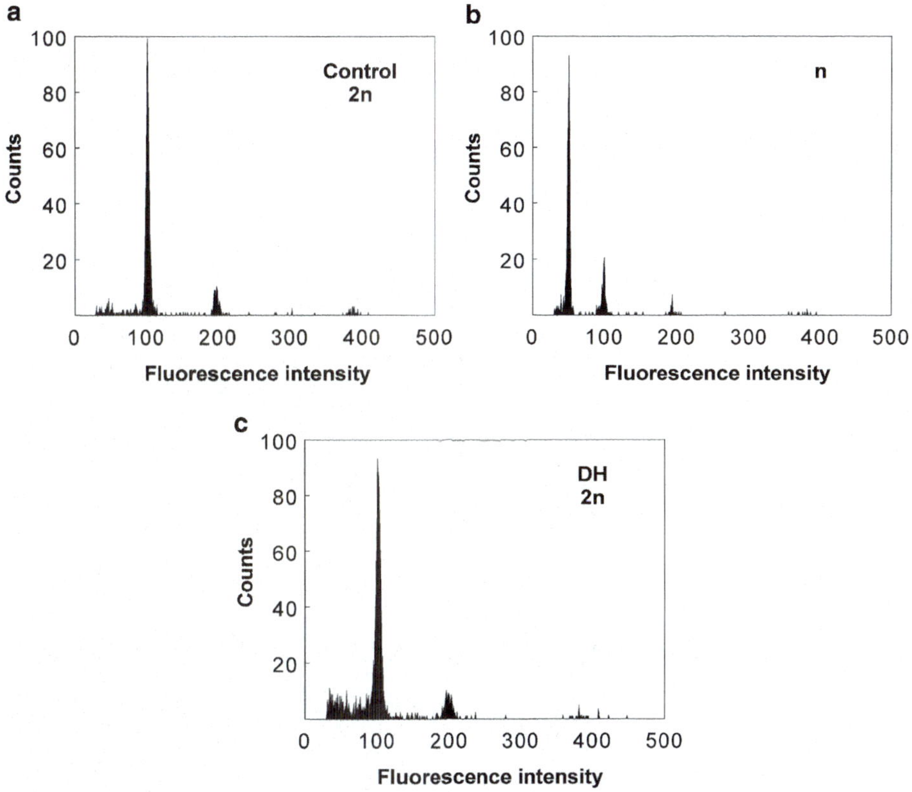

Fig. 9 DNA content of nuclei isolated from leaves of regenerated plants. The position of the main peaks indicate ploidy level. (**a**) Control diploid plant. (**b**) Haploid plant. (**c**) Dihaploid plant

Fig. 10 Regenerated plants from anthers of spring barley cv. Golden Promise. The picture shows the differences in appearance between a haploid plant (n), dihaploid plant (2n), and tetrahaploid plant (4n)

Fig. 11 Barley dihaploid lines derived from anther culture in the field trials

($FeCl_2 \cdot 6H_2O$); add 5 mL of a 10% ferric chloride solution per 100 mL of the 1% acetocarmine solution.

Two to three anthers from flowers in the middle section of the spike are placed on a glass slide and a drop of 1% acetocarmine solution is added. Cut with a razor and add a cover slip and squeeze the anthers with forefinger between two layers of filter paper. Microspores are observed under a microscope.

2. Stock solution of regenerated media 190-2

Solutions of macro- and micro-nutrients are better prepared as stock solutions of ten times the strength of the final medium (*see* Table 1). Stock solutions can be stored in a refrigerator at 4 °C. Vitamins are prepared as a100 times concentrated stock solution and stored in a freezer (−20 °C), or can be stored in a refrigerator for 1 month. Stock solutions of growth regulators are usually prepared 1 mg in 1 mL (add 10 mg plant hormone to 10 mL of distilled water), sterilized by filtration (0.22 mm filter) and stored in −20 °C.

3. Chromosome doubling

   The haploid set of chromosomes in microspore-derived plants becomes spontaneously doubled under these conditions. However, the percentage of doubling varies depending on the crop, including genotypes or hybrids. In barley approximately 75–80% regenerated plantlets spontaneously double, are fertile and only 15–20% plantlets are haploid. Haploid plants are usually smaller, have narrower leaves, and the plants are sterile. In general, colchicine is used for chromosome doubling at the whole plant level for barley (*see* **Note 4**).

4. Colchicine treatment

   Colchicine is used in the advanced tillering stage (developmental stages 15–23) where the plant has two to three tillers with several leaves. Haploid plants are removed from Jiffy peat or pots, and roots are trimmed to 5 cm, rinsed with water and completely immersed in a solution of colchicine concentration 0.1% for 5 h. After treatment, plants are rinsed in running water for 6 h, re-potted and for a few days allowed to grow at a lower temperature with high humidity to recover.

   To prepare a solution of 0.1% colchicine: add 1 g of colchicine to 1000 mL of water (Sigma C 3915, *see* product information), add 20 mL Dimethyl sulfoxide (Sigma D 2650, *see* product information) and ten drops of Tween 20.

   CAUTION: Colchicine is a HIGHLY TOXIC POISON.

Acknowledgments

This work was supported by the Ministry of Education, Youth and Sports, Czech Republic (Grant LO1204 from the National Program of Sustainability I.). We would like to thank the Gordon-Zeto Center for Global Education at Spelman College for supporting Tomas Vlcko and thus facilitating our collaboration. The authors are grateful to Ota Blahousek for the technical assistance.

References

1. Guha S, Maheshwari SC (1964) In vitro production of embryos from anthers of Datura. Nature 204:497
2. Bourgin JP, Nitsch JP (1967) Production of haploids nicotiana from excised stamens. Annales De Physiologie Vegetale 9:377–382
3. Clapham D (1973) Haploid Hordeum plants from anthers in-vitro. J Plant Breed 69:142–155
4. Hu H (1997) In vitro induced haploids in wheat. In: Jain SM, Sopory SK, Veilleux RE (eds) In vitro haploid production in higher plants: Volume 4—Cereals. Springer Netherlands, Dordrecht, pp 73–97
5. Li H, Devaux P (2005) Isolated microspore culture overperforms anther culture for green plant regeneration in barley (Hordeum vulgare L.). Acta Physiol Plant 27:611–619
6. Lantos C, Páricsi S, Zofajova A et al (2006) Isolated microspore culture of wheat (Triticum aestivum L.) with Hungarian cultivars. Acta Biologica Szegediensis 50:31–35
7. Reynolds TL (1997) Pollen embryogenesis. Plant Mol Biol 33:1–10

8. Bonet FJ, Azbaid L, Olmedilla A (1998) Pollen embryogenesis: atavism or totipotency? Protoplasma 202:115–121
9. Sunderland N, Evans LJ (1980) Multicellular pollen formation in cultured barley anthers. J Exp Bot 31:501–514
10. Chen X-W, Cistué L, Muñoz-Amatriaín M et al (2007) Genetic markers for doubled haploid response in barley. Euphytica 158:287–294
11. Lazaridou T, Sistanis I, Lithourgidis A et al (2011) Response to in-vitro anther culture of F 3 families originating from high and low yielding F 2 barley (Hordeum vulgare L.) plants. Aust J Crop Sci 5:265–270
12. Jacquard C, Nolin F, Hécart C et al (2009) Microspore embryogenesis and programmed cell death in barley: effects of copper on albinism in recalcitrant cultivars. Plant Cell Rep 28:1329–1339
13. Makowska K, Oleszczuk S, Zimny A et al (2015) Androgenic capability among genotypes of winter and spring barley. Plant Breed 134:668–674
14. Rodríguez-Serrano M, Bárány I, Prem D et al (2012) NO, ROS, and cell death associated with caspase-like activity increase in stress-induced microspore embryogenesis of barley. J Exp Bot 63:2007–2024
15. Caredda S, Doncoeur C, Devaux P et al (2000) Plastid differentiation during androgenesis in albino and non-albino producing cultivars of barley (Hordeum vulgare L.). Sex Plant Reprod 13:95–104
16. Devaux P, Pickering R (2005) Haploids in the improvement of Poaceae. In: Don Palmer CE, Keller WA, Kasha KJ (eds) Haploids in crop improvement II. Springer Berlin Heidelberg, Berlin, Heidelberg, pp 215–242
17. Kahrizi D, Mahmoodi S, Khaniki GB et al (2011) Effect of genotype on androgenesis in barley (Hordeum vulgare L.). Biharean Biol 5:132–134
18. Shen Y, Pan G, and Lübberstedt T (2015) Haploid Strategies for Functional Validation of Plant Genes. Trends in Biotechnology. 33, 611–620
19. Chu CC (1978) The N6 medium and its applications to anther culture of cereal crops. In: Proceedings of symposium on plant tissue culture, Science Press, Peking, p 45–50
20. Zhuang J, Xu J (1983) Increasing differentiation frequencies in wheat pollen callus. In: Hu H, Vega M (eds) Cell and tissue culture techniques for cereal crop improvement. Science Press, Beijing, p 431
21. Dolezel J, Greilhuber J, Suda J (2007) Estimation of nuclear DNA content in plants using flow cytometry. Nat Protocol 2:2233–2244

Chapter 5

Isolated Microspore Culture in Barley

Patricio Esteves and François J. Belzile

Abstract

Isolated microspore culture (IMC) is the most efficient way to produce large numbers of doubled-haploid (DH) barley plants in a short time. Yet, while IMC is more cost-efficient and less labor-intensive than anther culture, it is technically more complex and requires more experienced personnel if it is to yield its full potential. In part, this is because of multiple and important interactions that exist between factors at its many different phases, including genotype effects as well. When every phase is fine-tuned, the protocol that is presented below yields a useful number of DHs with almost all genotypes and can allow the production of up to 300 DH plants from a single F1 plant in just a few months.

Key words Barley, Doubled haploids, Isolated microspore culture, Pretreatment, Double layer induction medium, Plating density

1 Introduction

Regeneration of cultured haploid cells to yield doubled-haploid (DH) adult plants is a widely practiced method to assist breeding programs [1]. By producing DHs from a heterozygous parent, breeders can rapidly obtain homozygous lines. Thus, DHs have revolutionized breeding where they can be efficiently obtained. Moreover, they also attract considerable interest in genetics and genomics since they can be exploited to construct genetic maps, locate genes of agronomic and economic importance, identify markers for trait selection, and increase plant breeding efficiency [2, 3]. In barley, isolated microspore culture has been recognized as the most efficient method to produce DHs, e.g., surpassing the productivity of anther culture by a factor of 9.3× [4].

When exposed to stress and cultured in vitro, microspores can deviate from their developmental fate, the formation of pollen, to parthenogenetically form embryos instead. Even if microspores are haploid cells, an important fraction of the resulting embryos will have doubled their haploid genome, thus giving rise to doubled-haploid (DH) plants. Doubled-haploid plants are 100%

Wendy A. Harwood (ed.), *Barley: Methods and Protocols*, Methods in Molecular Biology, vol. 1900,
https://doi.org/10.1007/978-1-4939-8944-7_5,

homozygous and, when they are derived from the microspores of a heterozygous plant (e.g., an F1), each line represents a unique product of genetic recombination. In the context of a breeding program or genetic analysis, the fixation of lines is achieved in a single generation. Isolated microspore culture (IMC) is a highly productive method to obtain DH plants via androgenesis [5–7]. It allows for the isolation of huge numbers of uniformly treated microspores which rapidly develop into thousands of embryos many of which will develop into regenerated plants. In barley, IMC has been reported to offer the highest potential for DH production [8].

2 Materials

2.1 In Vitro Culture Solutions and Media

1. Disinfection solution: commercial bleach, 20%, freshly prepared, where a drop of Tween 20 is added just before use.
2. Deionized, distilled water, autoclave sterilized.
3. Mannitol 0.30 M, filter or autoclave sterilized.
4. Maltose 21% (w/v), filter sterilized.
5. Liquid M8M culture medium, filter sterilized.
6. Solid M8M induction culture medium, filter and autoclave sterilized, dispensed into 35 × 10 mm Petri dishes, 4 mL each.
7. Solid RMT regeneration medium, filter and autoclave sterilized, dispensed into 100 × 20 mm, 35 mL each.
8. Solid Germination medium, filter and autoclave sterilized, dispensed into glass tubes (25 × 150 mm) with translucent plastic caps, 20 mL each.

2.2 Culture Media Composition and Stock Solutions Organization

Four different types of culture media are included in this IMC protocol, and their composition is presented in Table 1. In our lab, solid induction and regeneration media are prepared starting from 25× stock solutions including Macronutrients. To avoid precipitation at this high concentration, the calcium dichloride salt ($CaCl_2 \times 2H_2O$) is included into the Micronutrients 25× stock solution. The Organic components of these media are also put together at 25×. The vitamins (thiamine-HCl, pyridoxine-HCl, and nicotinic acid), iron (FeEDTA), ascorbic acid, and arabinogalactans stock solutions are prepared each separately at 1000× and stored at 4 °C wrapped in aluminum foil to protect them from light. Stock solutions of plant hormones are prepared at a concentration of 1 mg/mL and stored in the same way. All these solutions are added separately when culture medium is being prepared.

2.3 Lab Equipment

A list of the necessary equipment for plant tissue culture operations must include certain essential elements, such as laminar air flow

Table 1
Culture media used in microspore culture (induction, plant regeneration, and germination phases) (mg/L)

Components	Liquid M8M induction medium	Solid M8M induction medium	RMT regeneration medium	Germination medium
Solid phase (autoclaved)				
Phytagel	–	3000	3000	3000
Liquid phase (filter sterilized)				
Macronutrients				
KNO_3	1900	1900	1900	1000
NH_4NO_3	165	165	165	1000
KH_2PO_4	170	170	170	300
$MgSO_4 \times 7H_2O$	370	370	370	71.5
$CaCl_2 \times 2H_2O$	440	440	440	–
$Ca(NO_3)_2 \times 4H_2O$	–	–	–	500
KCl	–	–	–	65
Micronutrients				
KI	0.83	0.83	0.83	0.75
H_3BO_3	6.2	6.2	6.2	1.6
$MnSO_4 \times 4H_2O$	22.3	22.3	22.3	4.9
$ZnSO_4 \times 7H_2O$	8.6	8.6	8.6	2.7
$NaMoO_4 \times 2H_2O$	0.25	0.25	0.25	0.2
$CuSO_4 \times 5H_2O$	0.025	0.025	0.025	0.076
$CoCl_2 \times 6H_2O$	0.025	0.025	0.025	0.05
$MnSO_4 \times H_2O$	–	–	–	4.9
Organic supplements				
Glutamine	500	500	500	–
Myo-inositol	300	300	300	100
Hydrolyzed casein	500	500	500	–
Sugar				
Maltose	60,000	60,000	30,000	20,000
Mannitol	32,000	32,000	–	–
Amino acids, vitamins, and plant hormones				
Glycine	–	–	–	2
Thiamine HCl	10	10	10	1
Nicotinic acid	1	1	1	5
Pyridoxine HCl	1	1	1	5
FeEDTA	43	43	43	43
Gum Arabic from acacia tree	50	50	–	–
(+)-Arabinogalactan	50	50	–	–
Ascorbic acid	25	25	–	–
Dicamba	0.60	0.60	–	–
Thidiazuron	0.30	0.30	–	–
Indole 3-acetic acid	–	–	0.1	–
Meta-topolin	–	–	0.1	–
pH	**5.8**	**5.8**	**5.8**	**5.8**

hood, autoclave, water demineralizer, refrigerator, freezer, water distillation unit, balances (electronic, and top pan type), magnetic stirrer/hot plate, pH Meter (digital), oven for dry-sterilizing, microwave oven, and media dispenser (for larger operations).

At the same time, culturing facilities such as a culture room (walk-in type) containing shelves with attached lighting devices and temperature control, as well as greenhouse, are also required.

Other than the standard facilities described above, a plant tissue culture lab wanting to perform IMC demands some particular equipment, such as:

- A refrigerated, swinging-bucket centrifuge capable of developing up to 1000 × *g* and to operate with 15- and 50-mL centrifuge tubes.
- A dissecting stereo microscope, with zoom and external or attached light source, with magnifying power up to 160×.
- A compound light microscope, bright field type, with a magnifying power of up to 1000×.
- A 4 °C refrigerator, preferably equipped with forced air circulation, to store the harvested spikes placed upright.
- A Waring blender mini container, two each of models MC1 and MC2, and a Heavy Duty Laboratory Blender with Digital Tachometer Jack (e.g., from The Lab Depot Inc., GA, USA).
- A vacuum pump, or an Aspirator attached to a water source (e.g., a Venturi-system aspirating device) to filter-sterilize liquids.
- An incubator facility (e.g., SHC10 Humidity Cabinets by Shel Lab).
- Availability of growth cabinets is highly recommended (e.g., of the type CONVIRON Model PGR15).

2.4 Lab Tools

1. Dissection scissors and forceps, autoclave sterilized.
2. Thirty-five millimeter diameter sterile cell culture dish.
3. Sixty millimeter diameter sterile cell culture petri dishes.
4. One hundred millimeter diameter sterile cell culture petri dishes.
5. Micro-pipettes, 0–20, 100–1000, and 10–5000 μL range, and corresponding filter-carrying, sterilized tips.
6. Thermoplastic flexible plastic film (e.g., Parafilm M®) 4″ × 125′.
7. Cell strainers (e.g., Falcon™), 100 μ pore, sterile.
8. Fifty milliliter polypropylene sterile (conical base) tubes.
9. Fifteen milliliter polypropylene sterile (conical base) tubes.

10. Glass sterile Pasteur pipettes with dropper bulbs.
11. Sterile disposable filter-sterilizing unit (pore size 0.2 μm) for large volumes.
12. Sterile disposable filter-sterilizing units (pore size 0.2 μm) for small volumes.
13. A hemocytometer.

2.5 Greenhouse and Growth Cabinets' Consumables

1. Multicell plastic trays, sizes 60 or 125 cm^3.
2. Plastic pots, 1 and 3.8 gallon sizes.
3. Commercial substrate (e.g., Grow Mix™).
4. Vermiculite, medium coarse.
5. Perlite.
6. Sand, medium-coarse.
7. Ceramic soil conditioner (e.g., Turface™).
8. Commercial substrate composed by peat, soil, and fertilizer (e.g., Grow Mix™).
9. Water-soluble 20-20-20 N-P-K fertilizer.

3 Methods

3.1 Culture Media Preparation

3.1.1 Induction Medium

Isolated microspores are plated onto an induction medium, called "M8M," which is used both in its liquid and solid forms. The M8M induction medium includes 50 mg/L of both arabic gum and arabinogalactan proteins [9], as well as 32 g/L of mannitol [10–12]. Thidiazuron (TDZ, *N*-phenyl-*N*′-1,2,3-thiadiazol-5-ylurea) and Dicamba (3,6-dichloro-*o*-anisic acid) are used as growth regulators, at 0.3 mg/L and 0.6 mg/L, respectively.

When preparing the M8M medium in its solid form, sterilization is performed separately on the two solutions, or components: a "solid phase" containing distilled water at two-thirds of the final volume and Phytagel (at 3.0 g/L) is autoclaved (at 121 °C and 1.05 kg/cm^2) for the amount of time required according to the volume of solution present in the container [13]. The other component, a solution made of macro- and micronutrients plus the organic compounds (vitamins, hormones, and sugar), is called the "liquid phase" and is filter sterilized (pore size 0.2 μm). The liquid phase is added to the autoclaved solid phase only once it has cooled enough so that its temperature is below 65–70 °C. The final mix of the two phases is dispensed at 3.5 mL per 35 × 10 mm Petri dish, and these are sealed with Parafilm™ before storing them in the dark at 4 °C. We find that it is not advisable to use solid induction culture medium that has been stored for more than 15 days. Thus, it is convenient to prepare it more often in a relatively small quantity each time.

As for the preparation of the liquid form of the M8M medium, corresponding aliquots are taken from stock solutions of macro- and micronutrients and other components, and diluted with the corresponding maltose quantity to obtain a final volume of 50 mL each time it is prepared. It is sterilized by filtration, and stored until use.

3.1.2 Regeneration Medium

The regeneration medium which we have developed is called "RMT," and its composition is shown in Table 1. As with the induction medium, it is prepared by mixing an autoclaved solid phase with a filter-sterilized liquid phase. Once the two have been combined, 35 mL of RMT medium is dispensed per Petri dish (100 × 20 mm), The RMT medium contains the same macro- and micronutrients as the induction medium, but its maltose concentration is reduced to 30 g/L, and no arabinogalactan proteins nor mannitol are included. The hormonal composition of the RMT medium is 0.1 mg/L 3-indolacetic acid (IAA) and 0.1 mg/L of meta-topoline [6-(3-hydroxybenzylamino)purine].

3.2 Starting the Donor Plants: Seed Disinfection and Germination

Spikes grown on healthy and vigorous donor plants are the best suited material to extract large numbers of viable and embryogenic microspores. Thus, care must be taken from the very beginning of plant cultivation (seed germination) to ensure optimal growth conditions.

Start the culture of the donor plants by surface-sterilizing the seeds by soaking them in 20% commercial bleach for 15 min, followed by rinsing/soaking them three times in sterilized distilled water (for 1 min each time, while shaking). Under a laminar flow hood, the sterilized seeds are then placed on a sterile (autoclaved) filter paper in 5 cm diameter Petri dishes. The filter paper is next wet with 2–3 mL of sterile distilled water, and the Petri dishes are stored in the dark at 25 °C for 2–3 days.

3.3 Donor-Plant Cultivation

Selected vigorous germinated seeds are first established in small pots (e.g., 60 or 125 cm^3 multicells trays) filled with a custom-prepared, pasteurized substrate mix composed of 5:2:1:1 soil: Grow-Mix™:perlite:sand. When plantlets reach the four to five leaf stage, they are transferred to larger pots (1 gallon) and maintained in these for a further 3–4 weeks. Finally, plants are ultimately transferred to 3- or 3.8-gallon plastic pots. These are filled with a 35:65 (in volume) mix of Turface™ and vermiculite (Figs. 1 and 2), in which plants are watered with a 20-20-20 fertilizer (0.25 g/L) solution, 5 out of 7 days. The 2 other days, only water is used.

Ideally, donor plants are grown in a growth cabinet (*see* **Note 1**). A photoperiod of 16 h of light (at an intensity of at least 350 μmol m^{-2} s^{-1} PAR) provided, i.e., by a combination of 14 Cool White Very High Output fluorescent tubes and 12 incandescent (standard, tungsten filament) 40 W bulbs per 1.5 m^2

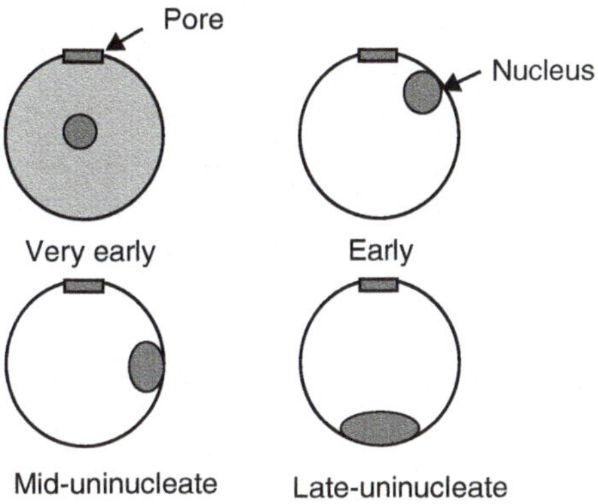

Fig. 1 Microspore maturation as defined by the relative position of the nucleus in relation to the pore. Mid- to late-uninucleate is the preferred stage [8]

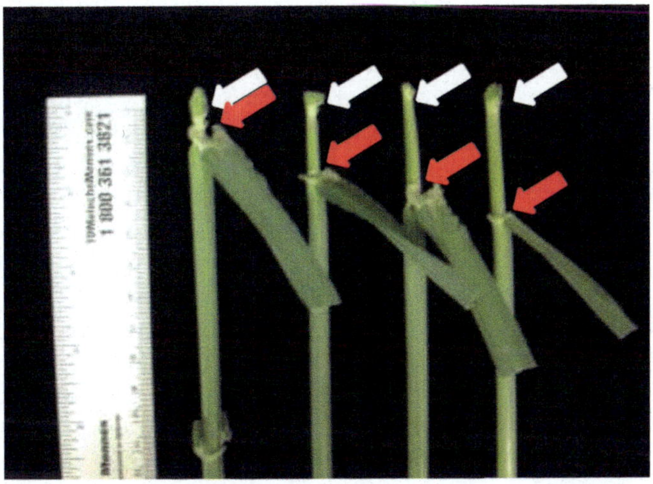

Fig. 2 Determination of the inter-ligule interval (in cm) at the time of harvest. White arrows shows the position of the ligule of the flag leaf (detached here), and red arrows the ligule of the second leaf

surface of the cabinet. Throughout the growth cycle, temperatures in the cabinet are set at 15/13 °C (day/night). Under these conditions, it takes about 40–48 days to reach the stage at which the harvest of the first stalks can be started (depending on the genotype). Prophylactic application of a systemic fungicide (such as Opus Team™, BASF) can be of help in order to ensure healthy, pathogen-free plants.

Vigorous donor plants will yield larger spikes (within the limits allowed by the genotype) and when these are harvested at the optimal stage these immature spikes will release a large number of microspores at the right developmental stage.

3.4 Harvesting of Immature Spikes

Harvesting the stalks containing spikes at the optimal stage, i.e., when a large number of the microspores are at the proper developmental stage to enter the pretreatment, is probably the most important factor to success in an IMC protocol. In the literature, the optimal harvesting stage is defined as the moment at which the majority of the microspores are at the mid- to late-uninucleate stage [14] (Fig. 1).

This moment can be determined by sampling the young spikes and taking out anthers from florets located at a position corresponding to a third of the length of the spike (measured from the bottom), and doing a squash—applying a drop of acetocarmine—to observe under a microscope the specific stage at which the majority of the microspores are found [15]. Conveniently, the developmental stage of microspores can be correlated with a more easily measured trait: the inter-ligule distance, i.e., the distance in cm between the ligules of the flag leaf and the previous leaf (Fig. 2) of the stalks at the time of harvest [16].

The objective of measuring this index is to correlate a standardized inter-ligule distance with the developmental stage of the microspores in the spikes. This will allow harvesting the stalks during the upcoming days at the optimal stage to enter the pretreatment. The inter-ligule distance associated with an abundance of mid- to late-uninucleate microspores is taken as a reference for the particular genotype growing under the particular environmental conditions.

As an alternate procedure, to determine the correct stage at which to harvest the spikes, we harvest them in sets of two to three spikes of similar inter-ligule distance for each of many such distances (e.g., of +2, +5, and +8 cm) for each genotype. We next isolate the embryogenic microspores from each one of these sets of immature spikes, and quantify their yield. This yield will vary from genotype to genotype (*see* **Note 2**). To compare the number of microspores obtained from different stages at harvest, we use a relative measure such as the number of embryogenic microspores/spike or the number of embryogenic microspores/total number of microspores. The best harvest stage (inter-ligule distance) is chosen on the basis of this measured relative yield (*see* **Note 3**). Figure 3 illustrates the extent to which the yield of embryogenic microspores can vary among different barley genotypes depending on the inter-ligule distance at harvest.

The chosen stalks are taken from the donor plant by cutting with scissors at a distance of at least two or three nodes below the node where the spike is inserted. Stalks are immediately wrapped in aluminum foil, and put into containers of distilled water, until taken to the lab, or conveniently stored at 4 °C until processed.

3.5 Pretreatment of Spikes

The pretreatment is a phase of stress which must be applied to the developing microspores in the harvested spikes. Its function is primarily to stop the progression of the cell cycle and, in a certain

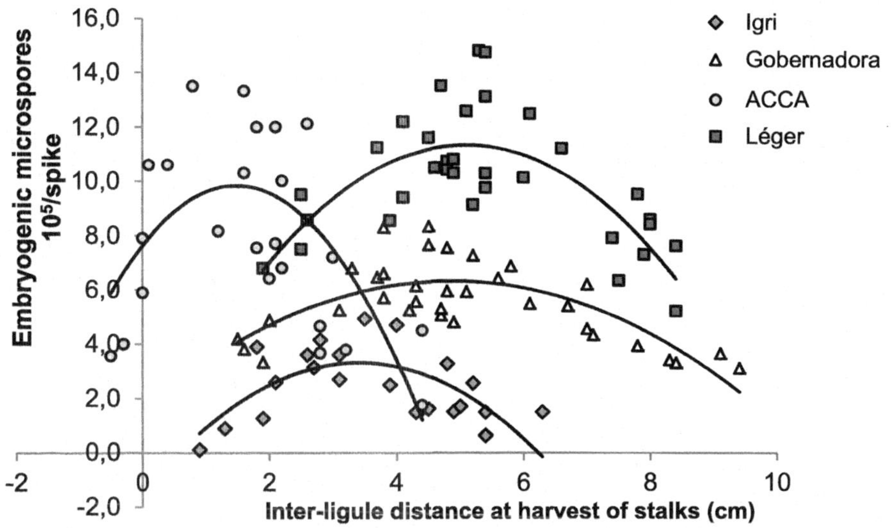

Fig. 3 Variation in the yield of embryogenic microspore between barley genotypes when stalks have been harvested at different inter-ligule distances

fraction of the population of microspores, to engage in the gametic-embryonic pathway. Many different types of pretreatments have been used by different researchers (heat, cold, pH, and osmotic stress). The efficiency of these various pretreatments can be markedly different.

In our lab, we have modified the procedure developed initially by Hoekstra et al. [17], and thus developed a pretreatment we call "M30" as it entails incubating the immature spikes in a sterile 0.3 M solution of mannitol stored at 26 °C for 48 h. When appropriate studies were performed, this M30 pretreatment showed itself to be more convenient than the traditional "cold pretreatment" (storing the stalks for 14 or 28 days at 4 °C in the dark in flower vases) [18, 19], or than the "14-14" pretreatment (storing the stalks for 14 days at 4 °C in the dark in flower vases, and exposing them to heat, at 32 °C, for 14 h) [20], in that it reduced the incidence of albinism among regenerated plants, and increased the proportion of isolated microspores that lead to the production of green plants [20, 21].

Ideally, only spikes extracted from freshly harvested stalks are put into the M30 pretreatment. However, this is not always practical as the harvest of a sufficient number of spikes to warrant microspore isolation may not always be possible on a single day. The top portion of many harvested stalks can be wrapped together in aluminum foil and kept upright with the cut portion of the stalk placed in water in an appropriate container (erlenmeyer flask, flower vase) at 4 °C in the dark, and stored for up to 2–3 days before entering the pretreatment. Reductions in the yield of isolated embryogenic microspores per spike are to be expected, but using a higher number of stalks can compensate for this.

To perform the M30 pretreatment, once the stalks at the optimal stage have been harvested, spikes are extracted from the boot, and surface disinfected (20% commercial bleach with a drop of Tween 20 for 15 min followed by three rinses with sterile, distilled water) in a laminar flow hood. Surface-sterilized spikes are then placed in groups of four to eight spikes in 100 × 20 mm Petri dishes containing 15–20 mL of a sterile mannitol solution (0.3 M), sealed with Parafilm™, and stored for 2 days at 26 °C in the dark.

3.6 Microspore Isolation and Plating

Upon completion of the pretreatment, microspores are isolated from the spikes using a Waring blender mini container (Model MC3, The Lab Depot Inc., GA, USA) containing 50 mL of cold (4 °C) mannitol 0.3 M. The whole operation is carried out in the laminar flow hood, under sterile conditions. The spikes are cut into 2–3 cm pieces before the cold mannitol is added to the chamber onto them. After carefully covering the container, the blender is operated for 15–25 s at low speed to homogenize the spikes. The debris are separated from the microspores by filtering through a 100-μm sieve (Falcon™ Cell Strainers, Sterile, VWR CAT N° 21008-950) placed on top of a 50-mL tube. Next, the micro-blender chamber is rinsed with another 50 mL of fresh 0.3 M mannitol and filtered through another 100-μm sieve on top of a second 50-mL tube to recover the remaining microspores. The two tubes are balanced, and centrifuged at 1500 rpm (300 × *g*) for 4 min using a centrifuge (e.g., an Eppendorf 5810R) equipped with a swing-bucket rotor. The pellet of both tubes is resuspended with a total of 50 mL of fresh mannitol, and the resulting 50-mL tube is recentrifuged under the same conditions.

This operation is repeated once again to wash the microspores, but after this the pellet is resuspended in only 10–12 mL of 0.3 M mannitol and the suspension of microspores is transferred to a 15-mL tube. Once again the resuspended microspores are centrifuged under the same conditions. Finally, to purify and select the viable and embryogenic microspores, a discontinuous gradient centrifugation is performed. Briefly, microspores in the final pellet are resuspended with 6–8 mL of cold (4 °C) maltose 21% (w/v) solution overlaid with 0.65 mL of 0.3 M cold mannitol. Then, a centrifugation at 900 rpm (60 × *g*) for 18 min is performed and, if the isolation has been successful, a resulting band of embryogenic and viable microspores is obtained (Fig. 4) (*see* **Note 4**).

The band consisting of purified microspores is collected from the maltose–mannitol interface using a sterile Pasteur-pipette with a rubber bulb, and transferred to a new 15-mL tube. Five to 10 mL of mannitol (0.3 M) is added and the resulting suspension is centrifuged at 1500 rpm for 4 min. Finally, the resulting pellet

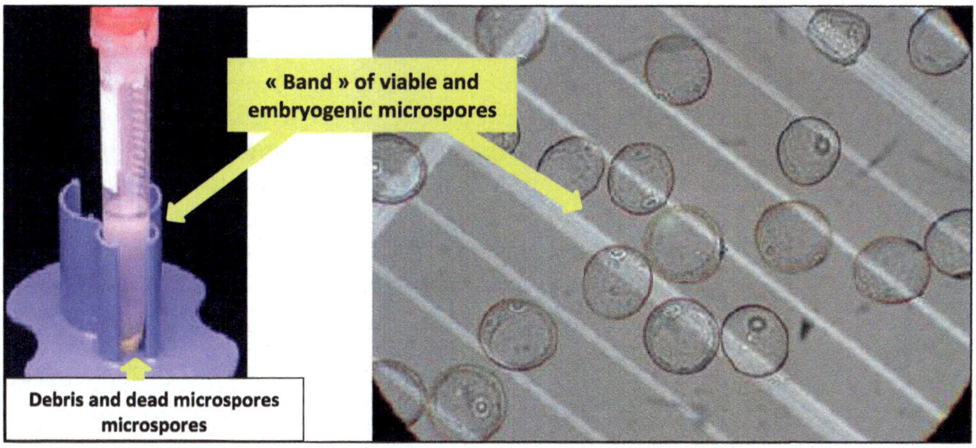

Fig. 4 Discontinuous gradient centrifugation yields a band of embryogenic microspores at the maltose-mannitol interface

will contain a purified preparation of viable and embryogenic microspores, which are resuspended in 100–200 μL of liquid M8M induction culture medium. A 10 μL aliquot is taken from the resulting resuspension and cells are counted using a hemocytometer (e.g., Thomas Scientific CAT N° 5971R30) under a light microscope. The final microspore density in the remaining suspension is adjusted to between 1 and 1.5 × 10^6 cells/mL by adding liquid medium to obtain the desired final cell density (*see* **Note 5**). The whole procedure for isolating embryogenic and viable microspores is schematized in Fig. 5a as a roadmap, and the time course for the entire process is illustrated in Fig. 5b.

Next, microspores are plated using a bi-layered induction culture medium, where each of the 35 × 10 mm Petri dishes containing a solid layer of M8M medium receives on its surface a 80–125 μL layer of liquid M8M medium containing the resuspended, freshly isolated microspores.

The Petri dishes containing the freshly plated microspores should not be shaken or even inclined, as this will change the surface of the dropped cell suspension, which in turn will alter the targeted density of the plated microspores.

The Petri dishes containing the plated microspores are sealed with Parafilm™ and stored at 26–28 °C, in the dark.

3.7 Microspore Embryogenesis

To control microspore development in culture, a dissecting stereo microscope, with zoom and external or attached light source and magnifying power up to 160× is needed. Using this device, after 7–10 days, cell divisions can be observed (Fig. 6), and, by the 15th to 18th day, masses of embryo-like structures (ELSs) will cover the medium surface (Fig. 7).

3.8 Transferring the ELSs

The size of the ELSs developing on the medium surface may vary depending on density. Thus, either when they have reached a size of approximately 1 mm, or when they have the normal anatomical structure (Fig. 6), they are ready to be transferred to the regeneration medium. The operation is performed in a laminar flow hood, assisted by a binocular stereo microscope (60–160×). A sterilized spatula is used to scoop up several masses of aggregated ELSs and to spread them evenly on the surface of the regeneration medium. Care must be taken to space the ELSs and to ensure that they are in contact with surface of the medium (Fig. 8). Regeneration plates are sealed with Parafilm™ after filling with ELSs.

Transferred ELSs need to be exposed to light, provided, e.g., by Standard™ 32 W bulbs and kept at room temperature (20–25 °C). Under these conditions, small green shoots

a

1) Cut 5-6 pre-treated spikes into 2-3 cm pieces, put them in the blender chamber with fresh ≈ 50 ml mannitol; blend for 20-25 sec (low speed)

2) Filter blended product with 100 μm filter, split into two 50 ml tubes

Centrifuge at 1500 rpm for 4 min

3) Throw away supernatant, re-suspend pellets with fresh mannitol and gather them into one 50 ml tube

Centrifuge at 1500 rpm for 4 min

4) Re-suspend pellet into 10-12 ml fresh mannitol (→ 15 ml tube)

Centrifuge at 1500 rpm for 4 min

5) Re-suspend pellet with 5-8 ml of cold 21% maltose; shake

Overlay suspension with 0.65 ml of 0.3 M cold mannitol

Centrifuge at 750 rpm for 18 min

6) Recover the resulting band, and add ≈ 10 ml of mannitol

Centrifuge at 1500 rpm for 4 min

7) Add 100 to 150 μl of liquid culture medium to the pellet and resuspend

8) Aliquot 10 μl: → hemocytometer / microscope

9) Adjust microspore density to 1.0 to 1.5 x 106 cells/ml

10) Plate microspores onto solid M8M induction medium (80-125 μl per plate)

Fig. 5 (**a**) Roadmap to embryogenic microspore isolation. (**b**) Time course of a typical isolated microspore culture experiment

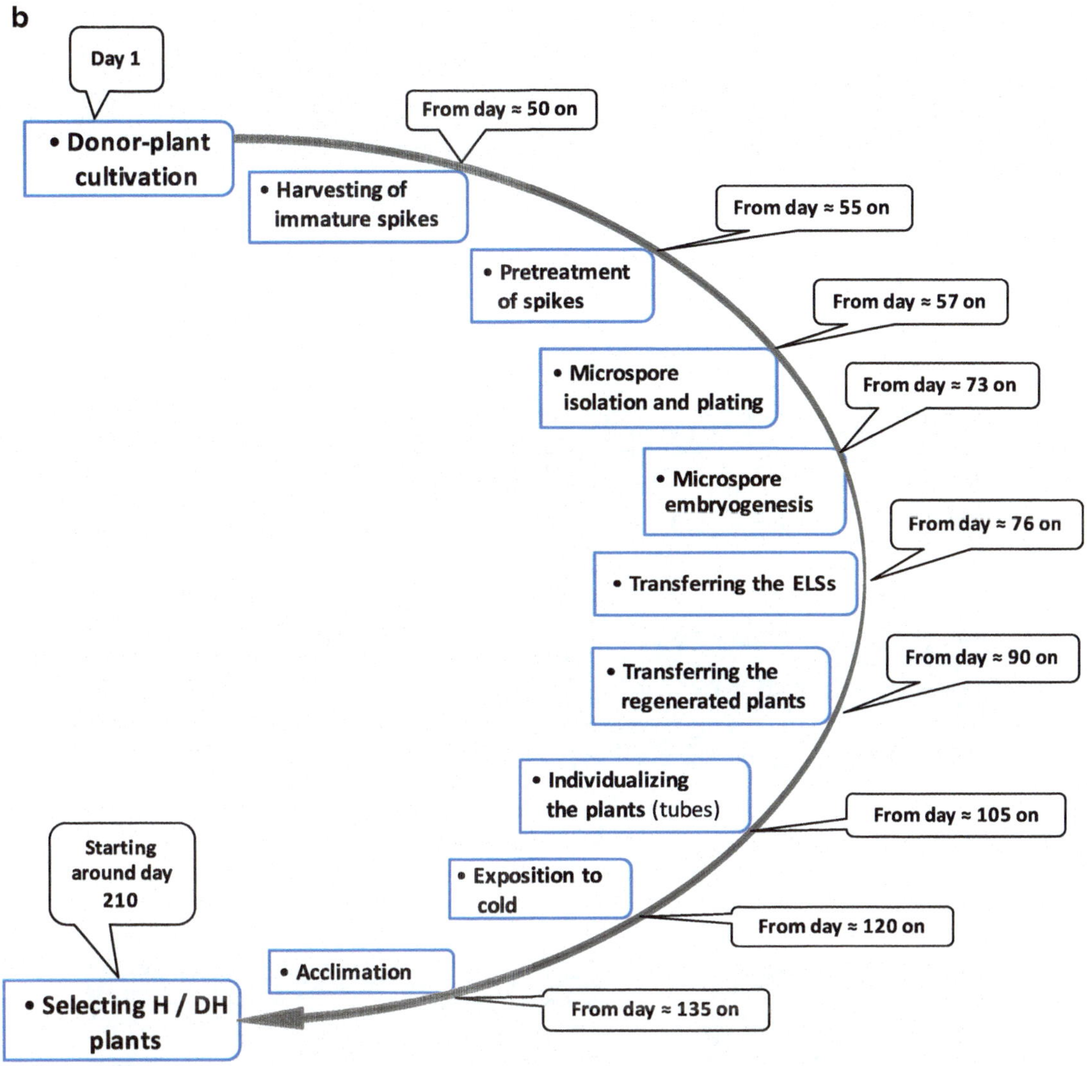

Fig. 5 (continued)

developing from the ELSs should start to appear 2–5 days after the transfer, and small plants should become visible to the naked eye after 10–12 days (Fig. 9) (*see* **Note 6**).

3.9 Plant Regeneration

ELSs derived from isolated microspore culture in barley give rise to bipolar structures. This is consistent with the "gametic embryogenic pathway" which has been described for plant regeneration in this system [22, 23]. Thus, small plants reaching a size of about 2–3 cm have normal roots developing synchronously with their aerial part, and can be transferred to germination medium, in 50 mL capacity tubes each filled with 20 mL of medium. This allows individualizing each plant from its derived population of

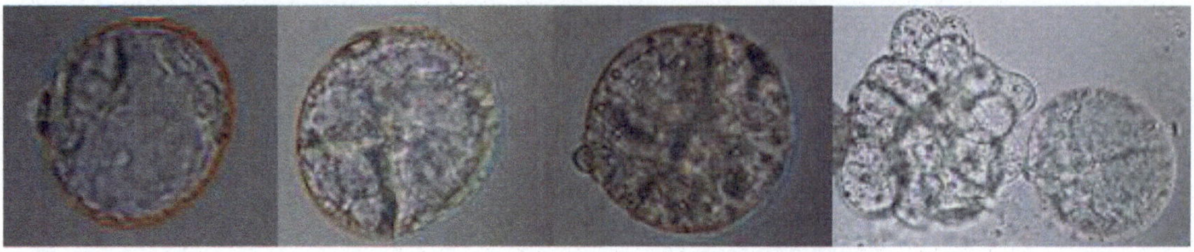

Fig. 6 The course of microspore embryogenesis: internal divisions become evident just a few days after plating, until multicellular masses arise. Days from plating (left to right) = 4, 6, 8 and 10

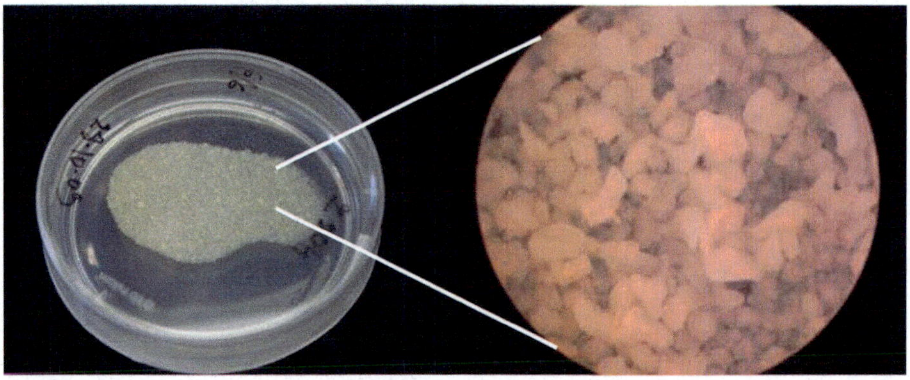

Fig. 7 Microspore embryogenesis is evident around 15–18 days after plating (left), and aggregated masses of ELSs are ready to be transferred to regeneration medium (zoomed; right)

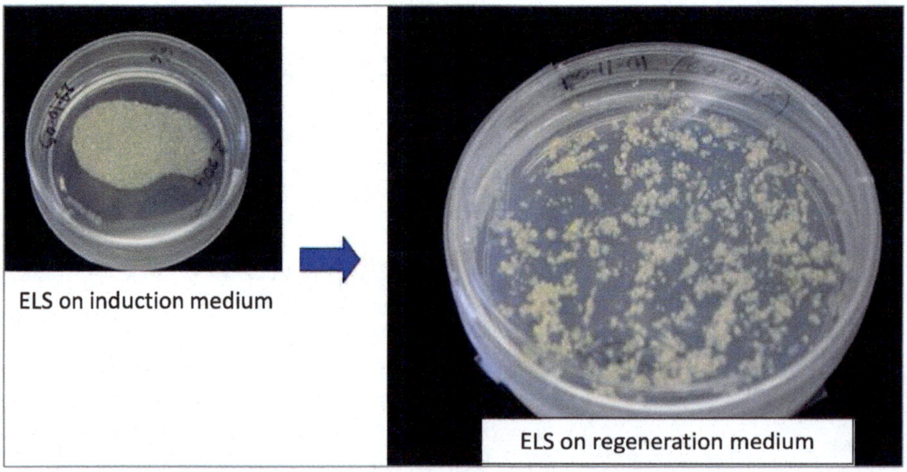

Fig. 8 The transfer of ELSs from the induction medium (left) is performed with a spatula and these are spread onto the surface of the regeneration medium (right)

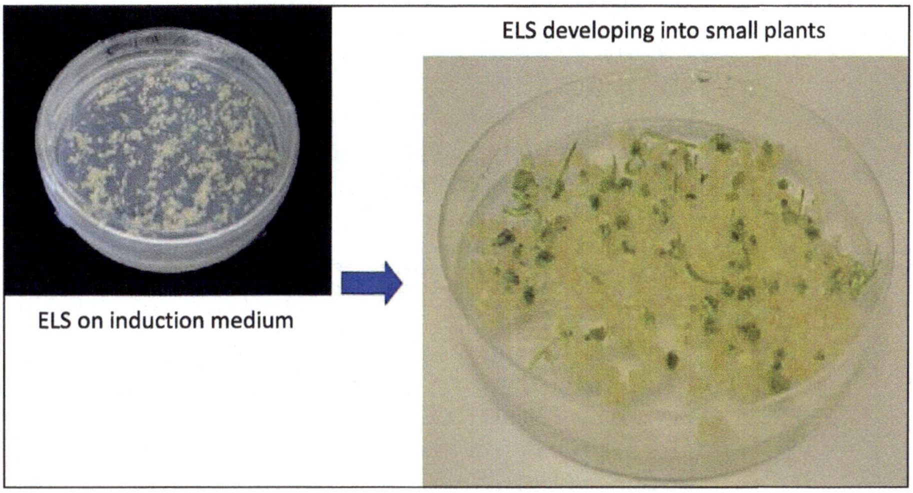

Fig. 9 Once transferred to the regeneration medium and exposed to light, small green plants begin to sprout from the ELSs

Fig. 10 Regenerated barley plants transferred to germination medium, and individualized in tubes

regenerated plants, and to handle it separately thereafter during ulterior acclimation and culture in the greenhouse. Each tube receiving a plant is capped with the corresponding plastic cap, and sealed with a band of Parafilm™ to avoid rapid desiccation of the medium.

Next, tubes and plants are exposed to light, under the same conditions described above for the transferred ELSs (Fig. 10). Once a good development of the root system is observed, usually

10–15 days after they have been transferred, plants are ready to enter the acclimation phase.

3.10 Acclimation

Acclimation of barley regenerated plants starts by giving them a 15-day treatment of cold (4 °C) with light (photoperiod of 16/24 h provided by the same type of fluorescent tubes described previously). In our case, this is performed in a cold chamber. Our observations indicate that this "cold treatment" helps to maximize the tillering of the plants, without causing losses due to physiological damage.

After a fortnight of exposure to cold, plants are ready to be transferred to pots. We use plug trays (e.g., of the type BM 24 Count Ellepot Tray -60 mm-, Blackmore, Kingsville, ON, Canada), filled with the same substrate that was described to grow the donor plants.

Acclimation of regenerated barley DH plants normally is not an issue: once the plants have formed conspicuous roots and shoots, they can be transferred to small pots and taken to the growth chamber under the conditions used for the cultivation of donor plants. We usually take care to remove every residue of agar from the roots, washing them generously with tap water, as the presence of sugar may favor the development of fungi or bacteria in the substrate. Every multicell tray carrying a set of plants is covered with a translucent, plastic cover, in order to maintain a high humidity during the first 2–4 days after the plants have been installed in the growth chamber.

3.11 Spontaneous Chromosome Doubling

Even if microspores are haploid cells, an important fraction of the resulting embryos developed in culture will have spontaneously doubled their haploid genome, thus giving rise to doubled-haploid (DH) plants. In barley, high frequencies of spontaneous doubling have been reported, at times in association with the type of pretreatment, genotype and culture conditions [24, 25]. Using the protocol described here, we have typically observed a frequency of about 75% of spontaneous doubling. It is thus unnecessary to use a colchicine treatment to double the chromosome complement of regenerated plants. Instead, we simply let the regenerated plants grow to maturity and, at that time, we discard sterile, haploid individuals.

4 Notes

1. To grow donor plants, the use of growth cabinets is highly recommended (e.g., of the type CONVIRON Model PGR15), as these allow the user to control both temperature and lighting conditions. In addition to their contribution to obtaining healthy donor plants, highly standardized growth

conditions can be considered a key factor to success, as this contributes substantially to the reproducibility of an adjusted, optimized protocol.

2. For obvious reasons, six-row genotypes will yield a larger number of microspores per spike than two-row genotypes. Nonetheless, the latter generally show a better response in IMC such that the reduced number of microspores does not impede the production of DH plants.
3. The optimal stage for harvesting the spikes is strongly influenced by genotype. Additionally, this index is also affected by the order of the spike (first, second, third . . .tenth, etc.), and by the growing conditions of the donor plant as well. Preliminary work exploring how various inter-ligule intervals at harvest impact the yield in isolated embryogenic microspores is recommended for each donor genotype before initiating any large-scale work. Another useful observable change associated with the optimal harvest stage is the very beginning of the thickening of the stalk in the absence of any lignification of the spike.
4. The size of the resulting embryogenic fraction of microspores is strongly influenced by three main factors: (1) the genotype, (2) the harvest stage, and (3) the type of pretreatment (e.g., its intensity and duration). For particular genotypes and culture conditions of the donor plants, experimenting on any modification of the variables of the pretreatment leading to an increase in the yield of embryogenic microspores per spike is always desirable.
5. It is strongly recommended that the final preparation of isolated embryogenic microspores is plated at a high density. We have been able to note that ELS formation increases considerably when a density of at least 10^6 cells/mL is plated. Once this is done, care must be taken not to shake or even incline the Petri dish. The idea is to not disturb in any way the plated drop of resuspended microspores, as an increase in the surface occupied by the liquid containing the microspores will result in a reduction in the density of the cells on the plate.
6. From the total population of microspores present in the spikes, only a relatively small fraction of them will engage in the embryogenic pathway in response to the pretreatment (e.g., $\leq$1%). Nonetheless, because a single microspore extraction from five to eight properly pretreated spikes can provide more than 500,000 viable, embryogenic microspores, the $\leq$1% fraction of responding microspores allows a sufficient number of embryo-like structures, or ELSs to be obtained. Given the appropriate culture conditions, these are capable of regenerating a few to several hundred haploid and doubled-haploid plants.

Acknowledgments

We wish to express our gratitude to Suzanne Marchand, Martine Jean, and Isabelle Clermont from the Département de phytologie of Université Laval for their help, support, and timely suggestions.

References

1. Weyen J (2009) Barley and wheat doubled haploids in breeding. In: Touraev A, Forster BP, Jain MS (eds) Advances in haploid production in higher plants. Springer Science+Business Media BV, Dordrecht, pp 179–187
2. Forster BP, Thomas WTB (2005) Doubled haploids in genetics and plant breeding. Plant Breed Rev 25:57–88
3. Chan SWL (2010) Chromosome engineering: power tools for plant genetics. Trends Biotechnol 28:605–610
4. Li H, Devaux P (2005) Isolated microspore culture overperforms anther culture for green plant regeneration in barley (*Hordeum vulgare* L.). Acta Physiol Plant 27:611–619
5. Devaux P, Kasha KJ (2009) Overview of barley doubled haploid production. In: Touraev A, Forster BP, Jain MS (eds) Advances in haploid production in higher plants. Springer Science +Business Media BV, Dordrecht, pp 47–63
6. Mirzaei M, Kahrizi D, Rezaeizad A (2011) Androgenesis and spontaneous chromosome doubling in *Hordeum vulgare* L. researches of the first international conference (Babylon and Razi Universities). Euphrates J Agric Sci 3 (9):248–252
7. Asif M (2013) Progress and opportunities of doubled haploid production. Series: SpringerBriefs in Plant Science 6, 75 pp
8. Li H, Devaux P (2003) High frequency regeneration of barley doubled haploid plants from isolated microspore culture. Plant Sci 164:379–386
9. Letarte J.E., Simion MM,·Kasha KJ. (2006) Arabinogalactans and arabinogalactan-proteins induce embryogenesis in wheat (*Triticum aestivum* L) microspore culture Plant Cell Rep 24: 691–698
10. Sorvari S, Schieder O (1987) Influence of sucrose and Melibiose on barley anther cultures in starch media. Plant Breed 99:164–171
11. Hoekstra S, van Zijderveld MH, Heidekamp E et al (1993) Microspore culture of *Hordeum vulgare* L.: the influence of density and osmolality. Plant Cell Rep 12:661–666
12. Wojnarowiez G, Caredda S, Devaux P et al (2004) Barley anther culture: assessment of carbohydrate effects on embryo yield, green plant production and differential plastid development in relation with albinism. J Plant Physiol 161:747–775
13. Burger DA (1988) Guidelines for autoclaving liquid media used in plant tissue culture. HortSci 23:1066–1069
14. Kasha KJ, Simion E, Oro R et al (2001) An improved in vitro technique for isolated microspore culture of barley. Euphytica 120:379–385
15. Devaux P, Pickering R (2005) Haploids in the improvement of Poaceae. In: Palmer CE, Keller WA, Kasha KJ (eds) Haploids in crop improvement II. Springer-Verlag, Berlin, pp 215–242
16. Wheatley WG, Marsolais AA, Kasha KJ (1986) Microspore growth and anther staging in barley anther culture. Plant Cell Rep 5:47–49
17. Hoekstra S, van Zijderveld MH, Louwerse JD et al (1992) Anther and microspore culture of *Hordeum vulgare* L. cv. Igri. Plant Sci 86:89–96
18. Huang B, Sunderland N (1982) Temperature-stress pretreatment in barley anther culture. Ann Bot 49(1):77–88
19. Jähne A, Lörz H (1995) Cereal microspore culture. Plant Sci 109:1–12
20. Esteves P, Belzile F (2014) Improving the efficiency of isolated microspore culture in six-row spring barley: I-optimization of key physical factors. Plant Cell Rep 33(6):993–1001
21. Esteves P, Clermont I, Marchand S et al (2014) Improving the efficiency of isolated microspore culture in six-row spring barley: II-exploring novel growth regulators to maximize embryogenesis and reduce albinism. Plant Cell Rep 33 (6):871–879
22. de Faria Maraschin S (2005) Androgenic switch in barley microspores. PhD thesis, Universiteit Leiden
23. Olmedilla A (2010) Microspore embryogenesis. In: Pua E-C, Davey MR (eds) Plant developmental biology—biotechnological

perspectives, vol 2. Springer-Verlag, Berlin Heidelberg, pp 27–44

24. Oleszczuk S, Sowa S, Zinny J (2006) Androgenic response to preculture stress in microspore cultures of barley. Protoplasma 228:95–100

25. Shim YS, Kasha KJ, Simion E et al (2006) The relationship between induction of embryogenesis and chromosome doubling in microspore cultures. Protoplasma 228:79–86

Chapter 6

TILLING in Barley

Matthias Jost, Miriam Szurman-Zubrzycka, Katarzyna Gajek, Iwona Szarejko, and Nils Stein

Abstract

TILLING (Targeting Induced Local Lesions IN Genomes), a popular reverse genetics approach in barley research, combines plant mutagenesis with efficient mutation detection for studying biological function of a specific gene. The high mutation frequency within a TILLING population principally enables the identification of induced variations in (almost) all genes of a given species (more precisely a given genotype of a species) of interest, which can be tested for their functional impact on morphological and/or physiological characteristics of the plant. Several TILLING populations induced by chemical mutagenesis were established for barley (Talame et al., Plant Biotechnol J 6:477–485, 2008; Gottwald et al., BMC Res Notes 2:258, 2009; Caldwell et al. Plant J 40:143–150, 2004) and showed the possibility for adapting protocols to develop further populations. This chapter describes a chemical mutagenesis protocol for barley seeds and two independent procedures for efficient single nucleotide polymorphism (SNP) detection in a large number of mutagenized plants either by slab-gel- or capillary gel-based electrophoreses on the LI-COR 4300 DNA Analyzer and the AdvanCE FS96 instruments, respectively.

Key words PCR, TILLING, LI-COR, SNP detection

1 Introduction

TILLING (Targeting Induced Local Lesions IN Genomes) is a general reverse genetics strategy that combines mutagenesis of a large population of plants with an efficient, high-throughput detection system for the identification of point mutations in principally any gene of interest in a given species [1]. Polymerase chain reaction (PCR) is used to target specifically nucleotide changes in any DNA fragment of interest, e.g., genes or regulatory elements. High dosage mutagenesis is applied to induce a maximally useful density of mutations (balance between efficiency of mutation detection and accumulation of lethal mutations) within a TILLING population. As a consequence, screening for mutations in the TILLING population will potentially allow detecting a series of induced alleles for a targeted locus, which then can be further analyzed for their impact

Wendy A. Harwood (ed.), *Barley: Methods and Protocols*, Methods in Molecular Biology, vol. 1900,
https://doi.org/10.1007/978-1-4939-8944-7_6,

on protein function ultimately resulting in morphological or physiological changes during plant development. The same strategy of mutation detection has been also applied for discovering naturally occurring variations in diverse germplasm collections, called EcoTILLING [2].

The TILLING approach consists of three steps: (1) establishment of a mutagenized population, (2) detection of mutations within a targeted sequence followed by (3) phenotypic analysis of the identified mutants [3]. The development of the population starts with chemical mutagenesis, introducing small lesions in the genome at high frequency. Frequently used substances are ethylmethanesulfonate (EMS), *N*-methyl-*N*-nitroso urea (MNU), or sodium azide (SA, NaN_3). EMS and MNU cause predominantly single base pair transitions from G/C to A/T through alkylation of the guanine residues [4]. SA also causes high frequency of point mutations—its mutagenicity depends on pH and its action is mediated through the production of an organic metabolite of azide [5]. EMS, MNU, and SA had been successfully applied in barley for establishing TILLING populations [6–9]. The protocol of chemical mutagenesis is species and genotype dependent and has to be empirically tested and optimized to find the right balance between mutation rate and sterility-, lethality rate [10].

Individuals grown from mutagenized seeds are referred to as the first mutant generation (M1) which has to be propagated through self-pollination to yield the actual TILLING population. This is required to purify for true breeding mutations and remove somatic mutational load which is not transmitted through the germline. The resulting second generation (M2) plants provide the source of DNA for mutation screening and for harvesting M3 seeds. These will be stored for future phenotypic progeny testing of mutant plants selected by TILLING.

The principle for the detection of induced SNPs in mutagenized plants relies on the formation of hetero-duplexes at mismatch positions between pairing polymorphic DNA fragments during PCR. This requires mixing of DNA from a mutant as well as a wild-type genotype to ensure also the identification of homozygous mutations. Usually this is achieved by pooling of different DNA samples from independent mutants of the same population, since, depending on mutation density, the chance of finding mutations at the same locus in two plants is low and it is usually required to screen by PCR between one or several thousands of M2 individuals for identifying independent mutations of the same gene. To reduce labor and cost, DNA of multiple plants [e.g., DNA from 8 individuals; 7] will be pooled which reduces the number of required PCR reactions by a similar factor. Two- or higher degree dimensional pooling of DNA allows furthermore the direct identification of individual mutants even from pooled template DNA [11]. The initially published TILLING procedure used denaturing high-performance liquid chromatography (DHPLC) for SNP detection in DNA pools by differential melting kinetics in homo- and heteroduplexes of DNA [1]. Subsequently the use of single-strand specific nucleases [e.g., Cel I, 12] were introduced to the method for cleaving single base pair mismatches in heteroduplex DNA fragments derived from PCR amplification [13]. The *CEL* I enzyme, which can be purified directly from celery juice extract, boosted the TILLING mutation detection providing a cost-efficient strategy that can be applied also in many small and less-equipped laboratories [11, 12]. Obtained digestion fragments visualized with an appropriate gel-based system enables the fast and cost-efficient detection of plants carrying mutation. PCR amplification followed by Sanger sequencing by using DNA of identified mutated plants allows the determination of the exact position and the status of zygosity of the mutant allele.

This chapter on TILLING in barley is reporting first the optimization of EMS-based chemical seed mutagenesis (Subheadings 3.1 and 3.4) followed by the description of two high-throughput and relatively low-cost mutation detection methods relying on heteroduplex-based mutation detection using either LI-COR slab-gel electrophoresis (Subheadings 3.8–3.13) or the AdvanCE 96 FS capillary gel electrophoresis system (Subheadings 3.14–3.17), respectively. This chapter is then closed by the describing strategies for down-stream de novo prediction of potential functional impact of the identified mutations (Subheading 3.19) as well as some introduction to phenotyping as a method of testing the effect of the identified variants (Subheading 3.20).

2 Materials

Prepare all solutions with ultrapure water (deionized, 18 MΩcm at 25 °C, abbreviated as ddH_2O in the text). The reagents reported for PCR may be replaced by alternative products from other suppliers and do not imply exclusivity for the success of the methods. Store solutions at room temperature (20–25 °C) or follow guidelines of the manufacturers for suitable storage of reagents.

2.1 Material for Seed Mutagenesis

1. Barley seeds (*see* **Note 1**).
2. EMS—ethyl methanesulfonate (*see* **Note 2**).
3. Distilled water—dH_2O.
4. 100 mM Na_2SO_3 (neutralizing solution).
5. Planting pots (e.g., square pots 13 cm × 13 cm × 13 cm).
6. Soil, vermiculite, and sand.

2.2 PCR Components

1. The Taq DNA Polymerase (Qiagen Cat No./ID: 201203) is routinely used for PCR. The product includes Taq polymerase [5 units/μL], ten times concentrated (10×) PCR buffer, 5× Q-solution (enhancer) and 25 mM $MgCl_2$ solution.
2. Dilute the dNTP Mix [10 mM] (ThermoFisher Scientific #R0192) with ddH_2O (1:2) to 5 mM working solution.
3. Dilute the primers to 10 mM. For mutation detection with AdvanCE FS96 instrument non-labeled primers are used. For mutation detection with LI-COR instrument use labeled primers (IRDye-700-labeled forward primer and IRDye-800-labeled reverse primer).

2.3 Gel Electrophoresis

1. Universal Agarose.
2. 0.5 M EDTA: Dissolve 93.06 g EDTA [M = 372.24 g/mol] in around 400 mL H_2O. Adjust to pH = 8.0 with 4 M NaOH. Bring to a volume of 500 mL and check the pH again (*see* **Note 3**).
3. 1 M Tris/HCl: Dissolve 121.14 g Tris [M = 121.14 g/mol] in 750 mL H_2O. Fill up the solution to 1 L final volume and adjust the pH to 8.0 with HCl.
4. 10× TBE buffer: Dissolve 108 g Tris base [M = 121.14 g/mol], 55 g boric acid, and 40 mL 0.5 M EDTA [pH = 8.0] in a total volume of 1 L. Dilute to 1× as working solution for gel preparation and for use as electrophoresis buffer.
5. 10× Loading Dye: 40 mL 100% glycerol, 40 mL 0.5 M EDTA [pH = 8.0], and 0.3 g bromophenol blue in a total volume of 100 mL.
6. TE-buffer (10 mM Tris–HCl, pH 8.0; 1 mM EDTA, pH 8.0): Mix 1 mL Tris/HCl and 500 μL EDTA (pH = 8.0) and fill the solution with ddH_2O up to 100 mL total volume.

2.4 Fragment Analysis with the Use of LI-COR 4300 DNA Analyzer (See Note 4)

1. *CELI*: Surveyor® Mutation Detection Kit, (Transgenomic, 706,025) or 1× CJE (Celery Juice Extract) (isolation of CJE and preparation of 1× CJE is described in [11]).
2. Mixture of ethanol and sodium acetate: Mix 50 mL ethanol (100% v/v, p.a. grade) and 500 μL of 3 M sodium acetate.
3. Ethanol 70% (50 mL): Mix 35 mL ethanol (100% v/v, p.a. grade) and 15 mL ddH_2O.
4. PRE-polymerization MIX: Dissolve 8.4 g urea [M = 60.06 g/mol], 2 mL 10× TBE buffer, 4 mL 30% acrylamide/bis-acrylamide solution and bring the total volume to 20 mL with ddH_2O.
5. TEMED (*N*,*N*,*N′*,*N′*-Tetramethylethylenediamine).
6. 10% APS (5 mL): Dissolve 0.5 g APS (ammonium peroxodisulphate) in 5 mL ddH_2O. Prepare 150 μL aliquots.

2.5 Fragment Analysis on the AdvanCE FS96 Instrument

Commercial Kits from the Company Advanced Analytical Technologies, Ankeny, USA, are used for capillary electrophoresis on the AdvanCE FS96 instrument.

1. Mutation Discovery 910 Gel Kit—(DNF-910T): Consisting of FA dsDNA Gel 35–1,500 bp (DNF-810-0500), Intercalating Dye (DNF-600-U030), 5X 930S dsDNA Inlet Buffer (DNF-355-0300), 5X Capillary Conditioning Solution (DNF-475-0100), Mutation Detection DNA Ladder (FS-SLR480-0001—not used for analysis), 35 bp and 5000 bp Markers (FS-SMK480-0003—not used for analysis) and a Mineral Oil Dropper Bottle (FS-SMO15).
2. dsDNA Cleavage Kit (DNF-480-3000) consisting of T-Digest Buffer—(FS-CLKTBF), dsDNA Cleaveage Enzyme—(FS-CLKENZ), and Dilution Buffer—(FS-CLKDBE).
3. Individual DNA fragments as size standards: NoLimits 100 bp, 250 bp, 500 bp, 750 bp, 1000 bp, 1500 bp, 2000 bp DNA Fragments [0.5 μg/μL], respectively (Thermo Fisher Scientific).
4. Capillary Storage Solution (GP-440-0100).

3 Methods

3.1 EMS-Based Chemical Mutagenesis

Remember that chemical mutagens may cause serious risks for your own health. Therefore, take proper safety precautions. Preparation of mutagen solutions and all steps of mutagenesis must be performed under the chemical fume hood. Use lab coat and disposable gloves. Collect all solutions containing the mutagen after use for decontamination and disposal in the appropriate waste containers.

3.2 Calibration of Optimal Dose of Mutagen

Different barley genotypes may respond with varying sensitivity to chemical mutagenesis. Before starting a large-scale TILLING population development it is recommended to calibrate for the optimal mutagen dose (mutagen concentration and duration of treatment) for the chosen genotype.

1. Presoak seeds in distilled water for 8 h (or overnight). To test four different mutagen doses, 1500 seeds are needed—300 seeds for each mutagen dose and 300 seeds for control.
2. Prepare four solutions with different EMS concentration—0.1%, 0.3%, 0.5%, and 1%. EMS should be dissolved in dH_2O. For 300 barley seeds 150 mL of mutagen solution is needed.
3. Before mutagenic treatment decant dH_2O from container with seeds and rinse the seeds in running tap water. It is important to completely remove water from the containers.

4. Prepare five glass containers and put 300 seeds in each of them. Remember to write on four containers the different mutagen dose (the fifth container mark as control).
5. Pour 150 mL of appropriate EMS solutions to the right containers. Pour 150 mL of dH_2O to the control container (*see* **Note 5**).
6. Place the containers on the shaker (<50 rpm) and perform mutagenic treatment at room temperature (20 ± 2 °C) for 16 h (*see* **Note 6**).
7. Decant the solutions from the containers. Be sure that the mutagen solutions are completely removed.
8. Add 150 mL of 100 mM Na_2SO_3 (neutralizing solution) to each container, mix, and incubate at room temperature for 5 min. Decant the solutions and repeat this step.
9. Rinse the seeds with running tap water at least three times (each time incubate in tap water for 10 min).
10. Perform germination tests and record growth and emergence reduction (the lethality of mutagenized plants should not exceed 50%).

3.3 Germination Tests

1. Prepare a mixture of moist soil and vermiculite in 2:1 ratio.
2. Prepare 15 square pots (13 cm × 13 cm × 13 cm). Remember to appropriately label all pots with the mutagen dose (three pots, as three repetitions, for each dose and three for the control).
3. Fill in the pots with mixture of soil and sand, leaving 5 cm space from the top.
4. Sow seeds to corresponding pots. Put 100 seeds in ten rows into each pot and cover it with 3 cm layer of moist sand.
5. Place the pots in growth chamber in controlled conditions (i.e., temperature 22 °C/20 °C—day/night, 16/8 photoperiod, light intensity 350–450 μmol m^{-2} s^{-1}) for 10 days. Remember to water the plants.
6. After 10 days count the number of emerged seedlings separately for each dose and each repetition and cut them at the sand level to measure the length of the first leaf. Based on three repetitions calculate an average number of emerged seedlings and an average length of the first leaf for each mutagen dose and the control.
7. Calculate the rates of reduction of germination and reduction of growth for each dose based on formulas:

   Germination reduction (%) = (1-average number of emerged seedlings for mutagen dose/average number of seedlings for the control) × 100

Growth reduction (%) = (1-average length of first leaf for mutagen dose/average length of first leaf for the control) × 100.

8. Evaluate the critical dose of the mutagen (the dose that leads to the 50–70% growth reduction at seedling stage). The mutagen dose used for TILLING experiment may be close to or a little below the critical dose (*see* **Note 7**).

3.4 Performance of Seed Mutagenesis and Maintenance of Mutagenized Population

1. Plan out the experiment. Evaluate the number of seeds for mutagenesis based on the preceding calibration experiment and the chosen mutagen dose (*see* **Note 8**).
2. Presoak seeds in dH_2O for 8 h (or overnight) in glass container.
3. Prepare an appropriate volume of EMS solution. For barley seeds it should be 0.5 mL/seed.
4. Decant dH_2O from container with seeds and rinse the seeds in running tap water. It is important to completely remove water from the container.
5. Perform mutagenic treatment. Pour into the container with seeds the calculated volume of EMS solution and incubate it on a shaker (<50 rpm) for 16 h at room temperature (20 ± 2 °C) (*see* **Note 6**).
6. Decant EMS solution from the container. Be sure that the solution is completely removed.
7. Add appropriate volume of 100 mM Na_2SO_3 to the container. The volume of neutralizing solution should be the same as the volume of EMS solution used for treatment. Mix and incubate at room temperature for 5 min. Decant the neutralizing solution and repeat this step.
8. Rinse the seeds with running tap water at least three times (each time, incubate in tap water for 10 min).
9. Plant M_1 seeds in pots containing soil mixed with vermiculite and grow them under optimal conditions (22 °C/20°C—day/night, 16/8 h photoperiod, light intensity 350–450 μmol m^{-2} s^{-1}) (*see* **Notes 9–11**).

3.5 Selection of TILLING Target

1. Select the target barley genes and determine their structure using bioinformatics tools that are based on annotated barley reference genome sequence [14]. Use, for example, Viro-BLAST [(http://webblast.ipk-gatersleben.de/barley/viroblast.php; 15] or *e!EnsemblPlants* (http://plants.ensembl.org/Hordeum_vulgare/Info/Index). The best target genes for TILLING are either single copy or belong to small gene families. Ideally they have no highly conserved paralogs in the genome and they have high conservation of encoded protein sequences among orthologs. Use the Monocots PLAZA 3.0

([http://bioinformatics.psb.ugent.be/plaza/versions/plaza_v3_monocots/; 16] database to identify the gene family and potential paralogs within the barley genome. PLAZA provides also information on gene ontology, including molecular function of a gene product, a biological process in which it is involved and a cellular component describing the physical location of the gene product.

2. Select TILLING amplicon. If the selected gene is 1.5 kb in length or shorter, the whole gene may be used as target for TILLING. If the candidate gene is longer, analyze the conservation of protein sequence in orthologous systems using, e.g., Pfam database [http://pfam.xfam.org/; 17]. Select the evolutionarily conserved fragment that increases the probability of a deleterious effect of missense mutation on the gene function [18]. Try to select fragments with no or a few introns, however remember that mutations in introns may also strongly affect the protein function (splice junction mutation may cause, e.g., intron retention and/or a premature STOP codon). Additionally, you may use CODDLE software (Codons Optimized to Discover Deletorious Lesions) (http://blocks.fhcrc.org/~proweb/coddle/) that is a tool for predicting the best gene fragment for TILLING analysis. CODDLE determines the regions in the protein sequence that are most likely to comprise the largest amount of deleterious lesions generated by G/C to A/T transitions.

3.6 PCR Optimization

We recommend performing a gradient PCR with a range of annealing temperatures to identify the optimal reaction conditions for the designed primer pairs. Use DNA of the mutant donor line, DNA of one eightfold pool, and a single DNA sample of the TILLING population as templates.

1. Prepare a reaction mix consisting of 2 μL 10× PCR buffer, 0.8 μL dNTP mix [5 mM], 0.6 μL of each primer [10 mM], 0.1 μL Taq polymerase [5 units/μL], and 1 μL DNA template [20 ng/μL]. Fill up the reaction mix with ddH_2O to a total volume of 20 μL per well (*see* **Note 12**). Remember that for mutation detection with AdvanCE FS96 instrument the unlabeled primers are used, whereas for mutation detection with LI-COR instrument use labeled primers (*see* **Note 13**).
2. Perform the PCR in a gradient thermocycler under following conditions: Initial denaturation for 5 min at 95 °C, then 35 cycles denaturation at 94 °C/45 s, 55–61 °C range of annealing temperatures/45 s and extension at 72 °C/60 s followed by a final extension step at 72 °C/5 min (*see* **Note 14**).
3. Prepare 1.5% (m/v) agarose gel dissolved in 1xTBE buffer containing 0.5 μg/mL ethidium bromide.

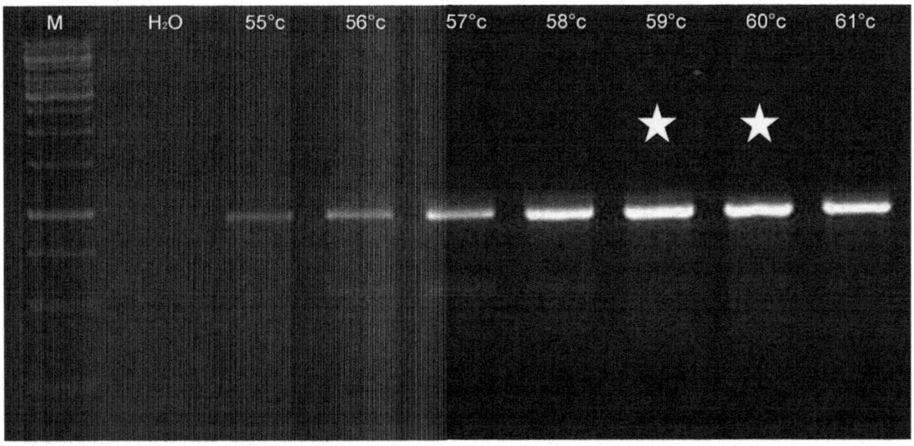

Fig. 1 PCR optimization by temperature gradient PCR. The efficiency of amplification was tested for a range of annealing temperatures and revealed the optimal conditions at 59 °C and 60 °C (labeled by asterisks)

4. Mix 2 μL of the PCR product with 2 μL Loading Buffer and run the electrophoresis [5 V/cm power supply for 1 h] to check if the expected specific fragments are obtained at any selected condition (Fig. 1, *see* **Note 15**).

3.7 Heteroduplex Formation

1. Prepare a master mix for 96 samples as described above (Subheading 3.6, **step 1**) (*see* **Note 16**).
2. Distribute the master mix into a 96-well PCR plate.
3. Transfer 1 μL [20 ng/μL] from the first plate of DNA pools of the TILLING population into the respective well of the prepared plate containing the PCR master mix.
4. Seal the plate, spin down briefly in a plate centrifuge (400 × *g*), and insert the plate to a PCR cycler.
5. Program the thermocycler to optimized run conditions and add the heteroduplex formation step at the end of the program after the final extension step. Heteroduplexes will form after an initial denaturation at 99 °C for 10 min, followed by a phase of decreasing incubation temperature at a rate of—0.3 °C/20 s, starting at 70 °C for 69 steps.
6. Prepare 1.5% agarose gel (w/v) in 1× TBE buffer including DNA staining solution.
7. Mix 2 μL PCR product with 2 μL of 1× loading dye and load the gel to check the success of PCR amplification.

3.8 Mutation Detection with LI-COR 4300 DNA Analyzer

Skip to Subheading 3.14 if you want to use the AdvanCE FA 96 system. The first step of the protocol provides two options for the digestion: Digestion by commercial *CEL* I enzyme (Subheading 3.9) or Digestion with Celery Juice Extract (CJE) containing *CEL* I endonuclease (Subheading 3.10).

3.9 Digestion by Commercial CEL I Enzyme

1. Transfer 7 μL of each sample to a new 96-well PCR plate using multichannel pipette.
2. Mix the Surveyor Enhancer S and the Surveyor Nuclease S at a ratio 1:1.
3. Add 1.2 μL of the mixture to each well.
4. Seal the plate, vortex, and spin down briefly in a plate centrifuge.
5. Incubate the plate at 42 °C for 15 min in a thermocycler.
6. Add 0.8 μL of the Stop Solution (supplied with the Kit) to each sample.

3.10 Digestion with Celery Juice Extract (CJE) Containing CEL I Endonuclease

1. Prepare a required volume of 0.1× CJE (2 mL/96-well plate) by diluting the 1× CJE solution with ddH_2O.
2. Add 20 μL of 0.1× CJE to each well of the PCR plate.
3. Seal the plate, vortex, and spin down briefly in a plate centrifuge.
4. Incubate the plate at 45 °C for 15 min in a thermocycler.

3.11 Ethanol Purification of Samples

1. Add 60 μL of 96% ethanol with 1% sodium acetate to each well containing 20 μL of digested products.
2. Seal the plate, vortex, and spin down for 20 min (4 °C, 2800 × *g*) in a plate centrifuge.
3. Remove the cover seal and decant the supernatant by turning the plate upside down on a stack of paper towels.
4. Add 30 μL of 70% ethanol to each well.
5. Seal the plate, vortex, and spin down for 15 min (4 °C, 2800 × *g*) in a plate centrifuge.
6. Remove the seal and decant the supernatant by turning the plate upside down on a stack of paper towels.
7. Insert the plate into the thermocycler and incubate the uncovered plate at 80 °C for 30 min (*see* **Note 17**).
8. Dissolve the pellets in 3 μL of stop buffer.
9. Proceed immediately to Subheading 3.12. or store samples in dark at 4 °C for no longer than 2 days or at −20 °C for no longer than 1 month.

3.12 LI-COR Electrophoresis

1. Thoroughly clean the inner sides of glass plates using ddH_2O and a paper towel.
2. Place the 0.2 mm spacers at the left and right edges of the larger glass plate and add the smaller one, generating a 0.2 mm deep space between both plates.
3. Set up the rail assemblies clamping the glass plates.

4. Prepare 6% denaturing polyacrylamide gel: add 20 μL of PRE--polymerization-MIX, 150 μL of 10% APS, and 15 μL of TEMED (*see* **Note 18**).
5. Pour the gel between the glass plates and remove any occurring air bubbles with a bubble hook (*see* **Note 19**).
6. Insert the comb from a flat side to form a brow of the gel, then set the casting plate and tighten the screws.
7. Wait an hour for complete polymerization of the gel.
8. Remove the casting plate and the comb.
9. Clean a brow of the gel using a syringe filled with ddH_2O.
10. Set the wells in the gel using a 48-well comb (or 64-well comb).
11. Clean the outer sides of the glass windows with ddH_2O.
12. Place the glass plates with a polyacrylamide gel into the LI-COR analyzer.
13. Set the upper and lower buffer containers, then fill them with 1× TBE buffer, place the electrodes, and connect them to the power supply.
14. Run pre-electrophoresis for 20 min (1300 V, 35 mA, 40 W).
15. In the meantime denature samples for 5 min at 94 °C.
16. When pre-electrophoresis is finished, disconnect the power source and the top electrode.
17. Rinse the wells using a syringe filled with electrophoresis buffer.
18. Load 1–1.5 μL of each sample and 0.5 μL labeled size marker to the gel.
19. Place back the top electrode, connect the power source.
20. Start the electrophoresis run (1300 V, 35 mA, 40 W, 4–5 h) (*see* **Note 20**).

3.13 Gel Image Analysis

1. Download the images from LI-COR 700 nm and 800 nm channels in order to detect plants with potential mutations (*see* Fig. 2).
2. Compare and mark pairs of additional signals per lane apart from the main bands of the PCR products on the images from 700 nm and 800 nm channels.
3. Determine the potential mutations by checking whether the sum of sizes of two bands which lie on the same lane on both images is equal to the size of the analyzed fragment.
4. Identify M_2 individuals carrying a potential mutation. 2D-pooling strategy allows direct identification of mutants since two pools carrying the same individual (only one possibility) should exhibit the same pattern of additionally observed fragments.

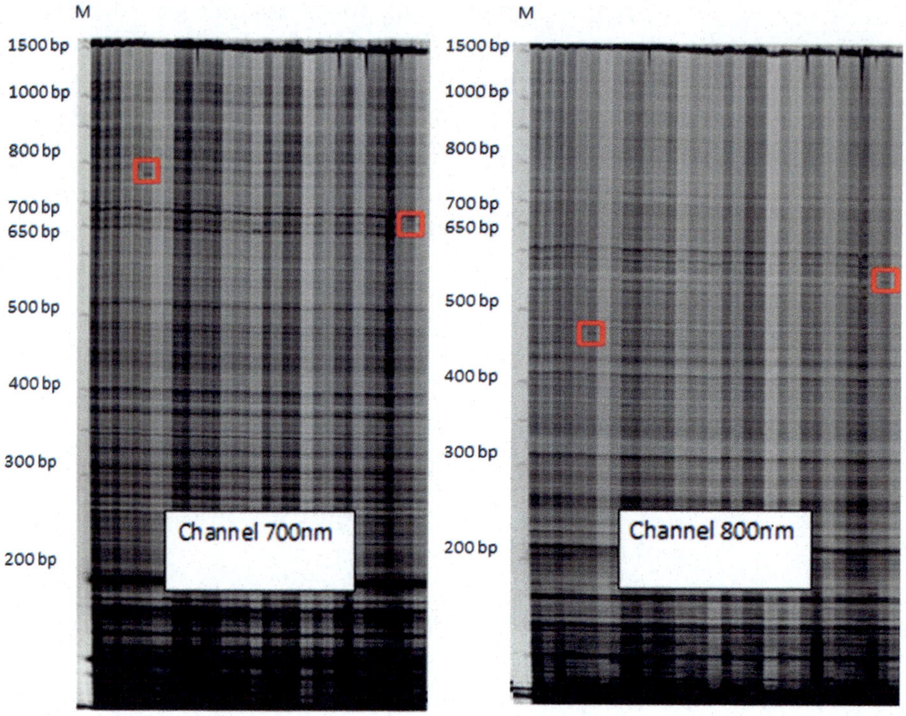

Fig. 2 *CelI*-based mutation detection using the LI-COR 4300 DNA Analyzer. Two potential mutations are highlighted by the red boxes on 700 nm and 800 nm channel images. For example, the CEL I enzyme digested the target sequence of one of individuals in pool 6 (lane 6) to two fragments—about 800 bp (visible in 700 nm channel) and 500 bp (visible in 800 nm channel). The size of fragments are estimated in comparison to the DNA Size Marker (M, visible in both channels) loaded on the left side of the gel

3.14 Mutation Detection with AdvanCE FS96 Fragment Analyzer

Skip the Subheadings 3.14–3.17 if you performed the mutation detection with the LI-COR 4300 DNA Analyzer and proceed to Subheading 3.18 for mutation confirmation.

3.15 Optimization of Heteroduplex Cleavage

Use the "dsDNA Cleavage Kit" to cleave DNA fragments carrying a mismatch.

1. Place two 12-well PCR strips into each of four tube/plate holding racks, respectively (4 racks = 4 incubation conditions, like shown in Fig. 3a).
2. Transfer 2 μL PCR product (*see* **Note 21**) of the first three rows (A1 to H3) from the PCR plate into the prepared 12 well strips (4 × 2), respectively.
3. The enzyme needs to be diluted at 1:125 in the digestion buffer. This is equivalent to 1.6 μL "dsDNA Cleavage Enzyme" and 198.4 μL "T-Digest Buffer" for one plate.
4. Mix 2 μL of the freshly diluted enzyme solution with the PCR products in the prepared strips by pipetting three times up and down.

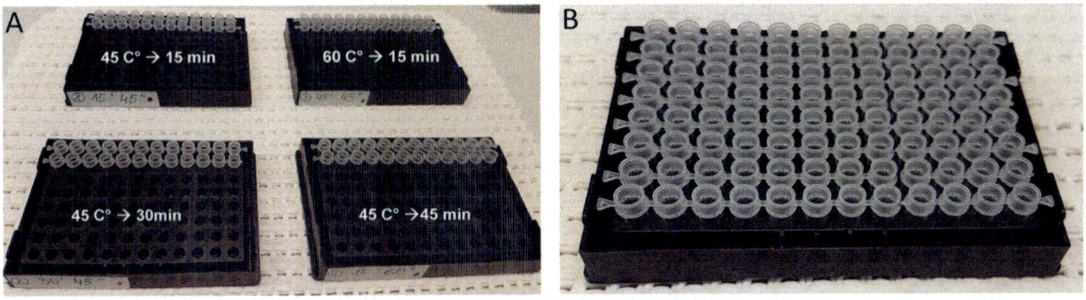

Fig. 3 Optimization of heteroduplex cleavage using four independent thermocycler settings. (**a**) Prepared racks carrying each two 12-well strips, respectively. (**b**) After incubation the samples can be combined and transferred to a 96-well PCR plate to process for analysis on the AdvanCE instrument (*see* Subheading 3.16)

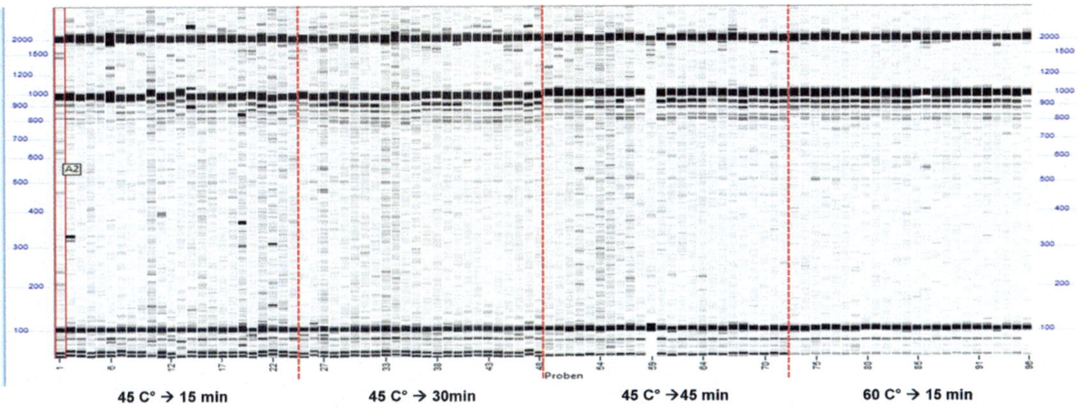

Fig. 4 Fragment analysis on Advance FS96 equipment. The figure shows an example of the same PCR products digested at four different temperature settings. The condition of 60 °C for 15 min (most right) revealed the best result (clear fragments and low background). The other three conditions have more background-digested products putatively caused by unspecific digestion

5. Incubate the reactions (always two strips) at 45 °C for 15, 30 or 45 min at 45 °C or 15 min at 60 °C, respectively, using, e.g., PCR cyclers.
6. Subsequently combine all strips together in one rack (Fig. 3b) and transfer the digested products into a new 96-well PCR plate.
7. Proceed with a test run on the AdvanCE FS96 system (*see* Subheading 3.16).
8. Analyze the result as described in the analysis part (*see* Subheading 3.17) and choose the optimal digestion conditions (less background, good contrast; *see* Fig. 4).
9. Prepare again the enzyme-mix (Subheading 3.15, **steps 3** and **4**) for the restriction digest and proceed with the optimized conditions on the entire plate.

3.16 Run on AdvanCE FS96

The following part of run setup is described for the AdvanCE FS96 instrument and the control software Advance System 1.01 (Advanced Analytical, Ankeny, USA). Please follow the instrument manual in case you have another instrument or software version.

1. Prepare DNA Ladder: Mix 100 μL 1× TE with a DNA fragment solution containing 3 μL of 100 bp, 250 bp, 500 bp NoLimits DNA fragments [0.5 μg/μL] and 5 μL of 750 bp, 1000 bp, 1500 bp, 2000 bp NoLimits DNA fragments [0.5 μg/μL], respectively. Transfer 4 μL of the ladder mix in an empty well of the sample plate (*see* **Note 22**).
2. Prepare the sample plate: Add 26 μL "Dilution Buffer" per well to each digested PCR product and mix carefully by pipetting up and down (*see* **Note 23**). Spin down briefly in a plate centrifuge (400 × *g*) to avoid air bubbles.
3. Prepare the marker plate: Add 5 μL [0.5 μg/μL] 100 bp and 5 μL [0.5 μg/μL] 2000 bp NoLimits DNA Fragments to 3 mL TE-buffer. Mix carefully and distribute 30 μL of the marker solution to each well of a 96 well PCR plate. Add a drop of mineral oil from the mineral oil bottle to each well to create an evaporation barrier (*see* **Note 24**).
4. Prepare the buffer plate (*see* **Note 25**): Dilute the "5× 930 dsDNA Inlet Buffer" to 1× and distribute 1000 μL of the dilution into each well of a 1 mL deep well plate (e.g., Costar 3958 96-Well Assay Blocks).
5. Prepare the gel matrix: Bring "Intercalating Dye" and gel to room temperature. Mix 5 μL of the "Intercalating Dye" with 100 mL of the dsDNA Gel solution (1:20,000 dilution) to prepare the gel matrix. Prepare the gel freshly before each run and calculate the amount according to number of plates you are going to analyze. Around 45 mL are required for each plate (*see* **Note 26**).
6. Prepare the Conditional Solution: Dilute the 5× Capillary Conditional Solution to 1× with H_2O.
7. Perform a "Full Conditioning" run (*see* **Note 27**): Open the control software by clicking on register card "system control" to select stage control for opening the instrument. Plate carrier of the instrument moves out. Insert the buffer plate in position 2 (front left *see* Fig. 5). Click on Full Conditioning and start the procedure.
8. Click on "One Sample Plate" to open the "Separation User Interface."
9. Edit Method: Perform capillary electrophoresis according to the following conditions: perform (i) Pre Run at 9 kV for 30 s, (ii) Marker Plate Injection at 5 kV for 20 s, (iii) Sample

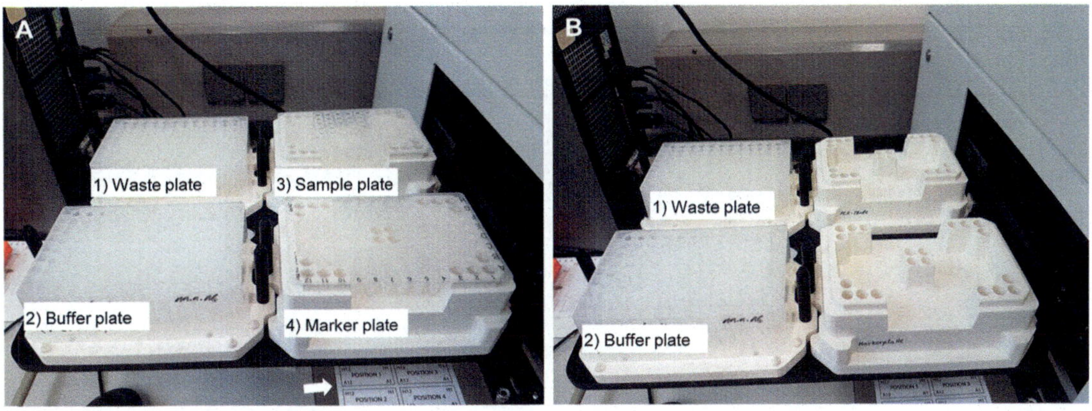

Fig. 5 Plate setup for run on AdvanceCE instrument. (**a**) Sample injection requires sample plate in slot 3 and the marker plate in slot 4. Take care of the required direction of the plate according to the scheme highlighted by the white arrow. (**b**) After injection is finished, the marker and sample plate have to be removed

injection at 7.5 kV for 45 s, and (iv) CE Separation at 9 kV for 90 min (*see* **Note 28**).

10. Click on "Load Sample Names file" to import your prepared sample table with the description of the samples, save data, and select directory for saving (Folder Location).
11. Open the control software by clicking on register card "system control" to select stage control for opening the instrument. Plate carrier of the instrument moves out.
12. Insert the prepared marker to position 4 and sample plate to position 3 (*see* Fig. 5).
13. Check if well A1 is located at the lower right site for each plate.
14. Start sample injection by clicking on "continuing." After 5–10 min, the injection is completed and the arm of the instrument moves out again. Remove the marker and sample plate. Click ok to start the run.

3.17 Data Analysis

1. Open the run file with the PROSize software (Advanced Analytical, Ankeny, USA).
2. Click on "Option" and open the "Set Global Configuration" and select under "Modus" "Mutation Analysis" and click on "Apply to All."
3. Click on advanced settings and change under the register card "Marker Analysis" the Marker settings for the lower and the upper DNA marker to 1000 RFU.
4. Change the color matrix to "White on Black" and optimize the contrast with the button in the upper control panel for your preferential view (*see* **Note 29**).

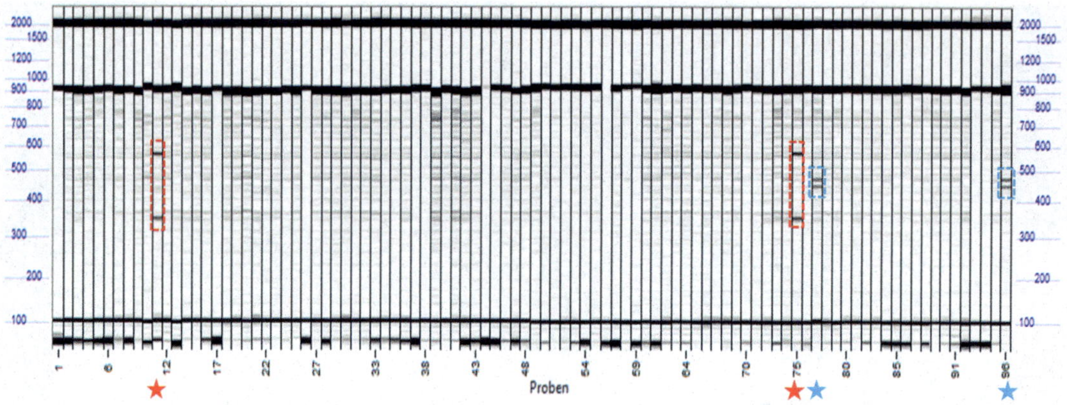

Fig. 6 Advance FS96-based detection of plants with mutation in the target fragment. The screenshot illustrates the result for a 96-well plate analysis shows two positive pools comprising a mutated plant (red and blue boxes and asterisks). Fragment of one plant in the pools of lane 11 and 75 (indicated by red stars) was digested by the restriction endonuclease in two fragments of 570 bp and 350 bp (indicated by red dashed boxes), respectively. A second mutated plant is included in pool of line 77 and 96 of which PCR product was digested into 490 bp and 480 bp fragments (indicated by blue stars). The product sizes were estimated in relation to the DNA size standard shown on the left and right side of the panel

5. Two pools with digested fragments of similar size enable the deconvolution to identify a single plant according to the 2D pooling strategy (Fig. 6). In case digestion fragments were observed only in a single pool, testing of all eight plants of the pool (*see* Subheading 3.18) could still lead to the identification of a mutated plant. In this case the mutant allele perhaps is present below the detection level in the pool of genotypes from the second dimension.

3.18 Mutation Confirmation

1. Amplify the target DNA fragment by PCR (*see* Subheading 3.6), under previously optimized conditions, from the candidate mutant plant and as well from the non-mutagenized wild-type genotype.
2. Send PCR fragments for Sanger sequencing or perform your own in-house Sanger sequencing.
3. Align the generated sequences with the original wild-type genotype and check sequence electrospherograms by help of an appropriate software (e.g., Sequencher 5.2.3, Gene Codes Corporation, USA) for sequence differences between the reference and mutant samples, respectively. A single peak is called as homozygous. Two overlying peaks at the same nucleotide position with approximately similar intensity indicate a mutation in a heterozygous state.

3.19 Bioinformatic Analysis of Mutations

1. Perform in silico translation for a mutant and a wild-type sequence in order to check the predicted effect of mutation.

Check if the mutation results in a synonymous or non-synonymous change, disruption or creation of a splice site, or the disruption of the open reading frame by introducing a premature stop codon. It may be done with the use of PARSESNP tool (http://blocks.fhcrc.org/blocks/uploads/proweb/).

2. If you identified many mutations in your gene of interest and some of them are at the heterozygous stage, you may need to prioritize which mutant lines should be used first for identification of homozygous mutants, seed increase and phenotyping. Analyze the conservation of the substituted amino acid residue in orthologous systems using Clustal Omega [http://www.ebi.ac.uk/Tools/msa/clustalo/; 19]. Mutations that lead to the change of the evolutionarily conserved amino acids in the functional domains may cause a deleterious effect to the gene function [18]. Some open source bioinformatic tools may be helpful to prioritize mutations, for instance SNAP^2 [https://rostlab.org/services/snap/; 20], HANSA [http://www.cdfd.org.in/HANSA/index.php?target=home; 21, 22], HOPE [http://www.cmbi.ru.nl/hope/input/; 23], or SIFT tool [Sort Intolerant From Tolerant; http://sift.jcvi.org/www/SIFT_seq_submit2.html; 24]. These programs assign the potential strength of identified mutations.
3. Silent mutations may also have impact on the protein. It is possible to assess a potential influence of a silent mutation on the splicing process with the use of ESEFinder 3.0 software (ExonicSplicingEnhancersFinder; [http://krainer01.cshl.edu/cgi-bin/tools/ESE3/esefinder.cgi?process=home; 25].
4. If a crystallographic structure model of the gene product of interest is available, it is possible to compare the 3D structure of the mutated and nonmutated proteins using, e.g., SWISS-MODEL [https://swissmodel.expasy.org/; 26]) or Phyre^2 homology-modeling approaches [http://www.sbg.bio.ic.ac.uk/phyre2/html/page.cgi?id=index; 27].
5. If you identified mutations in noncoding regions, check if they are located in the specific positions in introns, such as a polypyrimidine tract, a branch point, a donor- or acceptor-splice site. There is a high probability that such mutations may affect the splicing process and the function of the encoded protein.

3.20 Phenotyping

1. If the specific aspect of the plant phenotype is known that is controlled by the respective mutated gene, grow the M3 family of any identified mutant together with the wild-type genotype under controlled growth conditions (climate chamber or greenhouse) to inspect for expected differences in the trait of interest.

2. It is important to propagate from multiple individuals because of lethality effects of the mutation in a homozygous state or due to homozygosity in lethal background mutations which potentially could result in the loss of a family.
3. If the mutation was at a heterozygous state in M2, grow 16 plants of a given M3 progeny, genotype all and select homozygous and heterozygous individuals.
4. In case the function of the gene is unknown, several independent mutants should be analyzed to find multiple independent mutants sharing the same effect on the phenotype. This allows the differentiation of phenotypic effects caused by the mutant allele from potentially masking background mutations and avoids intensive backcrossing for background mutation removal.

4 Notes

1. We recommend to check the germination capacity of the seed stock selected for building the TILLING population before starting the mutagenesis experiment. It is important to discard any abnormal, injured, or diseased seeds.
2. Any other chemical mutagen with the potential of introducing single nucleotide lesions in the barley genome, such as MNU (*N*-methyl-*N*-nitrosourea) or sodium azide, may also be used for barley mutagenesis. The use of radiation or methods introducing larger lesions like deletions and translocations are not suited for developing a TILLING population because of the lack of compatibility with mutation detection by PCR.
3. EDTA will start to dissolve only close to the right pH value of 8.
4. Instead of LI-COR 4300 DNA Analyzer, other sequencers from the LI-COR company may also be used (e.g., 4200 LI-COR DNA Analyzer).
5. In the case of using other mutagens, calibration experiments for determining the appropriate doses should also be considered: for example, for MNU—0.5 mM, 1 mM, 1.5 mM, and 2 mM per 3 h (mutagen dissolved in dH_2O) and for sodium azide—1 mM, 1.5 mM, 2 mM, and 3 mM per 3 h (mutagen dissolved in a buffer containing KH_2PO_4 and H_3PO_4, $pH = 3$).
6. It is very important to perform mutagenesis at constant temperature conditions, because fluctuation of temperature may change the mutagenesis effectiveness.

7. A critical dose of mutagen leads to a low survival rate and a high degree of sterility of M_1 plants. In traditional plant breeding experiments, mutagen doses causing less than 30% growth reduction are usually applied. The aim of TILLING is to obtain a series of new alleles for every gene, therefore a very high density of mutations is needed and higher mutagen doses are usually used. Too low mutagen dose results in a low frequency of mutation and the necessity of development of a larger TILLING population which causes a higher follow-up cost, labor and time requirement. The somatic effects of chosen mutagen doses (germination and growth reduction rates) should be known in order to estimate the size of the M_1 generation required for creating a useful TILLING population.
8. We recommend to develop several thousands of M_2 plants in barley TILLING population. Each M_2 plant should be an offspring of an independent M_1 plant. This ensures a maximum of nonredundant mutations in the final population.
9. Described conditions are optimal for spring barley genotype. In case of winter genotypes it is recommended to vernalize 2-week-old seedlings at 4 °C for 8–10 weeks, then transfer them to the same conditions as spring genotypes.
10. Because of the chimeric structure of M_1 plants, the molecular mutation detection is performed in M_2 plants. It is important to carefully label all M_1 and M_2 plants and to create a seed bank for M_3 seeds (and for further generations).
11. It is very helpful to create a database containing information including phenotypes of plants in the TILLING population and characterization of any mutations found.
12. Optionally, 4 μL Q-solution can be added to the PCR reaction mix to improve the amplification of "recalcitrant" templates (e.g., high probability of secondary structures or GC-rich templates) according to manufacturer's information's (Qiagen, Hilden, Germany).
13. Fluorescently labeled PCR primers do not always perform in the same way as unlabeled ones, because the dye may change their conformations. To optimize the PCR performance it is recommended to mix labeled (*) and unlabeled primers in a ratio of 3*:2 for IRD700-labeled primers and in 4*:1 ratio for IRD800-labeled primers.
14. The range of annealing temperature needs to be adjusted to the melting temperatures of your designed oligonucleotides. The given example is suitable to melting temperatures of around 58–60 °C.
15. The success of the described mutant screen relies on a specific PCR amplification of the DNA fragment of interest. The PCR

fragment should have expected product size and an efficient amplification rate and no additional unspecific products. We recommend to sequence the PCR amplicon by Sanger sequencing to verify the amplified fragment reflects the targeted sequence and no secondary fragments.

16. Prepare the master mix for 100 samples to have a surplus to compensate for minor loss introduced by pipetting errors.
17. The plate should be incubated until all residual ethanol is evaporated and the samples are completely dry (30 min should be enough). Cover the plate using a box to avoid contamination and protect dyes.
18. Pour the gel immediately after addition of APS and TEMED.
19. You can prevent forming the air bubbles by tapping on the glass plates during pouring of the gel.
20. The time of electrophoresis depends on the PCR fragment length.
21. The manufacturer guideline recommends 25 ng/μL—100 ng/μL of uncut amplicon. Include some standard DNA controls at the gel electrophoreses to estimate the amount and adjust the volume of the PCR product used for the digest if required.
22. The DNA Ladder is used for calibrating the size of the analyzed DNA fragments. It is not necessary to run the DNA Ladder in each sample plate by using the same experimental conditions. The ladder can be imported in the ProSize software during the analysis step from previous runs.
23. A total volume of 30 μL/well is important for capillary injection.
24. Marker plate can be used for multiple runs and can be stored in −20 °C.
25. Buffer plate has to be refreshed every 2nd day.
26. Check before each run if sufficient gel is left and consider a surplus of gel for "Full conditioning" runs.
27. The device reports automatically after ten runs that a "Full Conditioning" is required. Perform the full conditioning also if the system was out of use for more than 10 days.
28. This is the configured run setup for our device after optimization. The setup configuration might be optimized individually.
29. We recommend to check multiple color and contrast configurations to increase the chance to identify mutated plants.

Acknowledgements

The work was supported by the European Regional Development Fund through the Innovative Economy for Poland 2007–2013, project WND-POIG.01.03.01-00-101/08 POLAPGEN-BD, task 22.

References

1. McCallum CM, Comai L, Greene EA, Henikoff S (2000) Targeted screening for induced mutations. Nat Biotechnol 18(4):455–457. https://doi.org/10.1038/74542
2. Comai L, Young K, Till BJ, Reynolds SH, Greene EA, Codomo CA, Enns LC, Johnson JE, Burtner C, Odden AR, Henikoff S (2004) Efficient discovery of DNA polymorphisms in natural populations by Ecotilling. Plant J 37 (5):778–786. https://doi.org/10.1111/j.1365-313X.2003.01999.x
3. Kurowska M, Daszkowska-Golec A, Gruszka D, Marzec M, Szurman M, Szarejko I, Maluszynski M (2011) TILLING: a shortcut in functional genomics. J Appl Genet 52(4):371–390. https://doi.org/10.1007/s13353-011-0061-1
4. Greene EA, Codomo CA, Taylor NE, Henikoff JG, Till BJ, Reynolds SH, Enns LC, Burtner C, Johnson JE, Odden AR, Comai L, Henikoff S (2003) Spectrum of chemically induced mutations from a large-scale reverse-genetic screen in Arabidopsis. Genetics 164(2):731–740
5. Gruszka DSI, Maluszynski M (2012) Sodium azide as a mutagen. In: Shu QY, Forster BP, Nakagawa H (eds) Plant mutation breeding and biotechnology. CABI International, Wallingford, pp 159–166
6. Talame V, Bovina R, Sanguineti MC, Tuberosa R, Lundqvist U, Salvi S (2008) TILLMore, a resource for the discovery of chemically induced mutants in barley. Plant Biotechnol J 6(5):477–485. https://doi.org/10.1111/j.1467-7652.2008.00341.x
7. Gottwald S, Bauer P, Komatsuda T, Lundqvist U, Stein N (2009) TILLING in the two-rowed barley cultivar 'Barke' reveals preferred sites of functional diversity in the gene HvHox1. BMC Res Notes 2:258. https://doi.org/10.1186/1756-0500-2-258
8. Caldwell DG, McCallum N, Shaw P, Muehlbauer GJ, Marshall DF, Waugh R (2004) A structured mutant population for forward and reverse genetics in barley (Hordeum vulgare L.). Plant J 40(1):143–150. https://doi.org/10.1111/j.1365-313X.2004.02190.x
9. Szarejko IS, Szurman-Zubrzycka M, Nawrot M, Marzec M, Gruszka D, Kurowska M, Chmielewska B, Zbieszczyk J, Jelonek J, Małuszyński M (2016) Creation of a TILLING population in barley after chemical mutagenesis with sodium azide and MNU. In: Jankowicz-Cieślak J, Tai TH, Kumlehn J, Till BJ (eds) Biotechnologies for plant mutation breeding, Protocols. Springer, Cham, pp 91–111
10. Till BJ, Reynolds SH, Greene EA, Codomo CA, Enns LC, Johnson JE, Burtner C, Odden AR, Young K, Taylor NE, Henikoff JG, Comai L, Henikoff S (2003) Large-scale discovery of induced point mutations with high-throughput TILLING. Genome Res 13 (3):524–530. https://doi.org/10.1101/gr.977903
11. Till BJ, Zerr T, Comai L, Henikoff S (2006) A protocol for TILLING and Ecotilling in plants and animals. Nat Protoc 1(5):2465–2477. https://doi.org/10.1038/nprot.2006.329
12. Oleykowski CA, Mullins CRB, Godwin AK, Yeung AT (1998) Mutation detection using a novel plant endonuclease. Nucleic Acids Res 26 (20):4597–4602. https://doi.org/10.1093/nar/26.20.4597
13. Till BJ, Burtner C, Comai L, Henikoff S (2004) Mismatch cleavage by single-strand specific nucleases. Nucleic Acids Res 32 (8):2632–2641. https://doi.org/10.1093/nar/gkh599
14. Mascher M, Gundlach H, Himmelbach A, Beier S, Twardziok SO, Wicker T, Radchuk V, Dockter C, Hedley PE, Russell J, Bayer M, Ramsay L, Liu H, Haberer G, Zhang X-Q, Zhang Q, Barrero RA, Li L, Taudien S, Groth M, Felder M, Hastie A, Šimková H, Staňková H, Vrána J, Chan S, Muñoz-Amatriaín M, Ounit R, Wanamaker S, Bolser D, Colmsee C, Schmutzer T, Aliyeva-Schnorr L, Grasso S, Tanskanen J, Chailyan A, Sampath D, Heavens D, Clissold L, Cao S, Chapman B, Dai F, Han Y, Li H, Li X, Lin C, McCooke JK, Tan C, Wang P, Wang S, Yin S, Zhou G, Poland JA, Bellgard MI, Borisjuk L, Houben A, Doležel J, Ayling S, Lonardi S,

Kersey P, Langridge P, Muehlbauer GJ, Clark MD, Caccamo M, Schulman AH, Mayer KFX, Platzer M, Close TJ, Scholz U, Hansson M, Zhang G, Braumann I, Spannagl M, Li C, Waugh R, Stein N (2017) A chromosome conformation capture ordered sequence of the barley genome. Nature 544:427–433 doi.org/10.1038/nature22043

15. Deng W, Nickle DC, Learn GH, Maust B, Mullins JI (2007) ViroBLAST: a stand-alone BLAST web server for flexible queries of multiple databases and user's datasets. Bioinformatics 23(17):2334–2336. https://doi.org/10.1093/bioinformatics/btm331
16. Proost S, Van Bel M, Sterck L, Billiau K, Van Parys T, Van de Peer Y, Vandepoele K (2009) PLAZA: a comparative genomics resource to study gene and genome evolution in plants. Plant Cell 21(12):3718–3731. https://doi.org/10.1105/tpc.109.071506
17. Finn RD, Coggill P, Eberhardt RY, Eddy SR, Mistry J, Mitchell AL, Potter SC, Punta M, Qureshi M, Sangrador-Vegas A, Salazar GA, Tate J, Bateman A (2016) The Pfam protein families database: towards a more sustainable future. Nucleic Acids Res 44(D1):D279–D285. https://doi.org/10.1093/nar/gkv1344
18. McCallum CM, Comai L, Greene EA, Henikoff S (2000) Targeting induced local lesions IN genomes (TILLING) for plant functional genomics. Plant Physiol 123(2):439–442
19. Sievers F, Wilm A, Dineen D, Gibson TJ, Karplus K, Li W, Lopez R, McWilliam H, Remmert M, Soding J, Thompson JD, Higgins DG (2011) Fast, scalable generation of high-quality protein multiple sequence alignments using Clustal omega. Mol Syst Biol 7:539. https://doi.org/10.1038/msb.2011.75
20. Bromberg Y, Rost B (2007) SNAP: predict effect of non-synonymous polymorphisms on function. Nucleic Acids Res 35(11):3823–3835. https://doi.org/10.1093/nar/gkm238
21. Acharya V, Nagarajaram HA (2012) Hansa: an automated method for discriminating disease and neutral human nsSNPs. Hum Mutat 33(2):332–337. https://doi.org/10.1002/humu.21642
22. Acharya V, Nagarajaram HA (2013) Response to: statistical analysis of missense mutation classifiers. Hum Mutat 34(2):407. https://doi.org/10.1002/humu.22250
23. Venselaar H, Te Beek TA, Kuipers RK, Hekkelman ML, Vriend G (2010) Protein structure analysis of mutations causing inheritable diseases. An e-Science approach with life scientist friendly interfaces. BMC Bioinformatics 11:548. https://doi.org/10.1186/1471-2105-11-548
24. Ng PC, Henikoff S (2003) SIFT: predicting amino acid changes that affect protein function. Nucleic Acids Res 31(13):3812–3814
25. Cartegni L, Wang J, Zhu Z, Zhang MQ, Krainer AR (2003) ESEfinder: a web resource to identify exonic splicing enhancers. Nucleic Acids Res 31(13):3568–3571
26. Arnold K, Bordoli L, Kopp J, Schwede T (2006) The SWISS-MODEL workspace: a web-based environment for protein structure homology modelling. Bioinformatics 22(2):195–201. https://doi.org/10.1093/bioinformatics/bti770
27. Kelley LA, Mezulis S, Yates CM, Wass MN, Sternberg MJ (2015) The Phyre2 web portal for protein modeling, prediction and analysis. Nat Protoc 10(6):845–858. https://doi.org/10.1038/nprot.2015.053

Chapter 7

Virus-Induced Gene Silencing (VIGS) for Functional Characterization of Disease Resistance Genes in Barley Seedlings

Lokanadha R. Gunupuru, Alexandre Perochon, Shahin S. Ali, Steven R. Scofield, and Fiona M. Doohan

Abstract

With the recent advances in sequencing technologies, many studies are generating lists of candidate genes associated with specific traits. The major bottleneck in functional genomics is the validation of gene function. This is achieved by analyzing the effect of either gene silencing or overexpression on a specific phenotypic or biochemical trait. This usually requires the generation of stable transgenic plants and this can take considerable time. Therefore any technique that expedites the validation of gene function is of particular benefit in cereals, including barley. One such technique is Virus-Induced Gene Silencing (VIGS), which evokes a natural antiviral defense mechanism in plants. VIGS can be used to downregulate gene expression in a transient manner, but long enough to determine its effects on a specific phenotype. It is particularly useful for screening candidate genes and selecting those with potential for disease control. VIGS based on Barley Stripe Mosaic Virus (BSMV) is a powerful and efficient tool for the analysis of gene function in cereals. Here we present a BSMV VIGS protocol for simple and robust gene silencing in barley and describe it to evaluate the role of the hormone receptor *BRI1* (*Brassinosteroid Insensitive 1*) in barley leaf resistance to *Fusarium* infection.

Key words Virus-Induced Gene Silencing (VIGS), Post-Transcriptional Gene Silencing (PTGS), Barley Stripe Mosaic Virus (BSMV), Cereal, Barley, Silencing, Leaf detachment assay, Disease, Functional genomics, *BRI1* (*Brassinosteroid Insensitive 1*)

1 Introduction

Virus-induced gene silencing (VIGS) is a simple and powerful tool that can be used to silence a gene in a relatively rapid time frame [1, 2]. VIGS is based on the natural phenomenon of RNA interference (RNAi), also known as post-transcriptional gene silencing (PTGS), that is induced as an antiviral response in plants [1]. PTGS is activated by the detection of double-stranded RNAs which are processed by the DICER enzyme into small interfering RNA (siRNA) molecules, 21–24 nucleotides in length [1, 3]. These

Wendy A. Harwood (ed.), *Barley: Methods and Protocols*, Methods in Molecular Biology, vol. 1900,
https://doi.org/10.1007/978-1-4939-8944-7_7,

siRNAs, in complex with ARGONAUTE proteins, form the RNA-induced silencing complex (RISC) that cleaves the target mRNA [4]. PTGS was first discovered in transgenic petunia plants in which the overexpression of chimeric chalcone synthase with the endogenous gene resulted in the reversible co-suppression of homologous genes [5, 6]. PTGS is now used widely to study gene function, either via the generation of stable RNA interference (RNAi) lines or by transiently expressing a fragment of the target gene using VIGS. The former has the advantage of the long-term availability of transgenic lines, while VIGS is more rapid, economic and eliminates the laborious transformation process. Indeed, VIGS is particular useful when used as a tool to screen and delineate the best target genes for a particular trait from a pool of candidate genes. The VIGS technique involves the insertion of a small fragment from the gene to be silenced into viral genome vectors. Subsequently corresponding viral RNAs are produced and applied to plants to induce the PTGS mechanism, which results in silencing of the target gene [7–9]. This technique has been applied in both monocots and dicots for the functional characterization of various genes involved in plant development and disease resistance [8–10].

Tobacco Mosaic Virus (TMV) derived VIGS vectors were the first gene silencing vectors employed for gene silencing in plants and these were used to silence the phytoene desaturase (*PDS*) gene in tobacco [8]. Potato Virus X was then developed which had a better silencing ability, and was used to silence genes in both *Nicotiana benthamiana* and *Solanum lycopersicum* [1, 11]. Thereafter, other gene silencing vectors were developed from various plant viruses [12–15]. The main disadvantages of these virus vectors were the narrower range of host plants that they could infect. For cereals, effective silencing vectors were developed from Barley Stripe Mosaic Virus (BSMV). BSMV is a positive-sense single strand RNA virus. Its tripartite genome consists of RNAs α, β, and γ [16]. The BSMV genome was modified as a gene silencing vector for use in barley and wheat [17].

BSMV vectors have been used as versatile tool to study disease resistance genes in wheat and barley. In wheat, Scofield et al. [18] reported the use of the BSMV system to silence the *lr21* (nucleotide binding leucine rich receptor) resistance gene in wheat, which conferred an increased susceptibility to leaf rust disease. BSMV-based silencing of *TaChlH* (magnesium chelatase subunit H) in wheat enhanced susceptibility to the causal agent of Septoria Tritici Blotch disease, *Zymoseptoria tritici* [19]. BSMV has also been used to confirm the role of genes in barley defense against pathogens. Silencing of barley *Sgt1* (suppressor of G-two allele of *Skp1*) and *Hsp90* (heat shock protein) compromised *Mla13*-mediated resistance toward the biotrophic mildew pathogen *Blumeria graminis* f. sp. *hordei* [20]. Silencing of barley chorismate synthase (*HvCS*), anthranilate synthase α subunit 2 gene (*HvASa2*), and chorismate mutase 1 (*HvCM1*) genes resulted in increased *B. graminis* penetration into epidermal cells of barley carrying mildew resistance locus a (*Mla*) [21]. BSMV-based silencing is not limited to seedlings but also can be used in adult plants to study the role of genes in disease resistance [18, 22–24].

The protocol we describe here has been successfully used in barley to study the role of *BRI1* in defense against *Fusarium culmorum*. This pathogen is responsible for yield loss in barley and other small-grains, attacking roots, stems and heads of cereal plants. In detached leaf assays, the silencing of *BRI1* using BSMV vectors in barley cv. Akashinriki enhanced *F. culmorum*-induced necrosis in leaves [22]. In this article we show the methodology used to silence the targeted gene in barley seedlings using BSMV vectors. The same protocol can be used to silence genes in barley heads and in wheat leaves and head tissue.

2 Materials

2.1 Plant and Growth Conditions

1. Barley seeds (cv. Akashinriki) (kindly provided by Dr. K. Sato, Barley Germplasm Centre, Okayama University, Japan).
2. Filter paper.
3. Circular Petri dish, 90 mm.
4. Leukopor (BSN medical, UK).
5. John Innes compost No-2.
6. Three inch diameter pots.
7. Plant growth room or chamber (22 °C, 16/8: light/dark, 70% humidity).

2.2 Amplification of Plant Gene Fragments and Cloning into BSMV Vectors

1. DNA extraction kit (E.Z.N.A.® HP Plant DNA Kit, USA, Cat. No. D2485-02).
2. *Escherichia coli* strain DH5α.
3. Platinum˙ *Taq* DNA polymerase (Thermo Fisher Scientific, USA, Cat. No. 10966026).
4. dNTPs.
5. PCR machine.
6. Agarose.
7. Solis BioDyne 1 Kb marker (Solis BioDyne, Estonia, Cat. No. 07-12-00050).
8. Safeview (NBS-Biologicals, UK, Cat. No. NBS-SV1).
9. QIAquick PCR purification kit (Qiagen, Germany, Cat. No. 28104).
10. NanoDrop™ 1000 Spectrophotometer (Thermo scientific, USA).

11. pGEM-T easy cloning kit (Promega, UK, Cat. No. A1380).
12. Plasmid vectors: pα42 (pα), pβ42.sp1 (pβ), pSL038-1 (pγ), pSL039-PDS (pγPDS) [2].
13. Luria-Bertani (LB) liquid broth (Sigma-Aldrich, USA, Cat. No. L3152) and LB solid media containing 1.2% (w/v) agar.
14. Ampicillin (100 mg/mL stock concentration).
15. Shaking incubators for cultures.
16. Plasmid isolation kit (Qiagen, Germany, Cat. No. 27106).
17. Restriction enzymes: PacI, Sma1, MluI, SpeI.
18. Ethanol.
19. Sodium acetate.
20. Tris–Borate–EDTA 10× buffer.
21. X-Gal (Sigma-Aldrich, Cat. No. B4252) and Isopropyl thio β-D-galactopyranoside (IPTG) (Sigma-Aldrich, USA, Cat. No. I6758).

2.3 Digestion of BSMV Plasmids and In Vitro Transcription

1. BSMV vectors (α, β, γ and modified γ) (Kindly provided by Prof. Steven R. Scofield).
2. Plasmid isolation kit (Qiagen, Germany, Cat. No. 27106).
3. Sodium acetate.
4. Ethanol.
5. EDTA.
6. mMessage mMachine T7 in vitro transcription kit (Ambion, USA, Cat. No. AM1344).

2.4 Inoculation of Plant RNA and Buffers

1. 5× GP buffer (0.5 M glycine, 0.3 M K_2HPO_4 dibasic).
2. FES buffer (100 mL 5× GP buffer, 22 mM sodium pyrophosphate decahydrate, 27.7 mM bentonite, 166 mM celite, bring final volume to 500 mL *see* **Note 1**).
3. Glycine.
4. K_2HPO_4 dibasic.
5. Sodium pyrophosphate decahydrate.
6. Bentonite (Aldrich, USA, Cat. No. 28,523-4).
7. Celite (Fluka, USA, Cat. No. 22141).

2.5 Production of Fusarium Spores

1. *Fusarium culmorum* (Strain FCF200, kindly supplied by Dr. Paul Nicholson, John Innes Centre, Norwich, UK).
2. PDA (Scharlau, Spain, Cat. No. 01-483-500).
3. Incubator.
4. Cork borer.
5. Mung beans.

6. Autoclave.
7. Glass ware.
8. Cheese cloth (from any grocery store).
9. Mung bean broth: 20 g Mung beans were boiled in 250 mL of distilled water for 20 min. The liquid was passed through a cheese cloth and diluted to 1 L with distilled water. 25 mL of the broth was transferred to 100 mL flasks and autoclaved for 15 min at 121 °C.
10. Centrifuge.
11. Tween 20.
12. Hemocytometer (Hycor Biomedical, USA, Cat. No. H2-LN-87144-PK).
13. Microscope (Olympus CH30, UK).

2.6 Gene Silencing Confirmation

1. Tungsten beads (Qiagen, Germany, Cat. No. 69997).
2. Tissue lyser II (Qiagen, Germany, Cat. No. 85300).
3. RNeasy Mini Kit (Qiagen, Germany, Cat. No. 74104).
4. Oligo(dT) 12-18 Primer (Invitrogen, USA, Cat. No. 18418012).
5. SYBR Green Master Mix (Takara, Japan, Cat. No. RR420A).
6. qPCR machine (Agilent Mx3000P, USA, Cat. No. 401512).

2.7 Detached Leaf Assay and Image Analysis of Infected Area

1. Petri dishes ($100 \times 100 \times 20$ mm).
2. Plant agar (Duchefa, Netherlands, Cat. No. P1001).
3. Benzimidazole (Sigma-Aldrich, USA, Cat. No. 194123).
4. Glass Pasteur pipette.
5. Leukopor® (BSN medical, UK).
6. Growth room.
7. Camera (Nikon coolpix p500, Japan).
8. IMAGE-J software (NCBI).

2.8 Other Consumables

1. Gloves.
2. Sterile culture tubes (Falcon, USA, Cat. No. 352057).
3. Micro-pipettes (Gilson, France).
4. Centrifuge tubes (Sarstedt, Germany, Cat. No. 72.695.500).
5. Gel imaging system (Fusion *Fx*, Vilber lourmat, Denmark).
6. Gel apparatus (Helixx Mupid-exU, USA).
7. Autoclaved ddH_2O.
8. Dry heating block (Thermo Fisher Scientific, USA, Cat. No. 11-473-70).

2.9 Software and Databases

1. Primer3 software (version 0.4.0; http://frodo.wi.mit.edu/primer3/).
2. Ensembl plants database (http://plants.ensembl.org/index.html).
3. GenBank database (https://www.ncbi.nlm.nih.gov/genbank/).

3 Methods

3.1 Overview of VIGS Metholology

1. Figure 1 shows an overview of the VIGS methodology. It overviews the steps involved from fragment selection through to the production of the RNA used to inoculate plants with the BSMV virus encoding the fragment that targets the specific gene for silencing.

3.2 Seed Germination and Plant Growth Conditions

1. Barley seeds were placed in 9 cm diameter Petri dishes containing two pieces of filter paper soaked with 6 mL of sterile double-distilled water (ddH_2O).
2. Plates were sealed with Leucopore tape and placed in the dark, for 2 days at 4 °C in order to stratify the seed, followed by 2 days incubation at 21 °C to facilitate germination.
3. Etiolated seedlings were transferred to 3 in. pots containing John Innes compost No 2; two seedlings per pot.
4. Plants were grown in a contained glasshouse where the temperature is 22 °C with a 16 h light/8 h dark photoperiod and 70% relative humidity.

3.3 Preparation of Fungal Conidial Production

1. Mycelium from glycerol stocks of *F. culmorum* FCF 200 (−80 °C) was subcultured on potato dextrose agar (PDA) plates and incubated at 25 °C for 7 days in the dark.
2. For conidia (asexual spore) production, three to four PDA plugs of 10 mm diameter were taken from these PDA cultures using a cork borer and these were used to inoculate 25 mL of Mung bean broth.
3. The flasks were kept in shaking incubator at 200 rpm, 25 °C for 5 days in dark.
4. The conidial suspension was passed through two layers of sterile cheese cloth into 50 mL centrifuge tubes and the filtrate was centrifuged at room temperature, $3220 \times g$ (5000 rpm). The liquid was discarded and the spore pellet was washed three times with sterile ddH_2O.
5. Conidia were resuspended in 0.02% (v/v) Tween20 and the concentration was determined using disposable hemocytometer (KOVA) slides.
6. The conidial concentration was adjusted to 1×10^6 spores/mL using 0.02% Tween20.

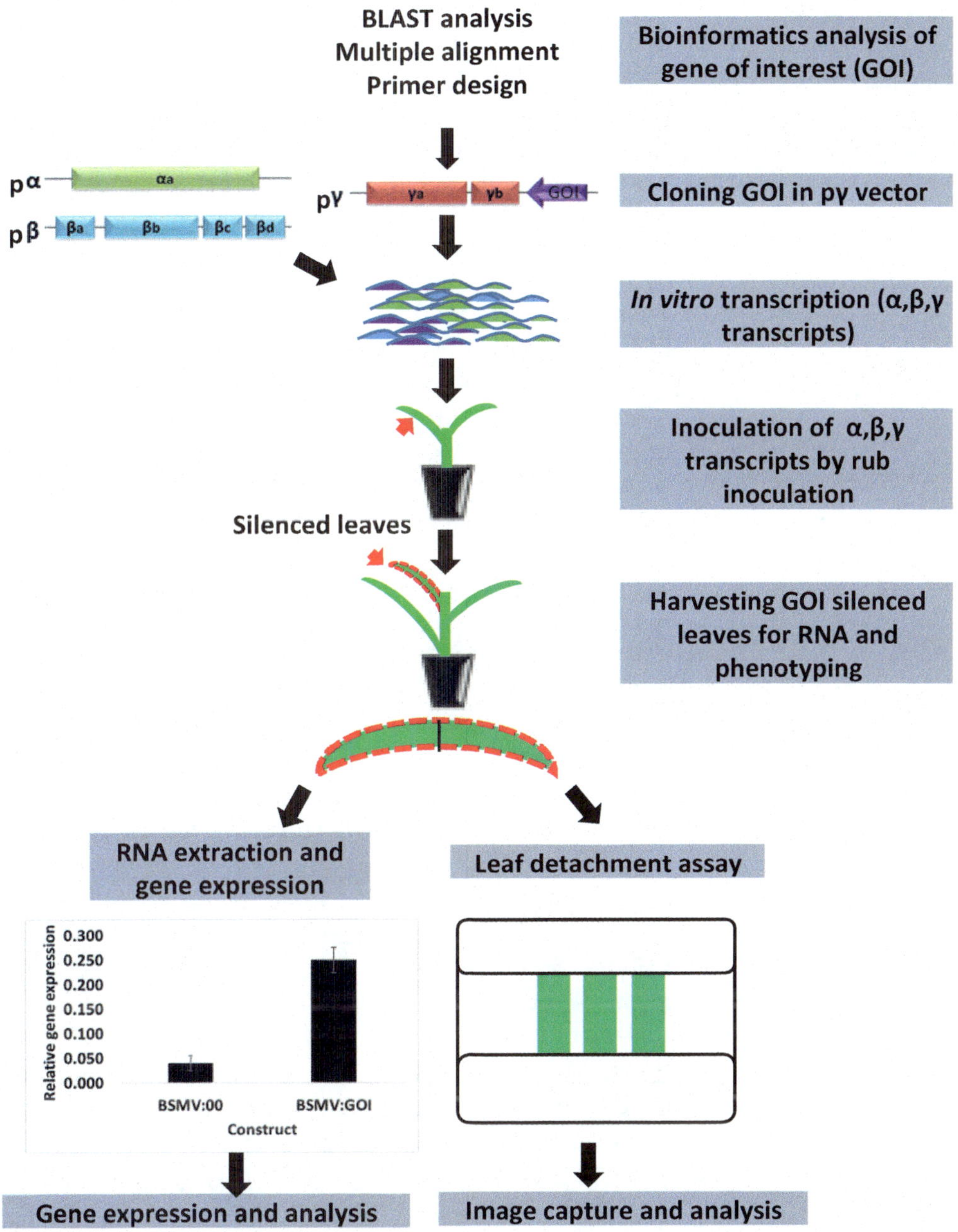

Fig. 1 Schematic representation of VIGS methodology and the detached leaf assay for the functional characterization of disease resistance genes in cereals. The tripartite genome (α,β,γ) of Barley Stripe Mosaic Virus (BSMV) was cloned into plasmids. A fragment of the gene of interest (GOI) was amplified from barley genomic DNA and subsequently cloned into the pγ vector. The plasmids carrying α, β, γ genomes were linearized and their respective transcripts were produced by in vitro transcription. Transcripts were suspended in FES buffer and rub-inoculated onto the first leaf of seedlings. The third leaf of inoculated plants was harvested and used for both gene expression studies and a detached leaf assay, in order to, respectively, evaluate the efficacy of gene silencing and role of the GOI in disease resistance. Empty vector treatment (BSMV:00) and gene of interest silenced plants (BSMV:GOI)

3.4 Selection of the Gene Target Regions for Both VIGS Fragments and VIGS Validation via qRT-PCR

1. A short exonic fragment of 120–500 bp is sufficient to initiate gene silencing by PTGS in plants [2]. Insert length influenced stability but not efficiency of VIGS and silencing is transient in most cases [25].
2. If the user aim is to silence an entire family of genes, then the gene silencing fragments need to be homologous to the most conserved exonic region of the gene family sequences. When a specific gene is the target, the most variable region should be used for gene silencing. The silencing fragments should be in the range of 120–500 bp in length for better stability in plants [18, 25, 26] and note that genes with a minimum homology of 85% are likely to be downregulated by VIGS [8, 26].
3. The gene fragment must target an exon. If amplifying the target sequence from gDNA extracts, it is also important to verify that the primers do not span introns. Therefore, it is important to compare genomic and mRNA gene sequence to locate introns.
4. At the gene fragment selection stage, one must try and ensure the fragment does not have any off target gene silencing potential. Thus it is important to analyze by BLAST analysis of the gene fragment sequence against the barley genome to determine its homology with other sequences (*see* **Note 2**).
5. The chosen fragment should not contain an Mlu1 restriction site, as this enzyme is later used to cut the vector into which the fragment is cloned. Also the silencing fragments should not contain the restriction sites that are used for cloning into γ vector.
6. The efficacy of VIGS was verified using quantitative reverse transcriptase polymerase chain reaction (qRT-PCR) analysis. At the gene fragment design stage, it is important to concurrently select the primers to be used for qRT-PCR in order to ensure that they do not overlap with the gene silencing fragment.
7. Its good practice to use two non-overlapping independent gene fragments to make sure a phenotype is due to the specific silencing of the target gene by using VIGS.
8. The empty vector γ was used as a negative control for the BSMV VIGS experiments and BSMV carrying a 185-bp fragment of the barley phytoene desaturase (*PDS*) gene which causes photo bleaching when silenced was used as positive control [17].

3.5 Amplification of Gene Silencing Fragments, Sequencing, and Digestion

1. Genomic DNA of barley cv. Akashinriki was isolated using the E.Z.N.A.® HP Plant DNA Kit and the concentration was measured using a Nano Drop. Concentration of the genomic DNA was then adjusted to 30 ng/μL with ddH_2O. Two independent, non-overlapping gene fragments of the barley *BRI1* (*Brassinosteroid Insensitive 1*) (Gen bank ID: AB109215.1)

Table 1
Primers used

Primer code	Primer sequence (5′ to 3′)	Application
BRI1-A for BRI1-A rev	CGATTAATTAAGCGGAGGCAGAAGAATGA CGACCCGGGGTCACCCTGGCCACTCAC	Cloning VIGS construct BSMV:BRI1-A
BRI1-B for BRI1-B rev	CGATTAATTAAGTGAGTGGCCAGGGTGAC CGACCCGGGTGGATGATGTGCGGAATG	Cloning VIGS construct BSMV:BRI1-B
BRI1:RT for BRI1:RT rev	CAACGATGCTCAAGGTGATG CCGGTGGTCATCTTCCTAAT	BRI1-specific qRT-PCR primers
RNAH:RT for RNAH:RT rev	GCACAGGGAATCGTCAAAGT TCAAAACAACACAACATCGAAGT	RNAH-specific qRT-PCR primers
pSL038 for pSL038 rev	TGATGATTCTTCTTCCGTTGC TGGTTTCCAATTCAGGCATCG	Sequencing primers for pSL038
M13 for M13 rev	CGCCAGGGTTTTCCCAGTCACGAC AGCGGATAACAATTTCACACAGGA	Sequencing primers for pGEM-T

were amplified from genomic DNA using gene-specific primers sets BRI1-A or BRI1-B (listed in Table 1). PacI and SmaI restriction site sequences were, respectively, added to the 5′ end of the forward and reverse primers.

2. PCR was performed in a 10 μL reaction containing 0.5 U Taq DNA polymerase and 1× PCR buffer (Invitrogen, UK), 1.5 mM $MgCl_2$, and 125 mM of each of forward and reverse primers (BRI1-A/B) with 1.25 units of Taq DNA polymerase with 30 ng barley genomic DNA. PCR was performed in a Peltier thermal Cycler DNA engine and the PCR program consisted of an initial denaturation step at 94 °C for 2 min, 30 cycles of denaturation 94 °C for 30 s, annealing at 60 °C for 30 s, extension at 72 °C for 45 s and a final extension step at 72 °C for 5 min.
3. 5 μL of PCR product was visualized on a 0.8% (w/v) agarose gel and remaining 45 μL of PCR product was purified using QIAquick PCR purification kit according to the manufacturer instructions, with the final product being eluted into 30 μL of sterile ddH_2O (*see* **Note 3**).
4. The concentration of DNA in the eluent was checked using the NanoDrop.
5. The fragments were cloned into the pGEM-T easy vector using the pGEM-T Easy cloning kit and following the manufacturer's instructions.
6. 10 μL of the ligation mixture was transformed into 20 μL of high efficient *E. coli* DH5α competent following the manufacturer's instructions. The transformed cells were spread on

LB-agar plates containing 100 μg/mL of Ampicillin, 20 mg/mL X-gal, and 100 mM of IPTG. The plates were incubated overnight at 37 °C.

7. The clones were screened for inserts via colony PCR, using a pinhead amount of cells suspended in 5 μL of ddH_2O as template. The PCR mix comprised of 10 μL reaction containing 0.5 U Taq DNA polymerase and 1× PCR buffer (Invitrogen, UK), 1.5 mM $MgCl_2$, and 125 mM of each of M13 forward and reverse primers (Table 1) in a total volume of 25 μL. The PCR program consisted of an initial denaturation step at 94 °C for 5 min, 30 cycles of denaturation at 94 °C for 30 s, annealing at 58 °C for 30 s, extension at 68 °C for 45 s, and a final extension step at 72 °C for 5 min.
8. The PCR product was visualized on a 0.8% agarose gel, and the positive clones were inoculated into 5 mL of LB broth supplemented with 100 μg/mL (w/v) of ampicillin, and cultures were incubated in an orbital shaker at 200 rpm, 37 °C, overnight.
9. Plasmids were extracted from the cultures using the QIAprep plasmid extraction kit according to manufacturer's instructions. The final elution was performed in a 30 μL of sterile ddH_2O.
10. pGEM-T clones carrying gene fragments were sequenced using M13 sequencing primers at Macrogen Inc. (Korea) (*see* **Note 4**).
11. Verified plasmids carrying gene fragments were double digested with PacI and SmaI. The digestion reaction consists of 2 μg of vector, 1× NEB cut smart buffer, 20 units of PacI and 20 units of SmaI and ddH_2O to a final volume of 50 μL. The reaction mixture was first incubated at 37 °C for 2 h and then at 25 °C for 2 h. Restriction digestion was performed on a dry heating block.
12. The digested product was run on a 1% (w/v) agarose gel, and the required insert was excised from the gel and purified using the QIAquick Gel Extraction Kit according to manufacturer's instructions. The final elution was performed in 30 μL sterile ddH_2O and eluent was stored at −20 °C (*see* **Note 5**).

3.6 Cloning of the Gene Silencing Fragments into the pSL038-1 that Encodes the BSMV Gamma Genome

1. Barley stripe mosaic virus (BSMV) has a tripartite RNA genome, and the three genome fragments from strain ND18 were cloned into ampicillin-resistant plasmids (a modified gamma-b genome) [17].
2. The BSMV gamma genome vector pSL038-1 was double digested with the restriction enzymes PacI and SmaI in same way as described above in Subheading 3.5, **step 11** (*see* **Note 6**).

3. The digested sample was purified using QIAquick Gel Extraction Kit by following the manufacturer's instructions, and the final elution was performed in 30 μL sterile RNase-Free water.
4. The purified VIGS fragments were ligated into the PacI and SmaI digested pSL038-1 vector. The ligation mixture comprised 1 μg of linearized vector, 0.5 μg of purified insert, 1× ligation buffer, 25 units of T4 ligase and the final volume was made up to 10 μL using sterile ddH_2O. The samples were incubated at 16 °C, overnight. All the ligation mixture was transformed into *E. coli* DH5α by following the manufacturer's instructions. The transformed cells were spread on LB-agar plates containing 100 μg/mL of Ampicillin. The plates were incubated at 37 °C, overnight.
5. The resulting colonies were screened for inserts by colony PCR as described in Subheading 3.5, **step** 7 with an exception that the primers used were specific to pSL038-1 (Table 1). The verified clones were sent for sequencing at Macrogen Inc. (Korea) using the pSL038-1 primers.
6. The verified clones containing either pSL038-1-BRI1-A (pγBRI1-A) or pSL038-1-BRI1-B (pγBRI1-B) were stored in 20% glycerol at −80 °C.

3.7 Preparation of BSMV Vector Plasmids

1. *E. coli* clones carrying ampicillin-resistant plasmids encoding either the BSMV RNA genomes α, β, γ (pα, pβ, and pγ), pγPDS [18], pγBRI1-A, or pγBRI1-B were cultured.
2. Glycerol stocks of each clone was streaked on LB-ampicillin (100 μg/mL, w/v) plates and incubated at 37 °C, overnight. A single colony was picked and inoculated in 5 mL LB broth supplemented with 100 μg/mL (w/v) of ampicillin and incubated overnight at 200 rpm, 37 °C.
3. Plasmid isolation was performed using Qiagen QIAprep Miniprep Kit as per the protocol; RNaseA was excluded from the buffer P1 in the first step and isolated plasmids were stored at −20 °C (*see* **Note** 7).

3.8 Linearization and Purification of BSMV Vector Plasmids

1. The plasmids pα, pγ, pγPDS, pγBRI1-A, and pγBRI1-B were linearized with the restriction enzyme MluI, while pβ was linearized with SpeI. All the restriction enzyme digestions were performed with NEB enzymes, following the manufacturer's instructions. The digestion reaction consists of plasmid DNA (2 μg), 1× NEB cut smart buffer, 20 units of either MluI or SpeI; the final volume was adjusted to 50 μL using RNase-free water (*see* **Note 8**).
2. Plasmid linearization was verified by visualizing the digestion products (5 μL) via agarose gel electrophoresis (1% w/v agarose gel). Once complete digestion was confirmed, enzymes were heat-inactivated by incubating the reaction at 65 °C for 20 min.

3. The linearized plasmids were purified by ethanol precipitation. The digested plasmids mixture was mixed with 0.1× volume of 3 M sodium acetate pH 5.2 and 2× volumes of 100% (v/v) ethanol. Precipitation was facilitated by incubating the mixture at −20 °C for 15 min.
4. The precipitation reactions were centrifuged at 16,200 × *g* (13,000 rpm) at 4 °C for 10 min. The supernatant was discarded and the pellet was washed with 750 μL of 70% (v/v) ethanol and centrifuged for 10 min at 16,200 × *g* (13,000 rpm), 4 °C.
5. The pellet was air-dried and dissolved in 20 μL of RNase-free water. The concentration of DNA was determined using the NanoDrop.

3.9 Preparation of Viral RNA by In Vitro Transcription

1. RNA was transcribed from the viral genomes using the purified linear plasmids as template and the mMessage mMachine™ T7 in vitro transcription kit, following manufacturer's instructions. Reactions contained linearized plasmid DNA (0.5 μg) 1× buffer, 5 units of in vitro transcription enzyme mix, 1× NTP/CAP and RNase-free water to a total volume of 20 μL.
2. The samples were incubated at 37 °C for 2 h and the quality of transcribed RNA was visualized by running 1 μL of each in vitro transcription reaction (diluted to 10 μL with RNase-free 6× gel loading buffer) through a 1% (w/v) agarose gel (Fig. 2). All the RNA samples were stored at −80 °C until further use (*see* **Note 9**).
3. For each plant to be treated, 0.35 μL of each RNA transcript (α, β, and relevant γ genome transcript) was required (based on the number of plants to be treated the reaction volume for in vitro transcription can be scaled up) (*see* **Note 10**).

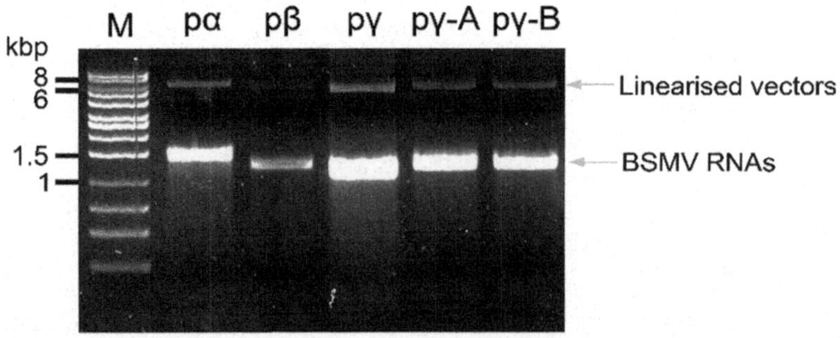

Fig. 2 Agarose gel electrophoresis of in vitro transcription products using linearized BSMV vectors pα, pβ, pγ, pγBRI1-A (pγ-A), and pγBRI1-B (pγ-B). DNase treatment was not performed on the samples; the linearized BSMV vectors can be seen above the RNAs, both highlighted with arrows. (M) Solis BioDyne 1 Kb marker

3.10 Inoculating Leaves with the BSMV Transcripts

1. Plants were inoculated with FES buffer containing a 1:1:1 transcript mixture of the α, β, and relevant γ genome transcript (γ: negative control; γPDS: positive control; γBRI1-A or γBRI1-B: targeting *BRI1* for silencing). FES buffer served as a mock treatment. For each plant, VIGS inoculum was prepared by mixing 9 μL of FES Buffer with 0.35 μL each of the α, β genome transcripts and the relevant γ-based genome transcript (*see* **Note 11**).
2. The VIGS inoculum was applied to the first leaf of 10-day-old seedlings. 10 μL of transcript mixture was pipetted onto the index. Keeping the leaf in between thumb and index finger, the transcript mixture was applied by pulling the finger tips gently (to prevent any major damage to leaf) from the base toward the tip and repeat this three times (Fig. 3 and *see* **Note 12**).

3.11 Validation of the Efficacy of the VIGS Platform Using PDS as a Positive Control Target Gene

1. During the infection BSMV moves systematically in seedling leaves and at 14 days post-viral inoculation, the visual symptoms of *PDS* gene silencing were evident on the third leaf of seedlings [26].
2. Silencing of *PDS* resulted in the photo bleaching of emerging leaves with a maximum photo bleaching observed in the third leaf (Fig. 4).

3.12 Confirmation of Gene Silencing by qRT-PCR

1. Gene silencing was quantified by qRT-PCR of RNA extracts, using primers specific to *BRI1* (BRI1:RT, Table 1) and relative to that of the RNA helicase housekeeping gene (RNAH:RT, Table 1) (*see* **Note 13**).
2. Total RNA was extracted from the VIGS-treated leaf segments at 14 days post viral infection. The frozen leaf samples were

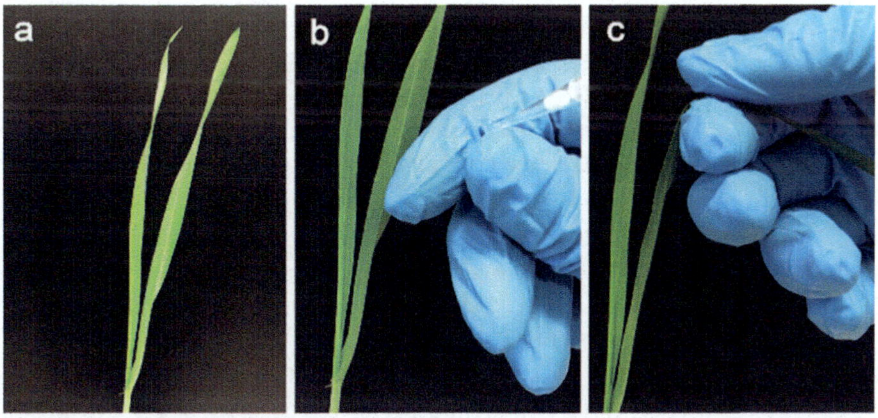

Fig. 3 Steps in the inoculation of barley cultivar Akashinriki with viral transcripts. (**a**) 10-day-old seedling. (**b**) Application of viral transcripts mixture in FES buffer to the index finger and thumb (10 μL). (**c**) Inoculation of viral transcripts mixture to the leaf by gently pulling from base to tip, keeping the leaf in between index finger and the thumb

Fig. 4 Silencing of phytoene desaturase (*PDS*) in barley seedlings resulted in photobleaching of newly emerging leaves. Codes: Mock (FES, VIGS buffer negative control), BSMV:00 (mock virus treatment), BSMV:PDS (positive control)

ground using a TissueLyser by adding two tungsten beads (1 min at 30 Hz). RNA was extracted from the ground leaf tissue using the RNeasy Mini Kit, according to the manufacturer's instructions (*see* **Note 14**).

3. cDNA was prepared from the extracted RNA following the protocol described by Ansari et al. [27] with the minor variation that the primer used was oligo $dT_{12–18}$ (Invitrogen) and the resulting cDNA was diluted to 200 μL using sterile ddH_2O.
4. PCR reactions contained 2.5 μL of diluted cDNA, 12.5 μL Premix Ex Taq™ (Takara, Japan), and 100 nM each of forward and reverse primers in a total volume of 25 μL. Independent PCRs were concurrently conducted for the housekeeping and target genes.
5. PCR reactions were conducted in a Stratagene Mx3000™ qRT-PCR machine (Stratagene, USA) and the program consisted of 1 cycle of 95 °C for 10 s, 40 cycles of 95 °C for 5 s, 60 °C for 30 s and 1 cycle of 95 °C for 60 s.
6. The cycle threshold (Ct) values obtained by qRT-PCR were used to calculate the relative gene expression using the formula $2^{-(\text{Ct target gene} - \text{Ct housekeeping gene})}$ (Fig. 5), as described previously [28].

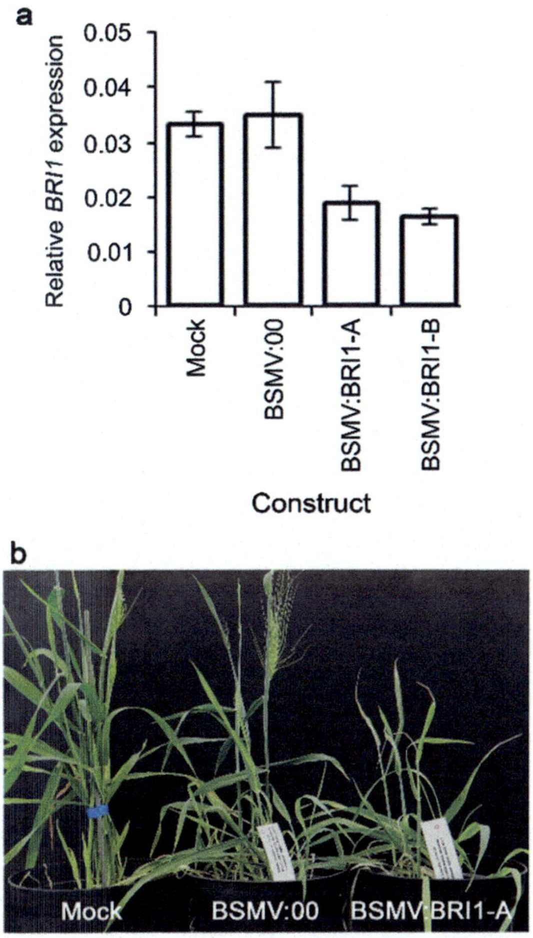

Fig. 5 VIGS of *BRI1* in barley cultivar Akashinriki. Plants were treated with Mock (FES VIGS buffer), VIGS constructs BSMV:00 (empty vector), or BSMV:BRI1-A/-B (two independent constructs targeting *BRI1* gene). (**a**) Graph showing the relative accumulation of *BRI1* transcripts, calculated using quantitative qRT-PCR ($2^{-(Ct\ BRI1\ -\ Ct\ RNA\ helicase)}$). (**b**) Silencing of *BRI1* resulted in stunting of plant growth

3.13 Using a Detached Leaf Assay to Determine the Effect of VIGS of BRI1 on Barley Response to the Pathogen Fusarium culmorum

1. At 14 days post-viral inoculation, the third leaf was harvested and an 8 cm segment was prepared for use in the detached leaf assay, and the remaining leaf material was flash frozen in liquid N_2 and stored at −80 °C prior to RNA extraction.
2. Square Petri dishes (100 mm × 100 mm × 20 mm) containing 1% (w/v) plant agar, pH 5.7 supplemented with 0.5 mM benzimidazole were prepared and a rectangle of agar (5 cm) was removed from the center of the plates to prevent excessive fungal growth.
3. Leaf segments were placed in each plate, the cut ends resting on the agar and the adaxial side facing upward. The removed agar

rectangle was cut into two strips. These strips were used to overlay, and thus hold in place and prevent dehydration, of the cut edges of the leaf segments (*see* **Note 15**).

4. The center of each leaf section was punctured with a glass Pasteur pipette. Wounds were treated with a 5 μL droplet containing either 0.02% Tween20 (mock treatment) or 5 μL of conidia of *F. culmorum* strain FCF 200 (1×10^6 spores mL^{-1} in 0.2% Tween20). Plates were taped with porous Leukopor® (BSN medical, UK). Plates were incubated in a controlled environmental room at 22 °C with a 16/8 h light/dark regime and 70% relative humidity (*see* **Note 16**).
5. For each virus/*Fusarium* treatment combination, a total of eight leaf sections were treated per technical replicate and there were at least three biological replicates of the experiment.
6. After 3 days leaves with symptoms were observed and photographed with a scale bar next to plates. The necrotic leaf area was measured using Image J software [29]. The *BRI1*-silenced leaf segments developed significantly larger necrotic lesions upon *F. culmorum* infection than did those treated with the negative control BSMV:00 viral treatment (Fig. 6).

4 Notes

1. FES buffer can be aliquoted into 50 mL volumes and autoclaved (121 °C for 20 min); these aliquots can be stored at room temperature under sterile conditions.
2. It is highly recommended to test potential off target gene silencing by qRT-PCR on a closely related gene that shares some homology (less than 85%) with the VIGS gene fragment of the targeted gene. This will prove that a potential phenotype due to VIGS is due specifically to the targeted gene.
3. PCR products were run on a gel to check specificity before purification with the QIAquick PCR purification kit.
4. It is highly recommended to sequence the plasmids to confirm the correct fragment was cloned into pGEM-T vector before subcloning into pSL038-1.
5. During the gel extraction of the gene fragment, it's important to avoid any residual ethanol from the wash buffer (Buffer PE) in the QIAquick column. Before the elution step, we recommend to let the column stand 10 min to evaporate any remaining ethanol. Residual ethanol in the eluate may interfere with the ligation of the gene fragment into pSL038-1.
6. Use pGEM-T with the gene fragment insert as a positive control in restriction digestion; the release of the fragment

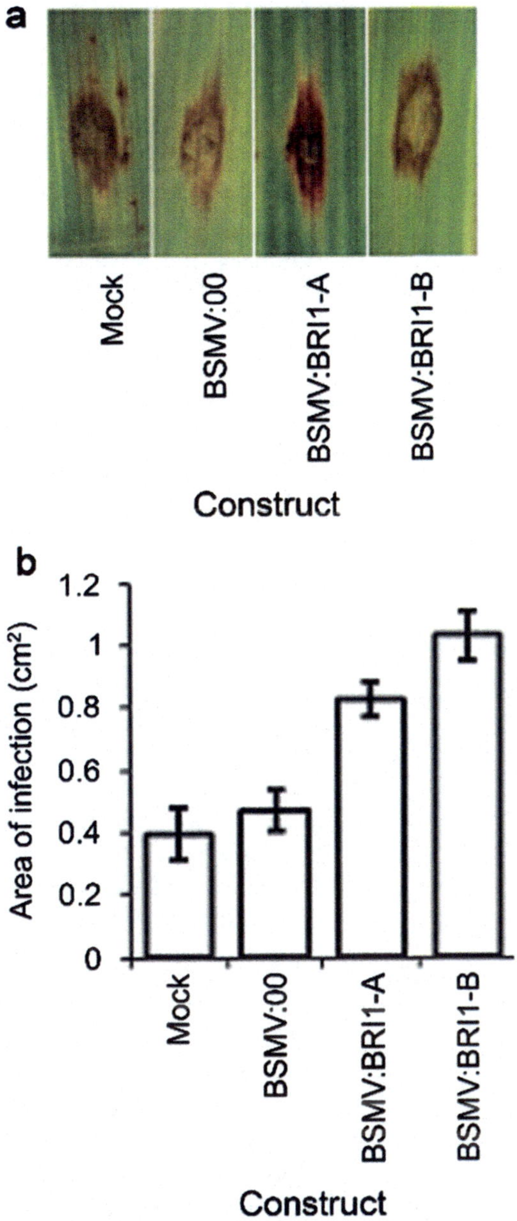

Fig. 6 Effect of VIGS of *BRI1* in the barley cultivar Akashinriki on *Fusarium culmorum* infection of seedling leaves as assessed in a detached leaf experiment. Plants were treated with Mock (VIGS FES buffer), VIGS constructs BSMV:00 (empty vector) or BSMV:BRI1-A/-B (two independent constructs targeting *BRI1* gene). (**a**) Phenotype of leaf segments showing necrosis upon infection with *F. culmorum*, at 3 days post inoculation. (**b**) Quantification of necrosis using ImageJ software

from this vector will act as a check to confirm that the double restriction worked.

7. For all the VIGS plasmids isolation, it is recommended to omit any RNase from your plasmid isolation protocol; residual RNaseA will interfere with the in vitro transcription process.
8. While performing the in vitro transcription reaction, always use RNase-free water and RNase-free filter tips to minimize any chances of RNase contamination. While running the agarose gel for RNA visualization, use only buffers prepared with DEPC-treated water and clean the gel tank with 3% (v/v) H_2O_2.
9. The agarose gel will indicate the quality of the RNA. A smear in the in vitro transcription product indicates degradation of the RNA. If the samples were not treated with DNase, there will be a band at approximately 6 Kbp (linearized vector) and a bright band of RNA at approximately 1.5 Kbp (Fig. 2).
10. As the only difference between the 1:1:1 transcript mixtures (α, β, γ) is the relevant γ genome transcript, therefore one needs more α and β transcript than of the specific γ constructs.
11. Before the application of viral transcripts, the plants should be well watered to aid recovery from the mechanical damage caused by the abrasive material present in FES buffer. The abrasive material causes damage to the leaf surface to facilitate entry of the viral transcripts into the plants. Be gentle on the leaves when applying the VIGS transcripts. It is important to compare FES treatment with BSMV:00 treatment to evaluate the influence of the viral infection on your disease assessment.
12. Always wear gloves before applying the viral transcripts and change the gloves between each treatment to prevent any cross contamination of the transcripts.
13. Housekeeping genes need to be checked for constitutive expression under the specific disease conditions and in the tissue being assessed.
14. Prior to investing effort in the phenotyping procedure, it is best to test the efficacy of the gene silencing via qRT-PCR.
15. In plates containing the detached leaves, it is advised to include a few mL of sterile water at the base to ensure high humidity in the plate, this is essential for the germination of fungal conidia.
16. It is better to avoid using a cultivar that is sensitive to the BSMV pathogen.

Acknowledgments

This work was supported by the Science Foundation Ireland research fund (14/IA/2508) and Department of Agriculture Research Stimulus Grant RSF 11/S/103.

References

1. Baulcombe DC (1999) Fast forward genetics based on virus-induced gene silencing. Curr Opin Plant Biol 2(2):109–113
2. Scofield SR, Brandt AS (2012) Virus-induced gene silencing in hexaploid wheat using barley stripe mosaic virus vectors. Methods Mol Biol 894:93–112
3. Fire A, Xu S, Montgomery MK, Kostas SA, Driver SE, Mello CC (1998) Potent and specific genetic interference by double-stranded RNA in Caenorhabditis elegans. Nature 391 (6669):806–811
4. Baulcombe D (2004) RNA silencing in plants. Nature 431(7006):356–363
5. Napoli C, Lemieux C, Jorgensen R (1990) Introduction of a chimeric chalcone synthase gene into petunia results in reversible co-suppression of homologous genes in trans. Plant Cell 2(4):279–289
6. Van der Krol AR, Mur LA, Beld M, Mol J, Stuitje AR (1990) Flavonoid genes in petunia: addition of a limited number of gene copies may lead to a suppression of gene expression. Plant Cell 2(4):291–299
7. Ding XS, Schneider WL, Chaluvadi SR, Mian MR, Nelson RS (2006) Characterization of a Brome mosaic virus strain and its use as a vector for gene silencing in monocotyledonous hosts. Mol Plant-Microbe Interact 19 (11):1229–1239
8. Kumagai M, Donson J, Della-Cioppa G, Harvey D, Hanley K, Grill L (1995) Cytoplasmic inhibition of carotenoid biosynthesis with virus-derived RNA. Proc Natl Acad Sci U S A 92(5):1679–1683
9. Ratcliff F, Martin-Hernandez AM, Baulcombe DC (2001) Technical advance: tobacco rattle virus as a vector for analysis of gene function by silencing. Plant J 25(2):237–245
10. Xu J, Schubert J, Altpeter F (2001) Dissection of RNA-mediated ryegrass mosaic virus resistance in fertile transgenic perennial ryegrass (Lolium perenne L.). Plant J 26(3):265–274
11. Faivre-Rampant O, Gilroy EM, Hrubikova K, Hein I, Millam S, Loake GJ, Birch P, Taylor M, Lacomme C (2004) Potato virus X-induced gene silencing in leaves and tubers of potato. Plant Physiol 134(4):1308–1316
12. Pflieger S, Blanchet S, Camborde L, Drugeon G, Rousseau A, Noizet M, Planchais S, Jupin I (2008) Efficient virus-induced gene silencing in Arabidopsis using a 'one-step' TYMV-derived vector. Plant J 56 (4):678–690
13. Huang C, Xie Y, Zhou X (2009) Efficient virus-induced gene silencing in plants using a modified geminivirus DNA1 component. Plant Biotechnol J 7(3):254–265
14. Igarashi A, Yamagata K, Sugai T, Takahashi Y, Sugawara E, Tamura A, Yaegashi H, Yamagishi N, Takahashi T, Isogai M (2009) Apple latent spherical virus vectors for reliable and effective virus-induced gene silencing among a broad range of plants including tobacco, tomato, Arabidopsis thaliana, cucurbits, and legumes. Virology 386(2):407–416
15. Zhang C, Bradshaw JD, Whitham SA, Hill JH (2010) The development of an efficient multipurpose bean pod mottle virus viral vector set for foreign gene expression and RNA silencing. Plant Physiol 153(1):52–65
16. Donald R, Jackson AO (1994) The barley stripe mosaic virus gamma b gene encodes a multifunctional cysteine-rich protein that affects pathogenesis. Plant Cell 6 (11):1593–1606
17. Holzberg S, Brosio P, Gross C, Pogue GP (2002) Barley stripe mosaic virus-induced gene silencing in a monocot plant. Plant J 30 (3):315–327
18. Scofield SR, Huang L, Brandt AS, Gill BS (2005) Development of a virus-induced gene-silencing system for hexaploid wheat and its use in functional analysis of the Lr21-mediated leaf rust resistance pathway. Plant Physiol 138 (4):2165–2173
19. Lee W-S, Rudd JJ, Kanyuka K (2015) Virus induced gene silencing (VIGS) for functional analysis of wheat genes involved in Zymoseptoria tritici susceptibility and resistance. Fungal Genet Biol 79:84–88
20. Hein I, Barciszewska-Pacak M, Hrubikova K, Williamson S, Dinesen M, Soenderby IE,

Sundar S, Jarmolowski A, Shirasu K, Lacomme C (2005) Virus-induced gene silencing-based functional characterization of genes associated with powdery mildew resistance in barley. Plant Physiol 138(4):2155–2164

21. Hu P, Meng Y, Wise RP (2009) Functional contribution of chorismate synthase, anthranilate synthase, and chorismate mutase to penetration resistance in barley-powdery mildew interactions. Mol Plant Microbe Interact 22 (3):311–320
22. Ali SS, Gunupuru LR, Kumar GS, Khan M, Scofield S, Nicholson P, Doohan FM (2014) Plant disease resistance is augmented in uzu barley lines modified in the brassinosteroid receptor BRI1. BMC Plant Biol 14(1):1
23. Perochon A, Jianguang J, Kahla A, Arunachalam C, Scofield SR, Bowden S, Wallington E, Doohan FM (2015) TaFROG encodes a Pooideae orphan protein that interacts with SnRK1 and enhances resistance to the mycotoxigenic fungus Fusarium graminearum. Plant Physiol 169(4):2895–2906
24. Walter S, Kahla A, Arunachalam C, Perochon A, Khan MR, Scofield SR, Doohan FM (2015) A wheat ABC transporter contributes to both grain formation and mycotoxin tolerance. J Exp Bot 66:2583–2593
25. Bruun-Rasmussen M, Madsen CT, Jessing S, Albrechtsen M (2007) Stability of barley stripe mosaic virus-induced gene silencing in barley. Mol Plant Microbe Interact 20 (11):1323–1331
26. Scofield SR, Nelson RS (2009) Resources for virus-induced gene silencing in the grasses. Plant Physiol 149(1):152–157
27. Ansari KI, Walter S, Brennan JM, Lemmens M, Kessans S, McGahern A, Egan D, Doohan FM (2007) Retrotransposon and gene activation in wheat in response to mycotoxigenic and non-mycotoxigenic-associated Fusarium stress. Theor Appl Genet 114(5):927–937
28. Livak KJ, Schmittgen TD (2001) Analysis of relative gene expression data using real-time quantitative PCR and the 2− ΔΔCT method. Methods 25(4):402–408
29. Abràmoff MD, Magalhães PJ, Ram SJ (2004) Image processing with ImageJ. Biophotonics Int 11(7):36–42

Chapter 8

Agrobacterium-Mediated Transformation of Barley Immature Embryos

Alison Hinchliffe and Wendy A. Harwood

Abstract

Barley transformation is an essential tool for a range of functional genomics studies as well as for future crop improvement. The demand for efficient crop transformation systems continues to grow, with new genome editing technologies adding to that demand. Here we describe an efficient and routine transformation protocol for the spring barley Golden Promise, based on Agrobacterium-mediated inoculation of immature embryos. This protocol has been widely used for overexpression and RNAi applications and more recently for CRISPR/Cas9 mediated genome editing. Average transformation efficiencies of 25% can be easily achieved.

Key words Barley transformation, *Agrobacterium tumefaciens*, Immature embryos, Genome editing, Genome engineering

1 Introduction

Over the last few years enormous progress has been made in the development of new, more precise tools for the engineering of plant genomes. Genome editing using CRISPR/Cas9 is the highest profile of these new techniques [1, 2]. Transformation is often the bottleneck in the application of new techniques as it is usually required to deliver the components for genome editing. It also remains essential for a range of gene validation studies. For these reasons efficient transformation is vital to allow progress in a wide range of research areas.

Transformation of barley has been reported using a range of techniques including biolistic methods [3] and Agrobacterium-mediated protocols [4, 5]. Currently the most efficient methods are Agrobacterium-mediated and these also tend to give higher quality transgenic events with lower numbers of inserted transgene copies [6]. A range of target tissues have been employed for barley transformation; however, by far the most popular are immature embryos. Immature embryos collected from high-quality donor

Wendy A. Harwood (ed.), *Barley: Methods and Protocols*, Methods in Molecular Biology, vol. 1900,
https://doi.org/10.1007/978-1-4939-8944-7_8, © Springer Science+Business Media, LLC, part of Springer Nature 2019

plants of the spring barley Golden Promise yield highly embryogenic callus that easily regenerates.

The preparation of constructs for transformation used to be time consuming but with new assembly techniques, in particular Golden Gate assembly, this is no longer a limiting factor as most steps can be automated [7]. The choice of selectable marker is important in any transformation protocol. In barley the most efficient selectable marker is the *hpt* gene conferring resistance to the antibiotic hygromycin [5]. Other selectable marker genes such as *bar* or *nptII* can also be used [8]. The availability of a range of selectable markers gives the opportunity for re-transformation of existing transgenic lines if required.

The principal remaining challenge in barley transformation is genotype dependence. Whereas the spring barley Golden Promise transforms with high efficiency, many modern varieties are recalcitrant to transformation. For successful transformation a variety needs both efficient regeneration and good susceptibility to Agrobacterium.

In this chapter we describe a high-throughput method for the transformation of Golden Promise using Agrobacterium inoculation of immature embryos and selection on hygromycin. This protocol has been successfully used to introduce over 1000 constructs to barley. These constructs include those for overexpression, for RNAi-mediated silencing, for promoter testing, and for CRISPR/Cas9-based genome editing. Large T-DNAs up to 40 kb in size can be introduced. The protocol has been transferred to several laboratories worldwide and routinely gives efficiencies of 25%.

2 Materials

2.1 Plant Material

Seeds of Golden Promise are sown every 2 weeks to provide a constant supply of immature embryos at the correct stage. The barley compost mix contains a 2:2:1 mix of Levington M3 compost:Perlite:Grit. The mix also contains a slow release fertilizer, Osmocote at the manufacturers recommended concentration. Seed is sown directly into trays with 24 5 cm compartments and then plants are transplanted into 12 cm pots once they reach a height of approximately 20 cm. Plants are grown in controlled environment rooms with a day temperature of 15 °C and a night temperature of 12 °C. Light intensity is at least 400 $\mu mol/m^2/s^1$ at the mature plant canopy level and is provided by metal halide lamps (HQI) supplemented with tungsten bulbs. Humidity is kept at 70% (*see* **Note 1**).

2.2 Stock Solutions

1. 100X Vitamin stock for callus induction (CI) medium: 100 mg/L thiamine HCl, 35 g/L myo-inositol, and 69 g/L proline. This stock should be filter sterilized and stored at 4 °C.
2. 100X Vitamin stock for transition (T) and regeneration media (R): 40 mg/L thiamine HCl and 10 g/L myo-inositol. This stock should be filter sterilized and stored at 4 °C.
3. $CuSO_4$ stock: 125 mg of $CuSO_4 \cdot 5H_2O$ dissolved in a total volume of 100 mL water, filter sterilized, and stored frozen.
4. Hygromycin: purchased as a sterile 50 mg/mL stock, divided into 1 mL aliquots, and stored frozen.
5. Timentin: 160 mg/mL stock made up in water, divided into 1 mL aliquots, and stored frozen.
6. Dicamba: 2.5 mg/L stock made up in water, filter sterilized, divided into 1 mL aliquots, and stored frozen.
7. 2,4-D: 2.5 mg/mL stock made up in 100% ethanol and stored at −20 °C.
8. BAP: 1 mg/mL stock made up in water with a few drops of 1 *M* NaOH and stored frozen (*see* **Note 2**).

2.3 Bacterial Medium

For culture of Agrobacterium, bacterial culture medium MG/L [9] is used, this contains: 5.0 g/L tryptone, 5.0 g/L mannitol, 2.5 g/L yeast extract, 1.0 g/L L-glutamic acid, 250 mg/L KH_2 PO_4, 100 mg/L NaCl, 100 mg/L $MgSO_4 \cdot 7H_2O$, and 10 μL biotin (0.1 mg/L stock). The media is adjusted to pH 7.2 with NaOH. For the preparation of solid media 15 g/L agar is added.

2.4 Plant Culture Media

1. Callus induction (CI): 4.3 g/L Murashige and Skoog plant salt base (Duchefa M0221), 30 g/L maltose, 1.0 g/L casein hydrolysate, 10 ml/L of CI vitamin stock, 2.5 mg/L Dicamba, 1.25 mg/L $CuSO_4 \cdot 5H_2O$, 3.5 g/L Phytagel. The media is adjusted to pH 5.8 with NaOH. Additional copper is omitted during cocultivation (*see* **Note 3**).
2. Transition (T): 2.7 g/L Murashige and Skoog modified plant salt base (without NH_4NO_3) (Duchefa M0238), 20 g/L maltose, 165 mg/L NH_4NO_3, 750 mg/L glutamine, 10 mL/L T/R vitamin stock, 2.5 mg/L 2,4-dichlorophenoxy acetic acid (2,4-D), 0.1 mg/L 6-benzylaminopurine (BAP), 1.25 mg/L $CuSO_4 \cdot 5H_2O$, 3.5 g/L Phytagel. The pH is adjusted to 5.8.
3. Regeneration (R): 2.7 g/L Murashige and Skoog modified plant salt base (without NH_4NO_3) (M0238 Duchefa), 20 g/L maltose, 165 mg/L NH_4NO_3, 750 mg/L glutamine, 10 mL/L T/R vitamin stock, 3.5 g/L Phytagel. The pH is adjusted to 5.8.

2.5 *Agrobacterium* Strain

Agrobacterium strain AGL1 is used for all transformation experiments.

2.6 Constructs

pBRACT vectors are appropriate for barley transformation [10]. Vectors should contain the *hpt* gene conferring hygromycin resistance under a 35 s promoter at the left border (LB) of the T-DNA. Vectors suitable for overexpression or RNAi applications are described at www.bract.org. New vectors for CRISPR/Cas9-based gene knockout are described in Chapter 14.

2.7 Equipment

1. Very fine, sharp forceps for embryo isolation.
2. Binocular dissecting microscope for embryo isolation.
3. 500 mL disposable filter units with 0.22 μm membrane.
4. Vacuum pump for filter sterilization.
5. Standard and deep 9 cm Petri plates.
6. Glass culture tubes for plantlet rooting or disposable plastic rooting tubes.
7. Micropore tape for sealing Petri plates.
8. Oven set to 60 °C for warming media.

3 Methods

3.1 Plant Culture Media Preparation

1. All plant culture media are solidified with phytagel. Prepare phytagel in Elga purified water at double the final concentration and sterilize by autoclaving. Store at room temperature until required.
2. Weigh all other media components, except for those listed as stock solutions above, to give double the final required concentration (*see* **Note 4**). Adjust pH to 5.8 and sterilize by filter sterilization inside a laminar flow hood. Connect disposable filter units to a vacuum pump to speed up filter sterilization.
3. Heat the filter-sterilized media and the phytagel to 60 °C in an oven.
4. Add all the required sterile stock components to the warm media and mix well. During all tissue culture steps following cocultivation, Timentin is added at 160 mg/L to control Agrobacterium growth and Hygromycin is added at 50 mg/L for selection of transgenic material.
5. Pour the warm, double concentration medium into the bottle containing the warm double concentration phytagel.
6. Mix well and pour plates as quickly as possible. 500 mL of CI medium will give 20 standard 9 cm plates. 500 mL of T or R medium will give 15 deep 9 cm plates.
7. Leave plates with lids tilted to dry before packing for storage at 10 °C (*see* **Note 5**).

3.2 Preparation of Agrobacterium

1. Select a single colony of Agrobacterium AGL1, containing the required vector, and inoculate 10 mL of liquid MG/L medium containing 25 μg/mL rifampicin and 50 μg/mL kanamycin.
2. Incubate at 28 °C with shaking at 180 rpm for 40 h.
3. Add 10 mL of sterile 30% aqueous glycerol to the bacterial culture and mix by inverting several times.
4. Prepare 400 μL aliquots of the glycerol/bacterial mix in 0.5 mL tubes and maintain at room temperature for 2 h mixing by inversion every 30 min.
5. Store the prepared standard inoculums at −80 °C ready for use.

3.3 Collection and Sterilization of Immature Seed

1. Barley spikes containing immature seed should be collected when the embryos are approximately 1.5 mm in diameter. The spikes will have awns that are just beginning to splay out (Fig. 1a) and look yellow at the tips.
2. A single seed from the middle of the spike should be removed and the size of the embryo checked before cutting the spike.
3. Remove seeds from the spike and remove the awns taking care not to damage the seed coat. Place into a screw topped jar (Fig. 1b).
4. Wash the seed with 70% ethanol for 2 min with occasional shaking. All subsequent steps are carried out in a laminar flow hood under sterile conditions.
5. Tip off the ethanol and wash twice with sterile water.
6. Sterilize the seed for 4 min in a 10% sodium hypochlorite solution after diluting 1 to 1 with sterile water.
7. Remove the sodium hypochlorite and wash four times with sterile water. Drain off all water after the final wash and leave sterile seeds in the screw top jar (Fig. 1b).

Fig. 1 Selection of immature seeds. (**a**) Spikes at the correct stage for immature embryo isolation. (**b**) Immature seeds following sterilization

3.4 Isolation of Immature Embryos

1. Isolate immature embryos from the sterilized seed in a laminar flow hood under a dissecting microscope. It is convenient to place a dark colored tile on the microscope base, clean well with 70% ethanol and to carry out all manipulation on the sterile tile.
2. Tip 10–20 sterile seeds onto the sterile tile. Position the first seed embryo side up (*see* **Note 6**) (Fig. 2a).
3. Hold the seed firm by piercing it with one pair of fine forceps (Fig. 2b).
4. With a second pair of fine forceps, place one blade under the seed coat close to where the seed is being held (Fig. 2c).
5. Pinch the seed coat and peel back over the position of the immature embryo (Fig. 2d).
6. Further expose the immature embryo by gently pushing back the seed coat to the right of the embryo (Fig. 2e).
7. Position the fine forceps over the embryonic axis of the immature embryo with one blade just under the root axis and one blade under the tip of the shoot axis (Fig. 2f).
8. Pinch off the embryonic axis and discard (Fig. 2g).
9. Squeeze the seed gently to release the immature embryo (scutellum) and flip it onto the forceps so that it is positioned scutellum side up (Fig. 2h).

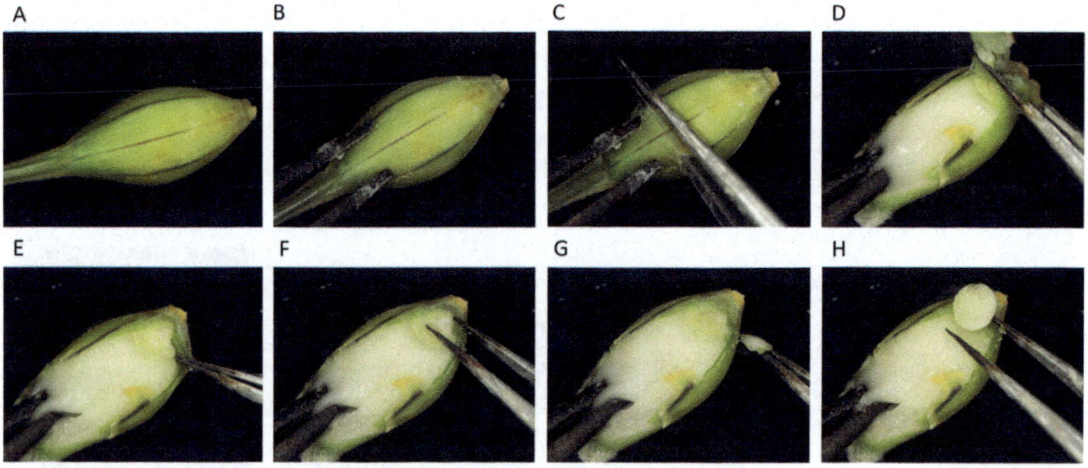

Fig. 2 Isolation of immature embryos. (**a**) Immature seed positioned for embryo isolation. (**b**) Seed held firm with forceps. (**c**) Second pair of fine forceps positioned to remove seed coat. (**d**) Seed coat peeled back to expose the embryo. (**e**) Seed coat pushed back on the right-hand side to further expose the embryo and embryo axis. (**f**) Fine forceps positioned to remove the embryonic axis. (**g**) Embryonic axis removed. (**h**) Embryo removed and flipped scutellum side up ready to transfer to CI medium

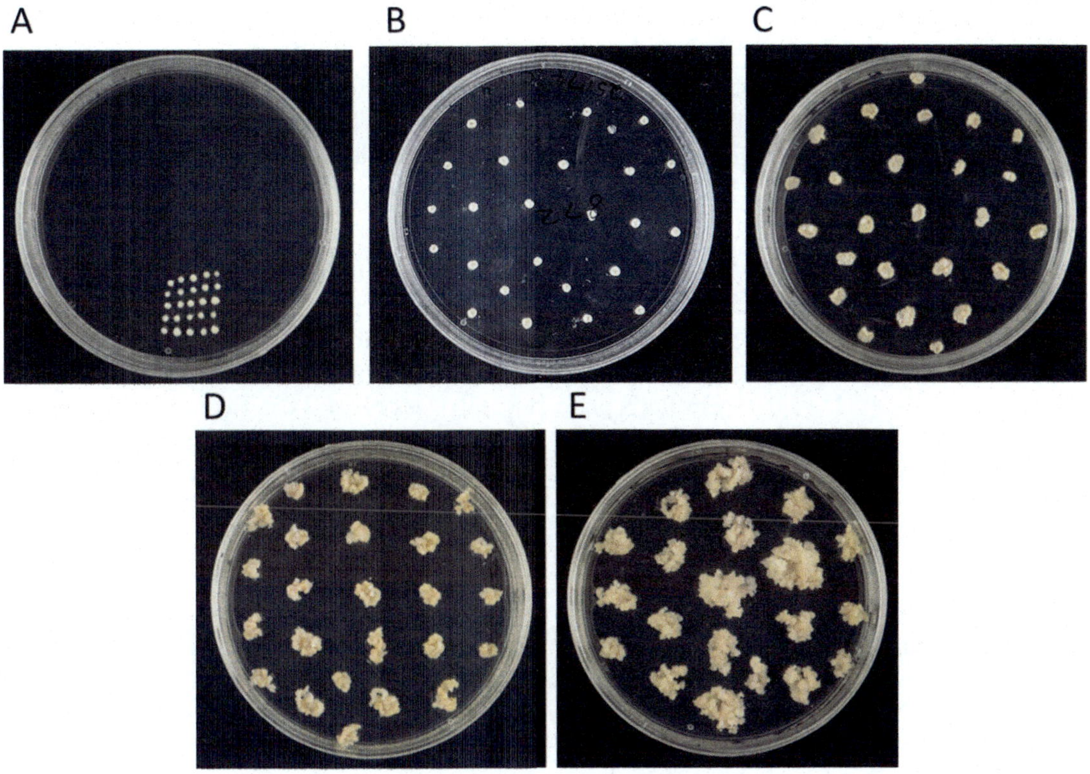

Fig. 3 Transformation and callus induction. (**a**) Immature embryos on CI medium after isolation arranged scutellum side up ready for Agrobacterium inoculation. (**b**) Embryo cocultivation with Agrobacterium. (**c**) Callus induction—selection 1. (**d**) Callus induction–selection 2. (**e**) Callus induction—selection 3

10. Place the embryo onto a prepared plate of CI medium without any antibiotics. Continue until 25 immature embryos have been isolated, place the embryos in a small block to one side of the 9 cm Petri plate (Fig. 3a) (*see* **Note 7**).
11. Continue until the required number of embryos have been isolated (*see* **Note 8**). Store plates of embryos at 23–24 °C in the dark overnight.

3.5 Preparation of Agrobacterium for Inoculation

1. Agrobacterium for inoculation of immature embryos is grown up overnight in the absence of antibiotics.
2. Remove one prepared Agrobacterium standard inoculum from the freezer and add to 10 mL sterile MG/L medium. Incubate overnight (approx. 18 h) at 28 °C with shaking at 200 rpm.
3. Use this full-strength culture for inoculation of immature embryos.

3.6 Agrobacterium Inoculation and Immature Embryo Cocultivation

1. Use 200 μL of the prepared Agrobacterium inoculum to inoculate each plate of 25 immature embryos.
2. Using a 200 μL pipette, drop one small drop of inoculum onto the surface of each of the 25 immature embryos. When all 25 embryos have been treated, tilt the plate so that excess Agrobacterium runs to the side.
3. Leave the plate for 5–10 min to dry (*see* **Note 9**).
4. After drying, transfer the immature embryos to a clean CI plate for cocultivation, placing embryos scutellum side down. For cocultivation the CI medium should not contain antibiotics or additional copper. Spread the immature embryos over the full 9 cm Petri plate (Fig. 3b) (*see* **Note 10**). Take care to place the embryos gently on the medium surface and do not touch the surface with the forceps as this may lead to excessive Agrobacterium growth.
5. Seal each plate with micropore tape and incubate at 23–24 °C for 3 days (*see* **Note 11**).

3.7 Callus Induction

1. After 3 days cocultivation transfer embryos to CI medium with the antibiotics Timentin and Hygromycin and with additional copper (selection plates). Seal plates with micropore tape and culture at 23–24 °C in the dark for 2 weeks. This step is referred to as selection 1 (Fig. 3c).
2. It is possible to check for expression of a reporter gene such as the *gus* gene at this stage. It is common to see relatively little GUS expression after cocultivation (Fig. 5a).
3. After 2 weeks transfer the immature embryos to a fresh CI selection plate (selection 2) Fig. 3d) and culture under the same conditions.
4. After a further 2 weeks transfer again to a fresh selection plate (selection 3) (Fig. 3e). At this stage it will be necessary to reduce the number of embryo-derived calli per plate. At each subculture the entire callus should be transferred and all callus from each individual immature embryo kept together.
5. Expression of the gus reporter gene can be checked after 4 weeks callus induction to monitor stable transformation events (Fig. 5b).

3.8 Regeneration and Rooting

1. After a total of 6 weeks callus induction, the immature-embryo-derived calli are ready to start regeneration. Gus reporter gene expression at this stage will show intense blue embryogenetic callus (Fig. 5c).
2. Transfer calli to transition medium (T) keeping the level of hygromycin and Timentin the same as during callus induction (Fig. 4a). Culture in the light at 24 °C. For the 1st week after

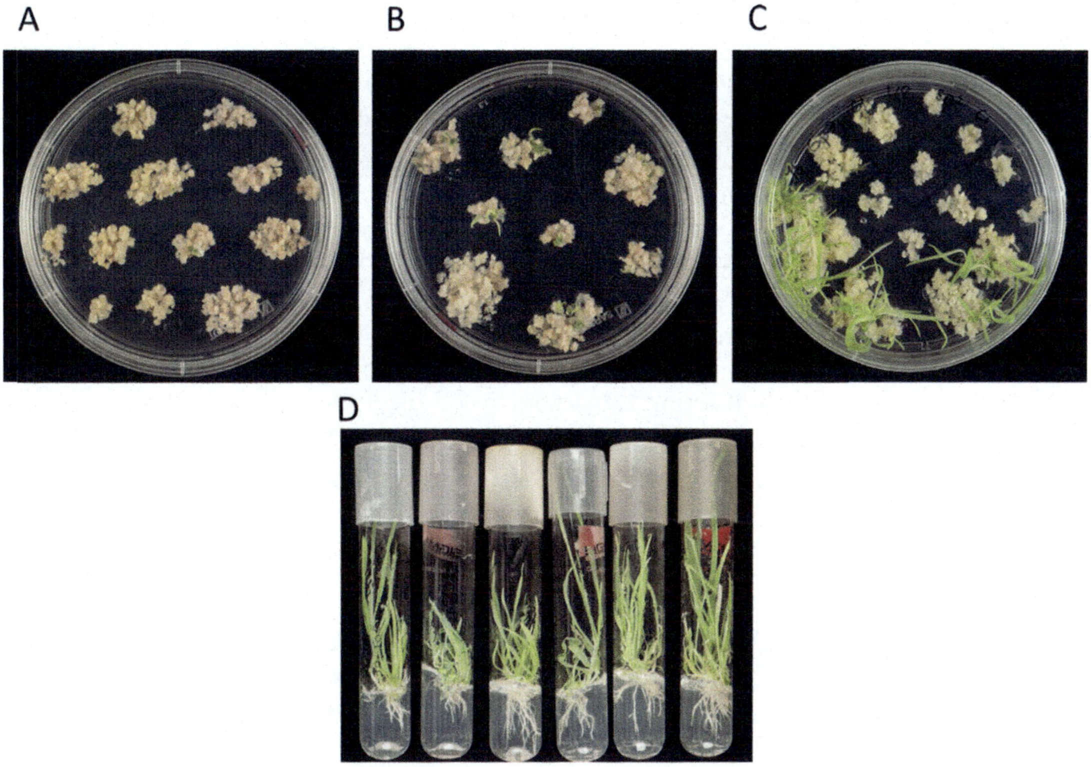

Fig. 4 Regeneration and rooting. (**a**) Callus on transition medium. (**b**) Early regeneration stage. (**c**) Late regeneration stage. (**d**) Plantlet rooting

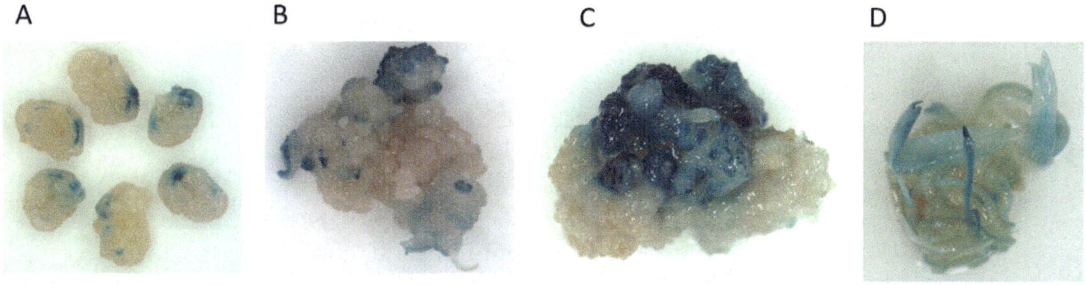

Fig. 5 GUS reporter gene expression. (**a**) Transient GUS expression following cocultivation with Agrobacterium. (**b**) GUS positive sectors after 4 weeks callus induction. (**c**) Large GUS positive region following 6 weeks callus induction. (**d**) GUS expression in regenerating shoots and roots

transfer to the light, cover the plates with a thin sheet of paper to reduce the light levels. The first green areas, indicating the initiation of transgenic shoots, should be visible after a few days in the light. At this stage the calli will grow rapidly, so further reduce the number of calli/plate during the transfer to T medium.

3. After 2 weeks on transition medium transfer all calli with green regions or developing shoots to regeneration medium (Fig. 4b). At this stage all growth hormones are omitted but hygromycin and Timentin remain at the same concentrations.
4. Leave regenerating calli on the same regeneration plates until small plantlets have shoots 2–3 cm in length and are forming roots (Fig. 4c). Regenerated shoots and roots can also be checked to monitor reporter gene expression (Fig. 5d).
5. Carefully separate out individual plantlets using fine forceps and transfer them to rooting tubes (Fig. 4d). When transferring to rooting tubes take care not to transfer callus as well. Rooting tubes contain CI medium without growth regulators or additional copper but still with the same levels of hygromycin and Timentin (*see* **Note 12**).

3.9 Transfer to Soil

1. Once plant shoots reach the top of the rooting tubes they can be transferred to soil. The same compost mix is used as described for the Golden Promise donor plants.
2. Gently remove plants from the culture tubes using a long spatula and the roots are washed under a running tap to remove all culture medium. Plants are then potted into moist compost in 5 cm pots. For the 1st week of growth they are covered with a propagator to maintain high humidity.
3. Grow plants to maturity under the same conditions as the Golden Promise donor plants (*see* **Note 13**).
4. The presence of the introduced genes can be confirmed by PCR and transgene copy number determined by qPCR (*see* **Note 14**).

4 Notes

1. The quality of the plants from which immature seed is collected is one of the most important considerations for success. If possible, plants should be grown under controlled conditions and should be kept free of pests and disease. Plants should not be sprayed, particularly when they are flowering.
2. All hormone, antibiotic, and copper stocks are kept frozen in small aliquots and one tube is removed from the freezer each time media is prepared. This reduces the risk of contamination and ensure that only fresh stocks are used.
3. Adding additional copper during callus induction and the early stages of regeneration leads to a significant increase in the amount of embryogenetic callus and subsequent levels of regeneration. It should not however be included during cocultivation with Agrobacterium.

4. If preparing a total volume of 500 mL of medium, then you will need 250 mL of double concentration phytagel and 250 mL of double concentration medium components.
5. It is convenient to wrap poured plates in cling film and then to place back into the original Petri dish bags for storage. All plates should be used within 2 weeks of preparation.
6. The embryo is located at the opposite end of the seed to the awn and on the side with the pigmented stripes, not the side with the groove.
7. Generally 100 immature embryos are isolated for a stable transformation experiment. If only a small number of plants are required, this number can be reduced to as few as 25.
8. A skilled operator can isolate 25 immature embryos in 10 min.
9. A maximum of three plates of immature embryos should be inoculated at once and left to dry before the first embryos are moved to a clean plate for cocultivation.
10. During transfer to the clean plate for cocultivation it is important to blot the immature embryos on a clean region of the original plate to avoid transferring too much Agrobacterium culture. However, do not overclean the embryos so that there is insufficient Agrobacterium left during cocultivation. This step may take practice so that there is very little Agrobacterium overgrowth after cocultivation but also good transformation efficiencies.
11. Immature embryos are usually isolated 1 day and inoculated the following day. However, it is possible to isolate and inoculate on the same day or even to store isolated immature embryos for 2 days before inoculation.
12. We have found that young plantlets grow well on CI medium. However, at this stage transgenic plantlets are robust and will also be fine rooting on regeneration medium without any growth hormones. Long forceps are useful for transferring plantlets to rooting tubes. Disposable plastic culture tubes can also be used for rooting instead of glass.
13. We prefer to grow transgenic barley in controlled environment rooms until it has set seed at which point we move plants to the glasshouse to mature for harvest.
14. Quantitative Real-time PCR analysis of large numbers of transgenic plants has revealed an average of 50% single copy events with another 25% of plants carrying two copies of the introduced gene.

Acknowledgments

We acknowledge support from the Biotechnology and Biological Sciences Research Council (BBSRC) via grant [BB/N019466/1] and grant [BB/P013511/1] to the John Innes Centre.

References

1. Lawrenson T, Shorinola O, Stacey N, Liu C, Østergaard L, Patron N, Uauy C, Harwood W (2015) Induction of targeted, heritable mutations in barley and *Brassica oleracea* using RNA-guided Cas9 nuclease. Genome Biol 16:258
2. Ricroch A, Clairand P, Harwood W (2017) Use of CRISPR systems in plant genome editing: towards new opportunities in agriculture. Emerging Topics Life Sci 1(2):169–182
3. Wan Y, Lemaux PG (1994) Generation of large numbers of independently transformed fertile barley plants. Plant Physiol 104:37–48
4. Tingay S, McElroy D, Kalla R, Fieg S, Wang M, Thornton S, Brettell R (1997) *Agrobacterium tumefaciens*-mediated barley transformation. Plant J 11:1369–1376
5. Bartlett JG, Alves SC, Smedley M, Snape JW, Harwood WA (2008) High-throughput *Agrobacterium*-mediated barley transformation. Plant Methods 4:22
6. Travella S, Ross SM, Harden J, Everett C, Snape JW, Harwood WA (2005) A comparison of transgenic barley lines produced by particle bombardment and *Agrobacterium*-mediated techniques. Plant Cell Rep 23: 780–789
7. Volpi e Silva N, Patron N (2017) CRISPR-based tools for plant genome engineering. Emerging Topics Life Sci 1:135–149
8. Harwood W, Smedley M (2009) Barley transformation using biolistic techniques. In: Jones H, Shewry P (eds) Transgenic wheat, barley and oats. Methods in molecular biology™ (Methods and Protocols), vol 478, Humana Press, pp 125–136
9. Garfinkel M, Nester EW (1980) *Agrobacterium tumefaciens* mutants affected in crown gall tumorigenesis and octopine catabolism. J Bacteriol 144:732–743
10. Smedley MA, Harwood WA (2014) Gateway-compatible plant transformation vectors. Methods Mol Biol 1223:3–16

Chapter 9

Methods for the Simple and Reliable Assessment of Barley Sensitivity to Abiotic Stresses During Early Development

Agata Daszkowska-Golec, Anna Skubacz, Marzena Kurowska, Michał Słota, Dorota Swiergolik, and Iwona Szarejko

Abstract

Physiological assays that facilitate screening for various types of responses to abiotic stresses are well established for model plants such as Arabidopsis; however, there is a need to optimize similar tests for cereal crops, including barley. We have developed a set of stress assays to characterize the response of different barley lines during two stages of development—seed germination and seedling growth. The assays presented, including the response to osmotic, salt, oxidative stresses, and exogenously applied abscisic acid, can be used for forward screening of populations after mutagenesis as well as for phenotyping of already isolated mutants, cultivars, or breeding lines. As well as protocols for stress treatments, we also provide methods for plant stress response evaluation, such as chlorophyll *a* fluorescence (ChlF) and image analysis.

Key words Abiotic stresses, Drought stress, Salt stress, Osmotic stress, Oxidative stress, Seed germination, Seedling development, High-resolution imaging, Barley, Assays

1 Introduction

The response to stress during the early stages of development, such as seed germination and seedling growth, significantly influences the further development of plants [1]. The ability to cope with adverse environmental conditions is determined by various factors, including genotype, physiology, stage of development, and the length of the exposure to stress conditions.

The majority of the knowledge that has been gathered in the field of seed germination and seedling growth under stress conditions comes from studies on Arabidopsis and rice for which an advanced set of phenotyping protocols, molecular tools, and technologies are available [2, 3]. Taking into account the importance of barley and the abiotic stresses, such as drought, that affect its yield, assays that allow phenotyping of plants under stress conditions can help to gain insight into the mechanisms of response to the stress. Selection of an appropriate treatment, including the concentration

Wendy A. Harwood (ed.), *Barley: Methods and Protocols*, Methods in Molecular Biology, vol. 1900,
https://doi.org/10.1007/978-1-4939-8944-7_9, © Springer Science+Business Media, LLC, part of Springer Nature 2019

of a stress factor, is problematic for cereals, since there are differences in sensitivity to the same factors between species or even cultivars of the same species. Among monocots, the most widely examined species is rice, which is considered to be a stress-sensitive cereal in comparison to the more tolerant species, e.g., barley [4].

The most commonly used protocols for the study of abiotic stress response in monocots are based on the evaluation of the kinetics of seed germination [5–7] or seedling growth in hydroponics conditions [8]. The observations of the kinetics, which are based on counting the number of germinated seeds every day of the experiment, enable a fast estimation of the genotype's sensitivity to the stress being analyzed. The most popular method is germinating seeds using filter paper that is soaked with MS (Murashige and Skoog) medium or water with or without a factor being studied [5]. The assays of the sensitivity to abiotic stresses during seedling growth involve most commonly the measurements of the first leaf and root of the seedlings that are grown in a hydroponic solution supplemented with a stress factor [8].

Here we describe protocols of the assessment of barley sensitivity during seed germination and seedling development to salt stress, evoked by NaCl, and osmotic stress, induced by mannitol. Taking into account that the crucial hormonal player in plant stress response is ABA, we also describe a protocol describing evaluation of the sensitivity to ABA during both stages of development. The presented protocols also include examination of barley seedlings in terms of tolerance to oxidative stress which causes induction of reactive oxygen species (ROS), and consequently the oxidative damage of various cell components: lipids, proteins, and DNA [9]. ROS include highly reactive singlet oxygen ($^1O_2$), superoxide anion radical ($O_2^{\cdot}$), peroxyl radical ($RO_2\cdot$), hydrogen peroxide (H_2O_2), and hydroxyl radical ($OH\cdot$). To evaluate plant response to the oxidative stress, we propose the treatment of seedlings with well-known oxidative stress inducers: hydrogen peroxide (H_2O_2) or methyl viologen (MV). Treatment with these two compounds leads to the same effect in the cell—the increased amount of H_2O_2, but in the case of MV treatment, the H_2O_2 is produced endocellularly. MV is thought to be a very effective electron acceptor that catalyzes the photo-reduction of O_2, thereby accelerating the production of (O_2^-) and H_2O_2 [10]. Leaf injury by MV treatment is often used to assay the resistance of plants to oxidative stress [11]. Compared with other ROS, H_2O_2 is a relatively long-lived molecule that is able to diffuse across cell membranes [12]. It is a highly reactive species, and therefore potentially harmful at high concentrations.

The measurements of the effect of oxidative stress could be done in different ways. In situ detection of H_2O_2 is based on staining the leaves with 3,3′-diaminobenzidine (DAB). DAB is oxidized by hydrogen peroxide in the presence of peroxidases and generates a dark brown precipitate [13]. Here, we propose using this staining method for evaluation of response to oxidative stress. The more brown discoloration on the leaves observed after DAB staining, the more sensitive to oxidative stress is the seedling. Another proposed method for the evaluation of response to oxidative stress is based on estimation of the lipid damage. The induction of oxidative stress leads to accumulation of ROS which degrade polyunsaturated fatty acids (PUFA), which in turn leads to the formation of typical products of lipid oxidation, including malondialdehyde (MDA). This compound is considered a marker for oxidative stress, as the content of MDA increases with the increased level of oxidative stress. The lower the MDA content, the more tolerant to oxidative stress is an analyzed genotype, due to the better enzymatic and nonenzymatic antioxidant defense system. Measuring of the MDA contents is based on the reaction with thiobarbituric acid (TBA). As a result, the colored products called thiobarbituric acid reactive substances (TBARS) are formed [14], which can be assayed spectrophotometrically.

In this chapter, together with detailed protocols of stress treatments, we provide methods for the evaluation of response to stress in barley based on chlorophyll *a* fluorescence (ChlF) and image analysis techniques. Taking into account that the most common plant response to stress is photoinhibition, it is crucial to investigate the photosynthetic performance in stressed plants. It can be measured using a fluorimeter, whose main advantage is the nondestructive and fast analysis of chlorophyll *a* fluorescence (ChlF). ChlF measurements are a very popular method for evaluating the influence of abiotic stress on photosynthesis. The determination of the parameters that are closely related to the functioning of PSII (photosystem II) and the photosynthetic electron transport chain is possible based on these measurements. Simultaneously, this technique allows the evaluation of plant vitality under unfavorable environmental conditions [15, 16]. Chlorophyll *a* fluorescence has been successfully used to evaluate the responses to stress in different species including barley [17, 18]. Parameters proposed to be analyzed in the presented protocols are very sensitive to abiotic stresses in plants. Performance Index (PI_{ABS}) is an important and multiparametric expression that contains information about three independent parameters—the density of the reaction centers that corresponds to the absorption flux, the quantum yield of trapping, and the probability that a trapped exciton will move an electron into the electron transport chain beyond Q_A [19, 20]. The maximum quantum yield of primary photochemistry ΦP_0 is another very sensitive parameter recommended to be evaluated when stress response is studied. The set of parameters referred to as phenomenological energy fluxes per excited cross section (CS) include five parameters that express different events within PSII and enable understanding of the negative effect of stress on the photosynthesis process. These parameters include ABS/CS—absorption energy flux per cross-section (CS); TR_0/CS—trapped energy flux per CS; ET_0/CS—electron transport flux per CS; DI_0/CS—dissipation energy flux per CS; and RC/CS—active reaction center per CS.

The image analysis techniques can strongly facilitate the laborious manual measurements of plant morphometric traits. The imaging of plant canopy using a stationary camera has an advantage of recording the phenotypic traits of a plant in a nondestructive manner. It allows the capture of multiple images of a plant canopy during the course of its growth providing the data on growth dynamics. Advanced image segmentation techniques allow for the assessment of more complex traits and provide high-quality quantitative data. The protocols of image analyses presented in this chapter comprise of the analyses of basic morphometric traits on raw images of plants (e.g., plant height, width, length of shoots) and more complex traits (e.g., the area of green pixels, color classification) employing image segmentation methods.

Summarizing, the presented assays of barley stress response are fast and easy to perform in an average laboratory, and do not require any advanced equipment. These advantages, taken together, indicate the possibility of a wide application of the described platform.

2 Materials

2.1 Media, Solutions, and Components Needed for Assays of Sensitivity to ABA, NaCl, and Mannitol Stresses During Seed Germination in Barley

Prepare all solutions using ultrapure water (prepared by purifying deionized water to attain a sensitivity of 18 M Ω cm at 25 °C) and analytical grade reagents. Prepare and store all reagents at room temperature (unless indicated otherwise).

1. *Abscisic acid solution*: Prepare 1.5 mL of 25 mM abscisic acid (ABA) solution in 99.8% ethanol. Weigh 9.91 mg abscisic acid (ABA) and dissolve it in 1.5 mL 99.8% ethanol. Transfer the completely dissolved solution to 1.5 mL Eppendorf tube. Keep at −20 °C in darkness.
2. *Salt solution*: Prepare 1 L of 1 M sodium chloride (NaCl) in water. Weigh 58.44 g sodium chloride and transfer to 1 L bottle. Add demineralized water to a volume of 1 L. Autoclave the prepared solution.
3. *Mannitol solution*: Prepare 1 L of 1 M mannitol in water. Weigh 182.17 g mannitol and transfer to 1 L bottle. Add demineralized water to a volume of 1 L. Autoclave the prepared solution.

4. *Components needed for assays of sensitivity to ABA and abiotic stresses during seed germination in barley*
   (a) Petri dishes, diameter 9 cm.
   (b) Whatman filters, Grade 1.

2.2 Media, Solutions, and Components Needed for Assays of Sensitivity to ABA, NaCl, and Mannitol Stresses During Seedling Development in Barley

Prepare all solutions using ultrapure water (prepared by purifying deionized water to attain a sensitivity of 18 M Ω cm at 25 °C) and analytical grade reagents. Prepare and store all reagents at room temperature (unless indicated otherwise).

1. *MS solution*
   (a) Prepare 1 L of macronutrient stock solution: dissolve 33 g NH_4NO_3, 38 g KNO_3, 8.8 g $CaCl_2 \times 2H_2O$, 7.4 g $MgSO_4 \times 7H_2O$, and 3.4 g KH_2PO_4 in H_2O. Each reagent should be dissolved *separately* in H_2O, then all solutions should be mixed and adjust to 1 L of H_2O.
   (b) Prepare 1 L of micronutrient stock solution: dissolve 6.2 g H_3BO_3, 22.3 g $MnSO_4 \times H_2O$, 8.6 g $ZnSO_4 \times 7H_2O$, 0.25 g $Na_2MoO_4 \times 2H_2O$, 0.025 g $CuSO_4 \times 5H_2O$, 0.025 g $CoCl_2 \times 6H_2O$, and 0.83 g KI in H_2O. Each reagent should be dissolved separately in H_2O, then all solutions should be mixed and adjust to 1 L of H_2O.
   (c) Prepare 100 mL of MS vitamin stock solution: dissolve 0.05 g Nicotinic acid, 0.05 g Pyridoxine HCl, 0.01 g Thiamine HCl, and 0.2 g Glycine in 100 mL H_2O.
   (d) Prepare 1 L of FeNaEDTA solution: dissolve 3.73 g FeNaEDTA in 1 L H_2O.

   Use 50 mL macronutrient stock solution, 1 mL micronutrient stock solution, 1 mL MS vitamin stock solution, 10 mL FeNaEDTA solution, 0.1 g myo-Inositol and make up with demineralized water to a volume of 1 L. Adjust pH to 5.8. Autoclave the prepared solution.
2. *Abscisic acid solution*: Prepare 1.5 mL of 25 mM abscisic acid (ABA) solution in 99.8% ethanol (as described in Subheading 2.1, **item 1**).
3. *Salt solution*: Prepare 1 L of 1 M sodium chloride in MS liquid medium. Weigh 58.44 g sodium chloride and transfer to 1 L bottle. Add MS solution to a volume of 1 L. Autoclave the prepared solution.
4. *Mannitol solution*: 1 M mannitol in MS liquid medium. Weigh 182.17 g mannitol and transfer to 1 L bottle. Add MS solution to a volume of 1 L. Autoclave the prepared solution.
5. *Components needed for assays of sensitivity to ABA and abiotic stresses during seedling development in barley*
   (a) Glass tubes (17 cm length) with cotton wool plug.

6. *Equipment needed for analysis of sensitivity to ABA and abiotic stresses during seedling development in barley*
   (a) Fluorimeter (PocketPEA fluorimeter, Hansatech Instruments Ltd., England).

2.3 Media, Solutions, and Components Needed for Assay of Sensitivity to Gradual Increase of Salt Concentration During Barley Growth in Pots

Prepare all solutions using tap water and analytical grade reagent. Prepare and store all reagents at room temperature (unless indicated otherwise).

1. *Salt solution*: Prepare 1 L of 1 M sodium chloride (NaCl) in water. Weigh 58.44 g sodium chloride and transfer to 1 L bottle. Add demineralized water to a volume of 1 L. Autoclave the prepared solution.
2. *Salt dilutions*: Prepare 1 L of 100, 200, and 300 mM NaCl dilutions. Add 100, 200, and 300 mL of 1 M NaCl stock solution, respectively, into individual bottles and fill up with water to a volume of 1 L. Mix gently the solutions.
3. *Components needed for assay of sensitivity to gradual increase of salt concentration during barley growth in pots*
   (a) Pots (8.5 × 8.5 × 9.5 cm) filled with mixture of soil and vermiculite (2:1).
4. *Equipment needed for evaluation of the sensitivity to gradual increase of salt concentration during barley growth in pots*
   (a) Fluorimeter (PocketPEA fluorimeter, Hansatech Instruments Ltd., England).

2.4 Media, Solutions, Components, and Equipment Needed for Assay of Sensitivity to Oxidative Stress During Seedling Development in Barley Using Mini-Hydroponic System and Detached Leaves

Prepare all solutions using ultrapure water (prepared by purifying deionized water to attain a sensitivity of 18 M Ω cm at 25 °C) and analytical grade reagents. Prepare and store all reagents at room temperature (unless indicated otherwise).

The plastic opaque containers used for hydroponic system have a capacity of 1 L. Therefore, the preparation of methyl viologen (MV) and hydrogen peroxide (H_2O_2) solutions are described for this volume. The solutions needed for determination of malondialdehyde (MDA) content are described for 30 mL (which was appropriate amount in conducted experiments).

1. *Half-strength Hoagland's solution*
   (a) Prepare 1 L of macronutrient stock solution: dissolve 28.76 g $NH_4H_2PO_4$; 25.28 g KNO_3; 59.04 g $Ca(NO_3)_2 \times 4H_2O$; and 61.62 g $MgSO_4 \times 7H_2O$ in H_2O. Each reagent should be dissolved separately in H_2O, then all solutions should be mixed and adjust to 1 L of H_2O.
   (b) Prepare 1 L of micronutrient stock solution: dissolve 2.86 g H_3BO_3; 1.81 g $MnCl_2 \times 4H_2O$; 0.22 g

$ZnSO_4 \times 7H_2O$; 0.08 g $CuSO_4 \times 5H_2O$; and 0.02 g $H_2MoO_4 \times H_2O$ in H_2O. Each reagent should be dissolved separately in H_2O, then all solutions should be mixed and adjust to 1 L of H_2O.

(c) Dissolve 1 g $FeC_6H_5O_7 \times 5H_2O$ in 100 mL H_2O. Store stock solutions at 4 °C.

Use 0.5 mL of $NH_4H_2PO_4$, 3 mL of KNO_3, 2 mL of $Ca(NO_3)_2 \times 4H_2O$, 1 mL of $MgSO_4 \times 7H_2O$, 0.5 mL of micronutrient stock solution, 1.25 mL of $FeC_6H_5O_7 \times 5H_2O$ and fill up with demineralized water to a volume of 1 L. Mix and adjust pH to 6.0 using 1 M HCl.

2. *Methyl viologen dichloride hydrate (MV) solution*

The molecular weight of methyl viologen dichloride hydrate is 257.16 g/mol for anhydrous basis and corrected for water content of the trihydrate form is 311.2 g/mol.
Methyl viologen stock solution: Prepare 1 mL of 100 mM methyl viologen solution in H_2O. Weight 31.1 mg methyl viologen and dissolve it in 1 mL H_2O. Transfer the completely dissolved solution to 1.5 mL Eppendorf tube. Prepare fresh solution before each experiment.

(a) For the treatment with 5 μM and 10 μM MV using a mini-hydroponic system:

- Prepare 1 L of 5 μM solution: add 50 μL of 100 mM stock solution and fill up to 1 L with a half-strength Hoagland's medium.
- Prepare 1 L of 10 μM solution: add 100 μL of 100 mM stock solution and fill up to 1 L with a half-strength Hoagland's medium.

(b) For the MV treatment of detached leaves use 10 μM of MV prepared as described above. Pour 15 mL of 10 μM of MV solution into conical tubes in which the treatment of detached leaves is performed.

3. *Hydrogen peroxide (H_2O_2) solution*

Use hydrogen peroxide solution, 50% wt. in H_2O, which is commercially available, a molecular weight 34.01 g, density 1.197 g/cm^3. A stock solution should be stored at 4 °C.

(a) For the treatment with 50 mM and 100 mM H_2O_2 in the mini-hydroponic system:

- Prepare 1 L of 50 mM H_2O_2 solution: add 2.84 mL of 50% wt. H_2O_2 stock solution and fill up to 1 L with a half-strength Hoagland's medium.
- Prepare 1 L of 100 mM H_2O_2 solution: add 5.68 mL of 50% wt. H_2O_2 stock solution and fill up to 1 L with a half-strength Hoagland's medium.

(b) For the treatment of detached leaves using H_2O_2 use 100 mM of H_2O_2 prepared as described above. Pour 15 mL of solution into conical tubes in which the treatment of detached leaves is performed.

4. *3-3′-diaminobenzidine (DAB) staining solution*

   (a) 0.1 mg/mL DAB solution: prepare 100 mL of solution by dissolving 10 mg DAB in 100 mL of 1× Tris-Acetate-EDTA buffer solution (pH 8.3), which is commercially available.

   (b) Mix by vortexing and adjust pH to 3.5 using 1 M HCl (*see* **Note 1**).

5. *Solutions needed for the determination of malondialdehyde (MDA) content*

   (a) 10% trichloroacetic acid (TCA) solution: prepare 30 mL of 10% TCA by adding 3 mL TCA solution and fill up to 30 mL with H_2O.

   (b) 0.25% thiobarbituric acid (TBA) solution in 10% TCA: prepare 30 mL of 0.25% TBA by dissolving 0.075 g TBA in 30 mL of 10% TCA.

6. *Components necessary for barley seed germination and growing*

   (a) Square Petri dishes 120 × 120 mm (Gosselin).

   (b) Whatman Filter Paper, Grade 1.

   (c) Pots (8.5 × 8.5 × 9.5 cm) filled with mixture of soil and vermiculite (2:1).

7. *The equipment needed for the mini-hydroponic treatment*

   (a) Mini-hydroponics setup consist of: a plastic opaque container of a 1 L capacity, a plastic opaque lid with 12 holes, air distributor with 12 outlets, air pipe with a non-return valve, air pump (Fig. 1).

8. *The equipment needed for the DAB staining*

   (a) Petri dishes, diameter 9 cm.

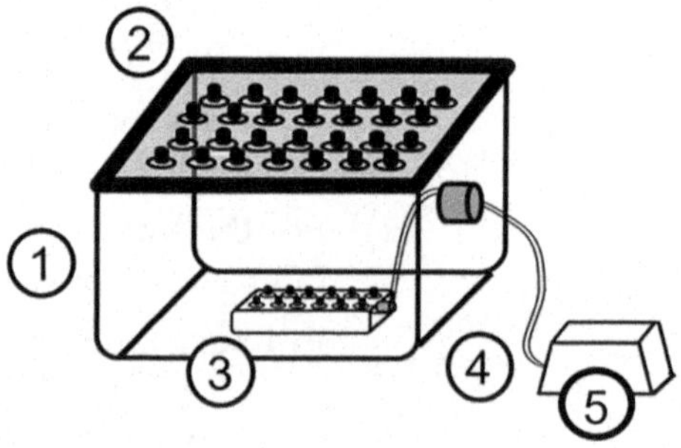

Fig. 1 A diagram of the mini-hydroponics setup consisting of: a plastic opaque container with a capacity of 1 L (*1*); a plastic opaque lid with openings (*2*); air distributor with outlets (*3*), air pipe with a non-return valve (*4*); air pump (*5*)

9. *The equipment needed for measuring the chlorophyll fluorescence*
   (a) Fluorimeter (PocketPEA fluorimeter, Hansatech Instruments Ltd., England).
10. *The equipment needed for the determination of malondialdehyde (MDA) content*
    (a) Spectrophotometer (e.g., Jena AG).

3 Methods

3.1 Assays of Sensitivity to ABA, NaCl, and Mannitol Stresses During Seed Germination in Barley

3.1.1 Plant Material Preparation

1. Sterilize seeds for 20 min using 1% sodium hypochloride.
2. Rinse the seeds three times using demineralized and autoclaved water. Each rinse should last 5 min.
3. Sow 30 seeds on each prepared Petri dish (*see* **Note 2**). Prepare 30 Petri dishes (diameter 9 cm) with two Whatman filter papers to set up three replicates for control and each of the treatments (Fig. 2).
4. Seal the Petri dishes using Parafilm.

3.1.2 Medium Preparation

In order to prepare appropriate dilution of reagents, use the formula:

$$v_1 = (c_2\, v_2)/c_1$$

where v_1 is required volume of stock solution, c_2 is concentration of prepared solution, v_2 is the volume of prepared solution, and c_1 is concentration of stock solution. The concentrations presented below are proposed to be used in protocols described in this chapter.

1. Prepare 15 mL of 10, 75, and 300 μM ABA dilutions. Add 6, 45, and 180 μL of 25 mM ABA stock solution, respectively, into conical tubes and fill up with water to a volume of 15 mL. Mix gently the solutions.
2. Prepare 15 mL of 150, 300, and 450 mM NaCl dilutions. Add 2.25, 4.5, and 6.75 mL of 1 M NaCl stock solution, respectively, into conical tubes and fill up with water to a volume of 15 mL. Gently mix the solutions.
3. Prepare 15 mL of 150, 400, and 550 mM mannitol dilutions. Add 2.25, 6, and 8.25 mL of 1 M mannitol stock solution, respectively, into conical tubes and fill up with water to a volume of 15 mL. Gently mix the solutions.
4. Prepare the Petri dishes containing two Whatman filters soaked with 5 mL of water or prepared dilutions of ABA, NaCl and mannitol.

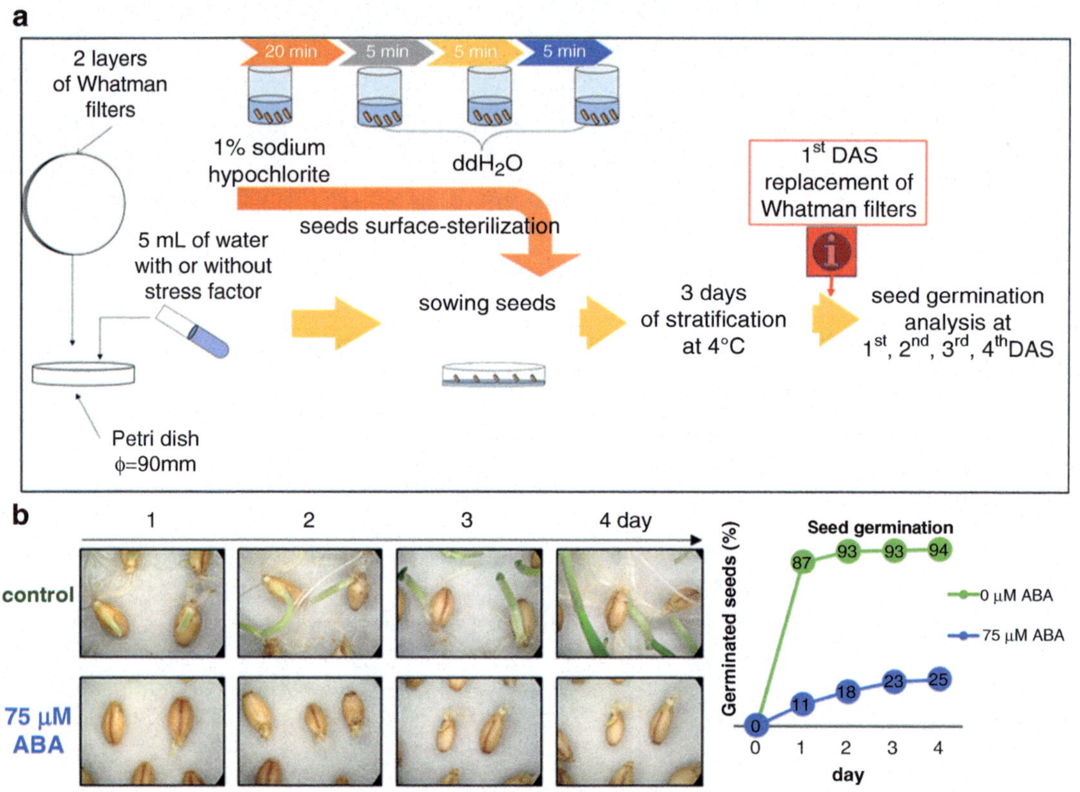

Fig. 2 The assay of sensitivity to ABA during seed germination in barley. (**a**) The workflow of the assay. (**b**) An example of results for one of the genotypes tested with the use of 75 μM ABA

3.1.3 Experiment Execution

1. Keep Petri dishes with sown seeds at 4 °C in darkness for 3 days in order to stratify the seeds.
2. After 3 days transfer Petri dishes to a growth room (22 °C, 16-h-light/8-h darkness, 200 μmol/m^{-2} s^{-1} illumination; *see* **Note 3**).
3. Analyze the number of germinated seeds on the 1st, 2nd, 3rd, and 4th day after stratification by manually counting and/or using the image analysis (Fig. 2; for detailed information *see* Subheading 3.5).

3.2 Assays of Sensitivity to ABA and Abiotic Stresses During Seedling Development in Barley

3.2.1 Plant Material and Experiment Setup Preparation

1. Prepare glass tubes (17 cm length) with a small cotton wool plug at the top, set up three replicates (each consisting of three tubes) for control and each treatment (*see* **Note 4**).
2. Soak cotton wool plugs with 3 mL demineralized and autoclaved water.
3. Sterilize 30 seeds (as described in Subheading 3.1.1).
4. Transfer single sterilized seeds into the glass tubes (*see* **Note 5**).
5. Cover the top with another glass tube and seal the junction using Parafilm (Fig. 3).

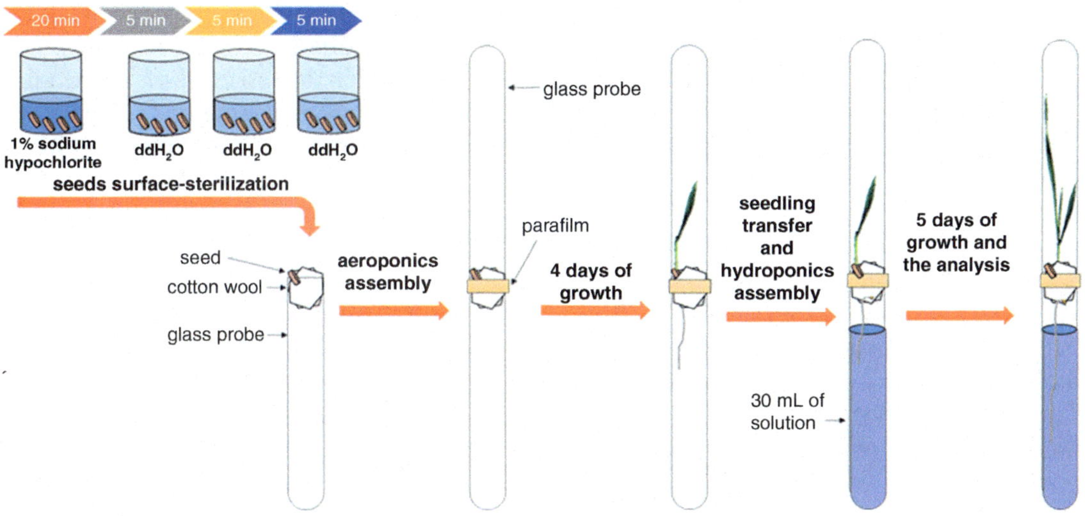

Fig. 3 The workflow for the assay of sensitivity to ABA and abiotic stresses during seedling development in barley

3.2.2 Medium Preparation

1. Prepare 90 mL of 10, 200, and 1000 μM ABA dilutions. Add 0.036, 0.72, and 3.6 mL of 25 mM ABA stock solution, respectively, into 100 mL cylinder. Add MS medium to a volume of 90 mL. Gently mix the solutions.
2. Prepare 90 mL of 100, 400, and 800 mM NaCl dilutions. Add 9, 36, and 72 mL of 1 M NaCl stock solution, respectively, into 100 mL cylinder. Add MS medium to a volume of 90 mL. Gently mix the solutions.
3. Prepare 90 mL of 200 and 500 mM mannitol dilutions. Add 18 and 45 mL of 1 M mannitol stock solution, respectively, into 100 mL cylinder. Add MS medium to a volume of 90 mL. Gently mix the solutions.

3.2.3 Experiment Execution

1. Transfer prepared aeroponics to a growth room (22 °C, 16-h-light/8-h darkness, 200 μmol/m^{-2} s^{-1} illumination) for 4 days. Keep the bottom part of aeroponic in darkness.
2. After 4 days open aeroponics, remove the cotton wool plugs with the seedlings and transfer 30 mL of MS medium or prepared dilutions of ABA, NaCl, and mannitol to the bottom glass tubes. For the 1000 mM mannitol treatment use directly 30 mL of 1 M mannitol stock solution.
3. Place the cotton wool plug with seedling into the bottom glass tube. Cover the top with another glass tube and seal the junction using Parafilm (Fig. 3).
4. Transfer prepared micro-hydroponics to the growth room (22 °C, 16-h-light/8-h darkness, 200 μmol/m^{-2} s^{-1} illumination) for 5 days (*see* **Note 6**).

5. After 5 days analyze the seedling growth by (i) measuring the length of the first leaf and the longest seminal root manually and (ii) image analysis (for detailed information *see* Subheading 3.5). Measure the chlorophyll *a* fluorescence (ChlF) using fluorimeter and then calculate parameters that express photosynthesis performance in stressed plants (for detailed information *see* Subheading 3.6). Perform ten measurements per treatment.

3.3 Assay of Sensitivity to Gradual Increase of Salt Concentration During Barley Growth in Pots

3.3.1 Plant Material Preparation

1. Sterilize 20 seeds for 20 min using 1% sodium hypochloride (as described in Subheading 3.1).
2. Sow the 20 seeds on Petri dish with two Whatman filter papers soaked with 5 mL of water (1st day of experiment).
3. Keep Petri dish with sown seeds at 4 °C in darkness for 3 days in order to stratify the seeds.
4. After 3 days transfer Petri dish to a growth room for 4 days (22 °C, 16-h-light/8-h darkness, 200 μmol/m^{-2} s^{-1} illumination).

3.3.2 Experiment Execution

1. Prepare pots with wet growth medium (mixed soil with vermiculite). Transfer seedlings to the pots (one seedling per pot). Measure the basic morphometric traits of the shoot using the image analysis (for detailed information *see* Subheading 3.5; precisely Subheading 3.5.3, **step 5**). Water each pot with 50 mL water every second day (up till 14th day of experiment). Keep the pots in a growth room (22 °C, 16-h-light/8-h darkness, 200 μmol/m^{-2} s^{-1} illumination).
2. After 15 days perform analyses including: chlorophyll *a* fluorescence using a fluorimeter (for detailed information *see* Subheading 3.6. The chlorophyll *a* fluorescence measurements and calculation of the most important parameters) and image analysis in order to measure morphometric traits of the plant shoot (for detailed information *see* Subheading 3.5; precisely Subheading 3.5.1, **step 8** and Subheading 3.5.3, **steps 5** and **8**). Perform ten measurements. On the 15th day of the experiment, split pots into two treatments: control (10 pots) and salt treated (10 pots). Water the control pots using 50 mL of water and the "salt treated" pots using 50 mL of 100 mM salt solution on the 18th and 20th day of the experiment.
3. After 7 days of growth in the presence of salt (22nd day of experiment) perform analyses both on control and salt-treated plants (as described in **step 2** of this paragraph). Water the control pots using 100 mL of water and the salt-treated pots using 100 mL of 200 mM salt solution, respectively, on the 25th and 27th day of experiment.

4. After 7 days of growth in the presence of 200 mM of salt (29th day of experiment) perform analyses both on control and salt-treated plants (as described in **step 2** of this paragraph). Water the control pots using 100 mL of water and the "salt treated" pots using 100 mL of 300 mM salt solution, respectively, on the 25th and 27th day of experiment.
5. After 7 days of growth in the presence of 300 mM of salt (36th day of experiment) perform analyses (as described in **step 2** of this paragraph).

3.4 Assay of Sensitivity to Oxidative Stress During Seedling Development Using a Mini-Hydroponic System and on Detached Leaves

Carry out all procedures at room temperature unless otherwise specified.

3.4.1 Plant Material and Medium Preparation, Execution, and Analysis of the Results of Oxidative Stress Treatment Using a Mini-Hydroponic System

Plant Material Preparation

1. Sterilize an appropriate number of seeds (as described in Subheading 3.1.1).
2. Sow 20 seeds on each prepared Petri dish (*see* **Note** 7). Prepare appropriate number of square Petri dishes (120 mm × 120 mm), containing two Whatman filter papers soaked with water, to set up three replicates for control and each of the treatments.
3. Seal the Petri dishes using Parafilm and keep at 4 °C for 1 day in darkness for seed stratification.
4. Germinate seeds in darkness at 22 °C for 2 days and transfer the Petri dishes to a growth room (22 °C, 16-h-light/8-h darkness, 200 μmol/m^{-2} s^{-1} illumination) for the next 2 days.

Experiment Execution

1. Cut a Whatman Filter Paper and cover the plastic opaque lid of the mini-hydroponics system. Pour water on the lid to make the paper wet. Make 12 holes in the paper using plastic tips.
2. Gently transplant 12 barley seedlings into holes.
3. Place lid with barley seedlings onto the container. Be sure that roots of the seedlings are submerged in the medium. Turn on the air pump. Run the experiment for 5 days.
4. Grow the plants in a growth room (22 °C, 16-h-light/8-h darkness, 200 μmol/m^{-2} s^{-1} illumination). Use three growth containers as replicates for control and for each of the treatments with MV or H_2O_2.

5. After 5 days of growth measure the chlorophyll *a* fluorescence using fluorimeter (for detailed information *see* Subheading 3.6). Analyze 12 plants per treatment per replicate.
6. Cut the leaf from each plant after 5 days of hydroponic growth and perform image analysis in order to quantify the effect of stress treatment. Estimate the ratio of damaged leaf area to the total leaf area of in case of MV treatment (for detailed information *see* Subheading 3.5; precisely Subheading 3.5.3, **step 8**). Check each plant for leaf chlorosis, necrosis, or other damage.
7. For the treatment using H_2O_2 pay attention to the leaf rolling phenomenon.

3.4.2 Experiment Execution and Analysis of the Results of Oxidative Stress Treatment of Detached Leaves

Plant Material Preparation

1. Sow one barley seed per pot filled with a soil/vermiculite mixture. Five plants should be set up per replicate for each genotypes analyzed. Three replicates should be performed.
2. Grow the plants in a growth room (22 °C, 16-h-light/8-h darkness, 200 μmol/m^{-2} s^{-1} illumination) for 14 days.

Experiment Execution

1. Pour 15 mL of sterile H_2O into 15 mL conical tubes for control and 15 mL of 10 μM MV or 100 mM H_2O_2 solution g for oxidative stress treatment.
2. Cut the first leaf from 14-old-day seedlings and submerge an individual leaf into a separate conical tube. Sample at least ten leaves from individual plants for each treatment combination.
3. Keep samples in a growth room (22 °C, 16-h-light/8-h darkness, 200 μmol/m^{-2} s^{-1} illumination) for 3 days.
4. Pour 15 mL H_2O to new conical tubes and transfer each leaf to a conical tube containing water for 2 days.
5. Pour 10 mL of DAB solution to each Petri dish (diameter 9 cm). Prepare an appropriate number of Petri dishes according to number of analyzed individual plants.
6. Transfer each leaf to a Petri dish containing DAB solution. Adjust the volume of DAB solution to ensure that leaves are immersed according to size of Petri dish. Depending on the number of analyzed genotypes prepare an appropriate amount of DAB solution. Wrap the Petri dishes with aluminum foil. Incubate samples for 24 h at 22 °C temperature.
7. Transfer samples into conical tubes containing 96% ethanol. Place conical tubes in a boiling water bath (95 °C) for 15 min (*see* **Note 8**).
8. Remove samples from conical tubes and perform image analysis.

Analysis of Results

1. In order to quantify DAB staining perform the image analysis using a white background under lighting. Quantify the DAB staining by estimating the area of the stain divided by the total area of the leaf (for detailed information *see* Subheading 3.5, precisely Subheading 3.5.3, **step** 7).
2. In order to measure malondialdehyde (MDA) content use five plants each from three replicates and follow the steps described below:
   (a) Weigh 0.2 g ± 0.01 g of leaf tissue and immediately homogenize in a mortar in 2 mL of 0.25% TBA prepared in 10% TCA.
   (b) Transfer the homogenate into 15 mL conical tubes.
   (c) Rinse the mortar with 2 mL of 0.25% TBA prepared in 10% TCA.
   (d) Transfer the homogenized tissue into 15 mL conical tubes.
   (e) Incubate samples at 95 °C for 30 min in a water bath.
   (f) Cool to room temperature in an ice bath for 10 min.
   (g) Centrifuge samples at 12,000 × *g* for 10 min at room temperature.
   (h) Transfer 2 mL of the supernatant into Eppendorf tubes.
   (i) Measure fluorescence at $\lambda = 532$ nm and $\lambda = 600$ nm using a spectrophotometer.
   Absorbance of the colored supernatant is measured at 532 nm and corrected for unspecific turbidity by subtracting the value of absorbance at 600 nm. Express the level of MDA concentration as nanomole per gram of fresh weight (FW) by using an extinction coefficient 155 mM^{-1} cm^{-1} [21, 22].

3.5 High-Resolution Imagining of Plants Grown Under Stress

All above described protocols engaged image analysis as the evaluation tool to assess plants response to stress. Imaging facilitates recording of the phenotypic traits of a plant in a nondestructive manner. As mentioned above we propose to use the image analysis in order to analyze (i) the seed germination under stress; (ii) seedling development under stress by measuring the length of the first leaf, the length of the longest seminal root, the width of the first leaf, the area of the first leaf; and (iii) the intensity of changes induced by stress, such as necrosis or chlorosis evoked by oxidative stress. It should be emphasized that there are common steps in imaging setup or imaging processing, regardless of the stress studied. Therefore, we describe here the pipeline of imaging analysis, starting from the imaging setup, through basic image processing to image segmentation. However, comments were added in order to underline steps that are specific for the type of stress analyzed.

3.5.1 Imaging Setup

1. For the purpose of high-quality plant imaging using a digital camera you may choose one of the available commercial "light tents" with a steady illumination or adapt a cabinet that will meet the indoor imaging prerequisites.
2. Cover the walls of the imaging chamber with an anti-reflex background for a more homogenous light saturation (e.g., a studio anti-wrinkle, white PVC background).
3. The light room should be equipped with four soft-boxes that are mounted at the ceiling corners, oriented to the center of the work area. The illumination setup should provide a broad ambient light to the imaged object.
4. Apply a daylight simulating, continuous illumination with a color temperature of 5400 K (the typical color temperature of the sun during a bright day at midday). You may use a 80 W daylight compact fluorescent light bulbs.
5. Preheat the lamps until the constant illumination is reached before the imaging process.
6. Adapt a setup to move plants vis-à-vis the stationary camera attached to a tripod. You may use a manual or automated turntable/rotating platform. You may adapt the setup for recording the images of root system growing in transparent pots and transparent substrate (*see* **Note 9**).
7. When performing the imaging operations on Petri plates (e.g., imaging of seeds germination) adapt the camera setup for a top-down view. You may use any of the commercial ceiling camera mounts.
8. The plants should be positioned in the middle of the turntable. The turntable should be moved in manual or automated manner relative to the camera to record the images from the front (0°), first side (90°), backside (180°), and second side (270°).
9. Use a color reference card (e.g., QPcard 202, QPcard AB; *see* **Note 10**) and a ruler for the calibration of the images.

3.5.2 The Image Processing

The image processing describes basic steps that should be performed in order to (i) calibrate the picture with a reference card to define the color components and (ii) establish the scale based on a ruler or a size marker. The latter step enables further measurements based on the established scale.

1. Create a directory of recorded images indicating the dates of the imaging time points, description of the genotypes and the position of the plant.
2. Perform the initial image processing steps applying the original images (RAW or TIFF files) of the plants using ImageJ software [23].

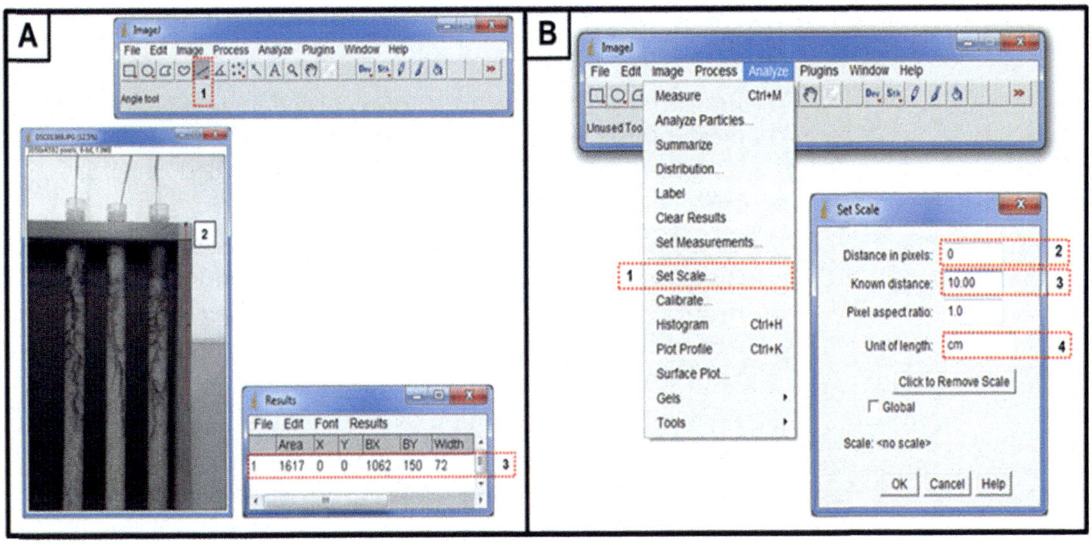

Fig. 4 The steps of raw image calibration using ImageJ software. To define the actual number of pixels aligned in a straight line between the start and end point of applied size standard draw a "*Straight Line*" parallel to the standard (e.g., ruler) and measure its length (**a**). To calibrate the image enter the measured pixel value of a known distance in a *Set Scale* menu using an *Analyze/Set Scale* command (**b**)

3. Calibrate the images using a color reference card and a ruler that were recorded in the picture.
4. Calculate the number of pixels per unit of measure by drawing a straight line (use a "*Straight Line*" shortcut on the ImageJ toolbar) along a selected distance on the size standard or a ruler (e.g., 10 cm). Determine the number of pixels using a command *Analyze/Measure* or using a keyboard shortcut *Ctrl+M* (Fig. 4a).
5. To calibrate the image use a command *Analyze/Set Scale* to enter the value of measured distance in pixels for a known distance (Fig. 4b). That will enable the conversion of the distance units with a given scale in automated manner. For the recorded image sequences (with steady camera and plant position) you may use the same scale throughout a whole image stack.
6. To standardize the image colorimetric scale define the color components of a color reference card using an available ImageJ plugin (*see* **Note 10**). That will determine the distribution of the clustered colors of similar values of a given color standard which will serve as reference for further analyses.

3.5.3 The Image Segmentation

The image segmentation has to be performed in order to measure both basic morphometric traits and advanced traits, such as projected shoot areas or the intensity of changes induced by stress, such as chlorosis or necrosis. The segmentation is a process of the separation of the regions of an image into distinct categories, which correspond to the recorded object and the background, or different objects.

1. Perform the image analyses of recorded images (RAW or TIFF files) of the plants using ImageJ software [23]. The initial image segmentation can be accomplished by using a HSV-segmentation method [24].
2. Convert the images from RGB (red/green/blue) to HSB (hue/saturation/brightness) color space using a "Color Transformer" plugin. This procedure will facilitate the subsequent step of image segmentation by defining the thresholds of each HSB channel.
3. Perform the segmentation by setting minimum and maximum threshold values for each single channel.
4. Additionally, the image cleaning process can be carried out using Erode/Dilate and Open/Close functions (go to *Process/Binary/Erode; Dilate; Open; Close-*). This step will contribute to reducing the noise in the image (e.g., isolated foreground pixels will be removed).
5. To analyze the basic morphometric traits of a plant shoot use a "*Straight Line*" function from the ImageJ toolbar. Draw a single line along the highest shoot or multiple lines along each shoot to determine the maximal or average height of a plant (Fig. 5). To attain the measurements in absolute values

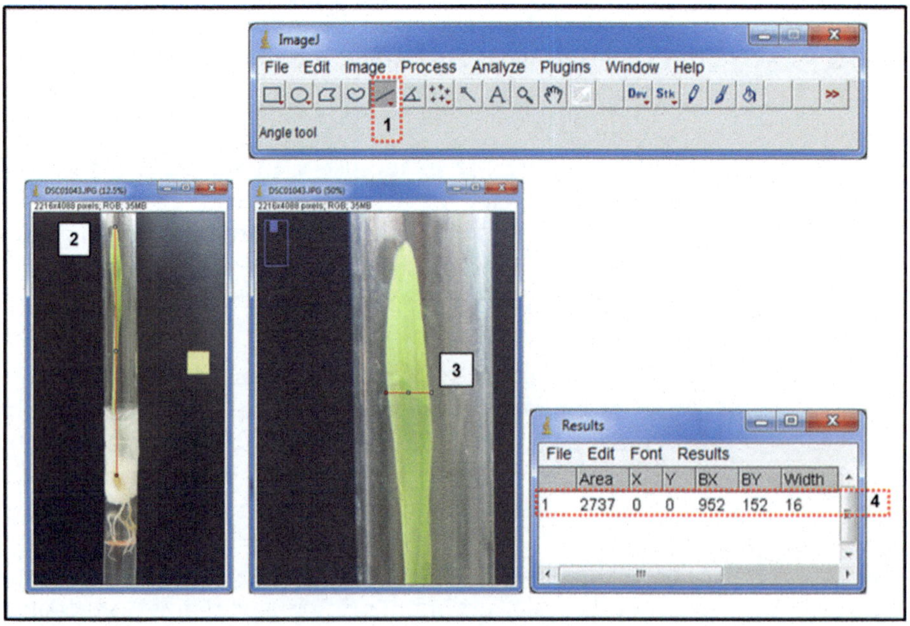

Fig. 5 The steps required for measuring plant shoot length and width using ImageJ software. Draw a "*Straight Line*" across the highest shoot of the plant and measure its length using a command *Analyze/Measure* or a *Ctrl +M* shortcut

(for previously scaled images) use a command *Analyze/ Measure.*

6. To analyze the whole area of a plant shoot, calculate the surface of green pixels (green pixel area) of the plant. The minimum and maximum threshold values for each single channel of a HSB color space of a scaled image should be selected to filter out the area of green pixels (Fig. 6). To attain the measurements in absolute values use a command *Analyze/Measure.*
7. For the analysis of the shoot color perform the color classification of the images of a plant shoot in RGB color space with the

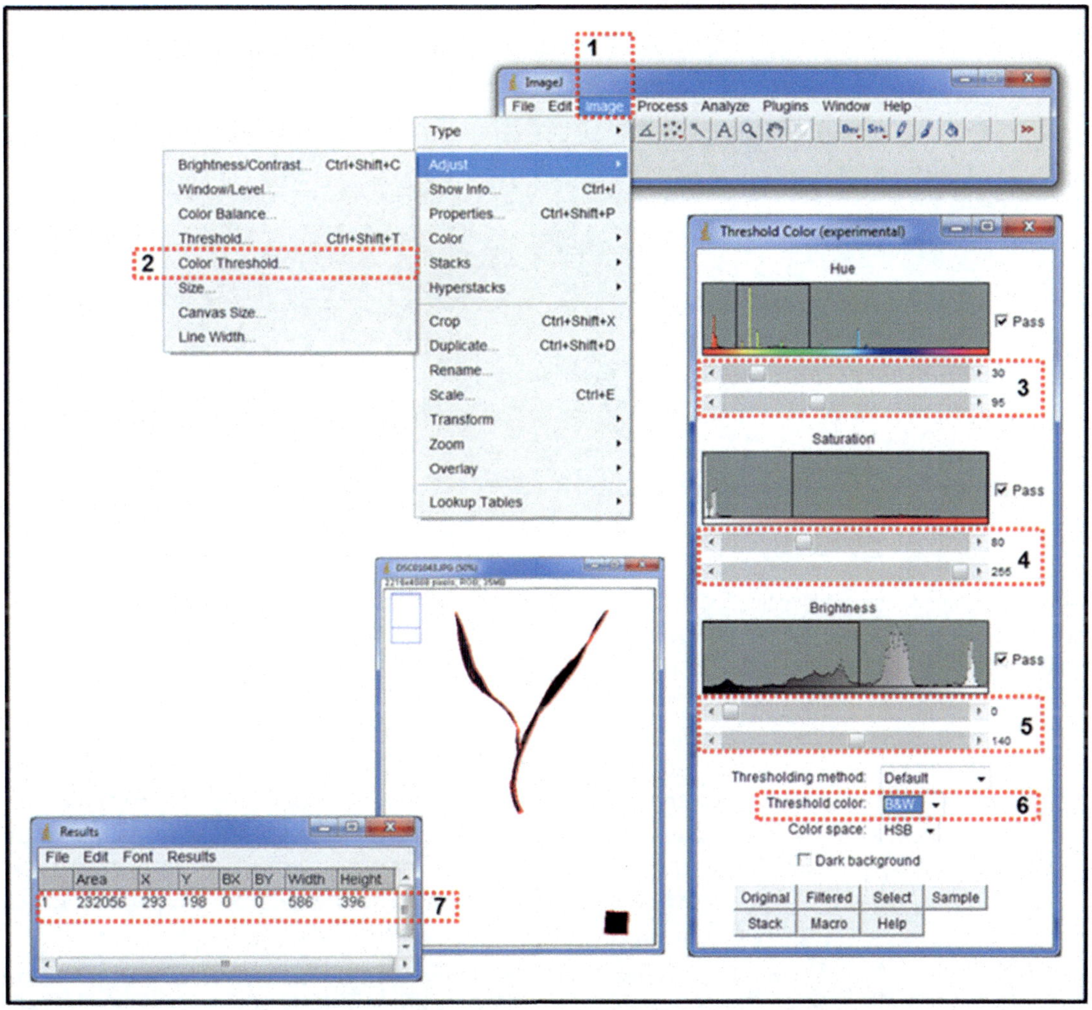

Fig. 6 The steps for image segmentation for green pixels of growing seedlings using the ImageJ software. The minimal and maximal threshold values for each channel of HSB color space should be defined for filtering the area of green pixels on the original image. Select a Threshold color as B&W where the black color will represent the filtered area of green pixels. Use a command *Analyze/Measure* or a keyboard shortcut Ctrl+M to measure the number of pixels in the filtered area

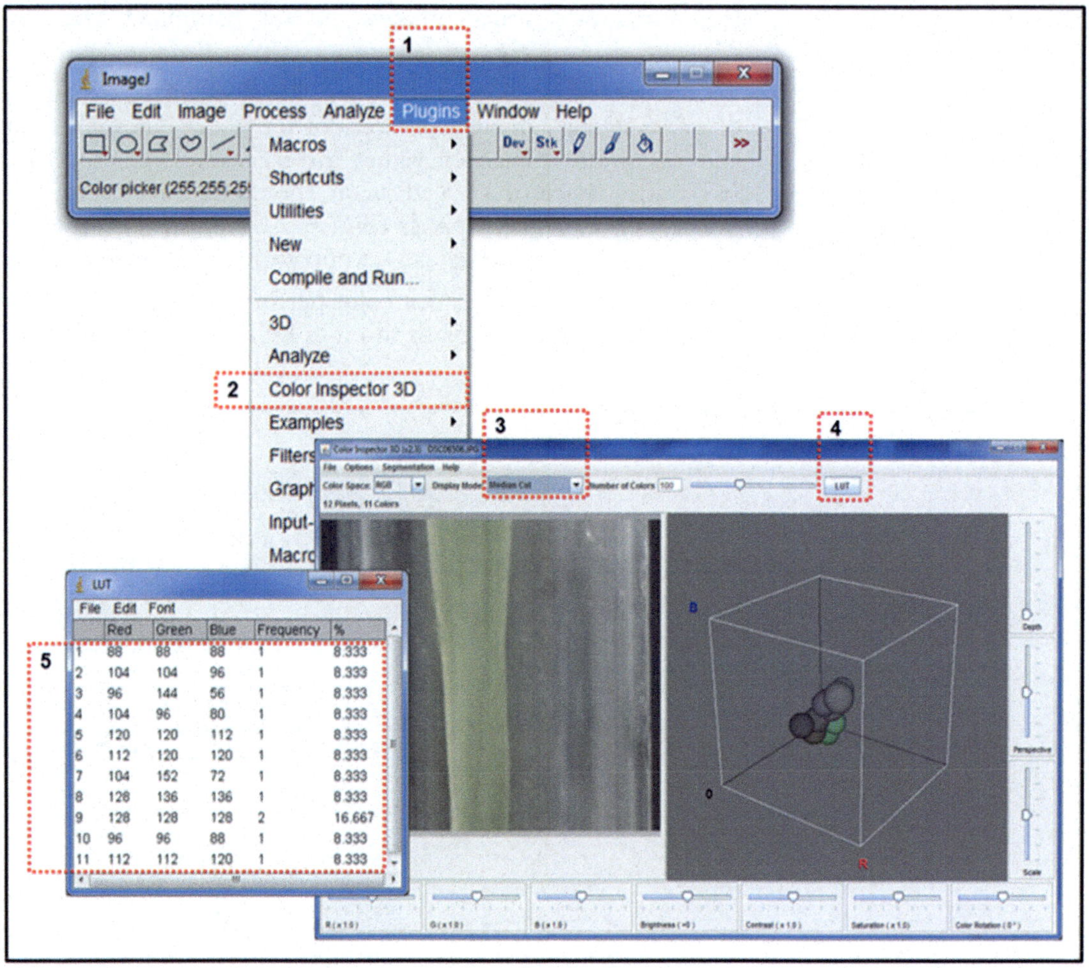

Fig. 7 The steps for color classification of a plant shoot using ImageJ software. Select the area of interest and run the Color Inspector 3D plugin available from the *Plugins* menu. Apply a "Median cut" filter to obtain a distribution of the clustered colors of similar values in the selected area

application of "median cut" filter available in color inspector 3D plugin menu (Fig. 7, *see* **Note 11**). The distribution of the clustered colors of similar values can be compared to the values measured for the control plants or on a color reference card. For the assessment of the statistical differences between the relative frequencies of each basic color components (red, green & blue), perform a chi-squared test.

8. The precise imaging of the changes on leaf blades may require performing the image segmentation process in two steps (Fig. 8). For the quantitative assessment of the intensity of the changes on the plant leaves (e.g., senescence, necrotic spots) you may repeat an image segmentation step by defining the threshold values of each channel of HSB color space to

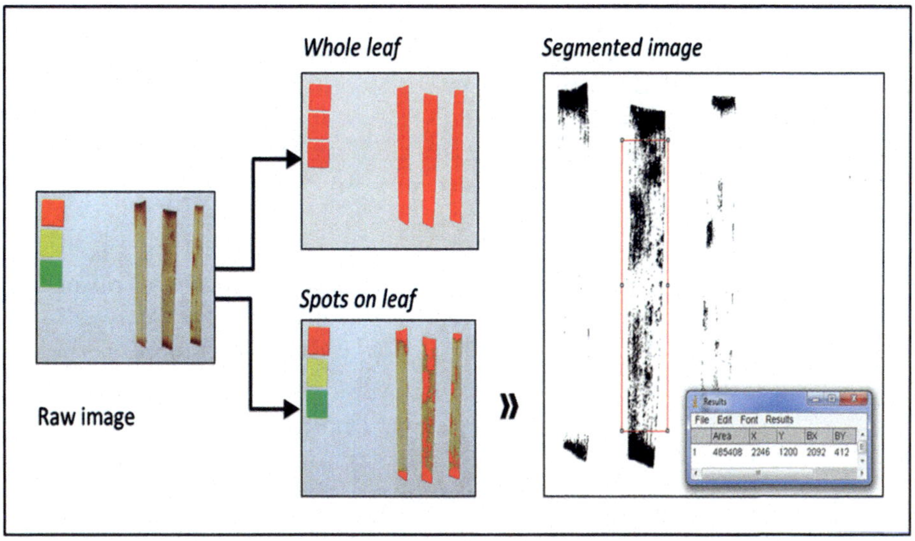

Fig. 8 The quantitative assessment of the intensity of the changes on the leaves after DAB staining using the ImageJ software. The minimal and maximal threshold values for each channel of HSB color space should be defined for the filtering of the area of whole leaf and visible changes separately. A color reference (red) of known dimensions (13–13 mm) serves as the reference for the calculation of the absolute values of the total area of whole leaf and identified spots

distinguish a "true green" area of a shoot from discolorations (e.g., necrosis, chlorosis, yellow fragments). The percentages of the areas of identified discolorations on leaf fragments to the total green pixel area of leaf can be calculated and the statistical differences between the analyzed genotypes can be determined using a Student's *t*-test.

9. The following steps of each measurement can be recorded as a macro script in order to enable the repetition of the analysis in an automated manner.

3.6 The Chlorophyll a Fluorescence Measurements and Calculation of the most Important Parameters

The measurements of chlorophyll *a* fluorescence can be performed using a fluorimeter, e.g., Plant Efficiency Analyser (PocketPEA fluorimeter, Hansatech Instruments Ltd., England). The protocol of measurements presented below is relevant to PocketPEA in accordance with the official manual (http://www.hansatech-instruments.com/products/introduction-to-chlorophyll-fluorescence/continuous-excitation-chlorophyll-fluorescence/pocket-pea/).

3.6.1 Chlorophyll a Fluorescence (ChlF) Measurements

1. Before the measurements, the leaves need to be dark adapted for 30 min. Place the leafclip dedicated for PocketPEA in the middle part of the leaf and close the shutter plate in order to exclude the light from the leaf sample during dark adaptation.
2. After 30 min of dark adaptation put the optical part of the switched on PocketPEA fluorimeter onto the ring of the

leafclip. Open the leafclip shutter. Click the bottom "Measure" on the fluorimeter and expose leaf sample to a pulse of saturating light at an intensity of 3500 μmol m^{-2} s^{-1} with a wavelength of 627 nm. The parameters measured are automatically saved in fluorimeter and can be further downloaded on a PC using Bluetooth connection. The fluorimeter is able to store up to 200 recordings.

3.6.2 Processing of Chlorophyll a Fluorescence (ChlF) Data

1. In order to deposit data from PocketPEA you need to install software dedicated for this instrument—PeaPLUS (Hansatech Instruments Ltd., England) and configure the Bluetooth connection.
2. From the main menu of PeaPlus software select in toolbar "PocketPea," then from the submenu select "Select Bluetooth" (select the number of COM port synchronized with PocketPEA in your PC). Then from the same submenu select the command "Download All Recordings."
3. Export your data to *.han file: using the toolbar select "File" and then from the submenu select "Export to HAN."
4. Open *.han file in Biolyzer software (Fluoromatics Lab, Switzerland) and from the main menu select "Calculations" then from submenu select "Calculate JIP-test parameters table."
5. Results of calculations of raw data save as *.csv (comma separated values) file and open it further in a Microsoft Office Excel datasheet.

3.6.3 Calculating Chlorophyll a Fluorescence (ChlF) Parameters Crucial for the Evaluation of Stress Response

1. Calculate parameters such as (i) maximum quantum yield of primary photochemistry (ΦP_O), (ii) performance index (PI_{ABS}), and (iii) phenomenological energy fluxes per excited cross section: ABS/CS—absorption energy flux per cross-section (CS); TR_0/CS—trapped energy flux per CS; ET_0/CS—electron transport flux per CS; DI_0/CS—dissipation energy flux per CS; RC/CS—active reaction center per CS, based on data obtained after processing records from PocketPEA with the use of PeaPLUS and Biolyzer software in Excel.
2. Perform statistical analysis adequate to your data set. We recommend for this purpose STATISTICA (data analysis software system), www.statsoft.com. StatSoft, Inc.

4 Notes

1. Cover the tube with aluminum foil as DAB is light-sensitive. The DAB solution should be prepared fresh. A complete dissolution of DAB is possible only at pH = 3.5.

2. To facilitate analysis, put the six seeds in five rows on each Petri dish.
3. Replacement of the solutions and Whatman filters on 1st day after stratification prevents overdrying the Petri dishes.
4. We recommend to set up a higher number of tubes if the germination potential of seeds is low.
5. Use a glass tube rack to stabilize the tubes when you set up the experiment. Insert the seed between a glass tube facet and a cotton wool plug.
6. Replacement of solutions after 2 days of treatment prevents contamination and ensures air circulation.
7. Place not more than 20 seeds per Petri dish to avoid root damage during seedling transfer.
8. This treatment will bleach out the chlorophyll but leave the brown precipitate formed by the DAB reacting with the hydrogen peroxide. The time of boiling can be adjusted by ± 5 min depending on the appearance of the leaves. They should be completely deprived of chlorophyll.
9. A single or a group of the tubes assembled in a narrow flask can be placed on a turntable (as described in [25]). That enables the imaging of the seedlings growing in transparent tubes filled with a solid substrate from different angles. The standard practice comprises of the imaging of four subsequent side views (0°, 90°, 180°, 270°) to visualize the growth of shoot and root in a non-invasive manner.
10. A color reference card contains mounted color chips. The normalization of the photographs (in RAW image format) with the reference card in the images enables the proper interpretation of the recorded colors in any imaging setting.
11. The color analysis of the recorded images requires the Color Inspector 3D plugin (available online at https://imagej.nih.gov/ij/plugins/color-inspector.html) for ImageJ software. A Color Inspector 3D plugin enables the display of the distribution of the image colors within a 3D color space. The spheres inside the RGB diagonal cube represent the relative frequency of each color interval of the sampled area where the volume is related to the frequency of occurrence of that color on the sampled area. The color classification of the color reference/control plants in the RGB color space should be conducted with the application of the "*Median cut*" filter available in the Color Inspector 3D menu.

Acknowledgments

This work was supported by the European Regional Development Fund through the Innovative Economy for Poland 2007–2013, project WND-POIG.01.03.01-00-101/08 POLAPGEN-BD "Biotechnological tools for breeding cereals with increased resistance to drought," task 22; National Science Centre, Poland, project SONATA 2015/19/D/NZ1/03441.

References

1. Nakashima K, Yamaguchi-Shinozaki K (2013) ABA signaling in stress-response and seed development. Plant Cell Rep 32:959–970. https://doi.org/10.1007/s00299-013-1418-1
2. Almeida DM, Almadanim MC, Lourenço T, Abreu IA, Saibo NJ, Oliveira MM (2016) Screening for abiotic stress tolerance in rice: salt, cold, and drought. Methods Mol Biol 1398:155–182. https://doi.org/10.1007/978-1-4939-3356-3_14
3. Dilukshi F, Schroeder DF (2016) Role of ABA in Arabidopsis salt, drought, and desiccation tolerance. Abiotic and biotic stress in plants—recent advances and future perspectives. Dr. Arun Shanker (Ed.), InTech, Rijeka. doi: https://doi.org/10.5772/61957
4. Munns R, James RA, Läuchli A (2006) Approaches to increasing the salt tolerance of wheat and other cereals. J Exp Bot 57:1025–1043. https://doi.org/10.1093/jxb/erj100
5. Munns R (2005) Genes and salt tolerance: bringing them together. New Phytol 167:645–663. https://doi.org/10.1111/j.1469-8137.2005.01487.x
6. Munns R, James R, Sirault X, Furbank R, Jones H (2010) New phenotyping methods for screening wheat and barley for beneficial responses to water deficit. J Exp Bot 61:3499–3507. https://doi.org/10.1093/jxb/erq199
7. Peleg Z, Apse M, Blumwald E (2011) Engineering salinity and water-stress tolerance in crop plants: getting closer to the field. Adv Bot Res 57:405–443. https://doi.org/10.1016/B978-0-12-387692-8.00012-6
8. Tavakkoli E, Rengasamy P, McDonald G (2010) The response of barley to salinity stress differs between hydroponic and soil systems. Funct Plant Biol 37:621–633. https://doi.org/10.1071/FP09202
9. Del Río LA (2015) ROS and RNS in plant physiology: an overview. J Exp Bot 66:2827–2837. https://doi.org/10.1093/jxb/erv099
10. Cornic G, Bukhov NG, Wiese C, Bligny R, Heber U (2000) Flexible coupling between light-dependent electron and vectorial proton transport in illuminated leaves of C_3 plants. Role of photosystem I-dependent proton pumping. Planta 210:468–477. https://doi.org/10.1007/PL00008154
11. Yoshimura K, Miyao K, Gaber A, Takeda T, Kanaboshi H, Miyasaka H, Shigeoka S (2004) Enhancement of stress tolerance in transgenic tobacco plants overexpressing *Chlamydomonas* glutathione peroxidase in chloroplasts or cytosol. Plant J 37:21–33. https://doi.org/10.1046/j.1365-313X.2003.01930.x
12. Bienert GP, Schjoerring JK, Jahn TP (2006) Membrane transport of hydrogen peroxide. Biochim Biophys Acta 1758:994–1003. https://doi.org/10.1016/j.bbamem.2006.02.015
13. Thordal-Christensen H, Zhang Z, Wei Y (1997) Subcellular localization of H_2O_2 in plants. H_2O_2 accumulation in papillae and hypersensitive response during barley-powdery mildew interaction. Plant J 11:1187–1194. https://doi.org/10.1046/j.1365-313X.1997.11061187.x
14. Velikova V, Yordanov I, Edreva A (2000) Oxidative stress and some antioxidant systems in acid rain-treated bean plants. Protective role of exogenous polyamines. Plant Sci 151:59–66. https://doi.org/10.1016/S0168-9452(99)00197-1
15. Bresson J, Vasseur F, Dauzat M, Koch G, Granier C, Vile D (2015) Quantifying spatial heterogeneity of chlorophyll fluorescence during plant growth and in response to water stress. Plant Methods 11:23. https://doi.org/10.1186/s13007-015-0067-5
16. Filek M, Łabanowska M, Kościelniak J, Biesaga-Kościelniak J, Kurdziel M, Szarejko I, Hartikainen H (2015) Characterization of

barley leaf tolerance to drought stress by chlorophyll fluorescence and electron paramagnetic resonance studies. J Agron Crop Sci 201:228–240. https://doi.org/10.1111/jac.12063

17. Kalaji HM, Bosa K, Kościelniak J, Żuk-Gołaszewska K (2011) Effects of salt stress on Photosystem II efficiency and CO_2 assimilation of two Syrian barley landraces. Environ Exp Bot 73:64–72. https://doi.org/10.1016/j.envexpbot.2010.10.009
18. Oukarroum A, El Madidi S, Strasser RJ (2016) Differential heat sensitivity index in barley cultivars (*Hordeum vulgare* L.) monitored by chlorophyll a fluorescence OKJIP. Plant Physiol Biochem 105:102–108. https://doi.org/10.1016/j.plaphy.2016.04.015
19. Zivčák M, Brestič M, Olšovská K Slamka P (2008) Performance index as a sensitive indicator of water stress in *Triticum aestivum* L. Plant Soil Environ 54:133–139
20. Kalaji HM, Oukarroum A, Alexandrov V, Kouzmanova M, Brestic M, Zivcak M et al (2014) Identification of nutrient deficiency in maize and tomato plants by in vivo chlorophyll a fluorescence measurements. Plant Physiol Biochem 81:16–25. https://doi.org/10.1016/j.plaphy.2014.03.029
21. Woo HR, Kim JH, Nam HG, Lim PO (2004) The delayed leaf senescence mutants of Arabidopsis, ore1, ore3, and ore9 are tolerant to oxidative stress. Plant Cell Physiol 45:923–932. https://doi.org/10.1093/pcp/pch110
22. Kacienė G, Žaltauskaitė J, Milčė E, Juknys R (2015) Role of oxidative stress on growth responses of spring barley exposed to different environmental stressors. J Plant Ecol 8:605–616. https://doi.org/10.1093/jpe/rtv026
23. Rasband WS (1997–2015) ImageJ. U. S. National Institutes of Health, Bethesda, MD. http://imagej.nih.gov/ij/. Accessed 31 Oct 2016
24. Walter A, Scharr H, Gilmer F, Zierer R, Nagel KA, Ernst M, Wiese A, Virnich O, Christ MM, Uhlig B, Jünger S, Schurr U (2007) Dynamics of seedling growth acclimation towards altered light conditions can be quantified via GROWSCREEN: a setup and procedure designed for rapid optical phenotyping of different plant species. New Phytol 174:447–455. https://doi.org/10.1111/j.1469-8137.2007.02002.x
25. Slota M, Maluszynski M, Szarejko I (2016) An automated, cost-effective and scalable, flood-and-drain based root phenotyping system for cereals. Plant Methods 12:34. https://doi.org/10.1186/s13007-016-0135-5

Chapter 10

Preparation of Barley Roots for Histological, Structural, and Immunolocalization Studies Using Light and Electron Microscopy

Marek Marzec and Michael Melzer

Abstract

Microscopic investigations of biological objects are an integral part in plant research and most fields of life sciences. They allow the description of morphological, histological, and structural aspects of individual cells or tissues. Based on various cell biological tools and methods it is possible to characterize different plant genotypes or study their adaptation to changing environmental conditions. In combination with antibodies raised against specific antigens and epitopes light and electron microscopy enable investigation of the function of single genes/proteins in plant growth and development or their role related to abiotic or biotic stresses.

Here, we describe sample preparation of barley roots for cell biological investigations using light and electron microscopy, to characterize morphological, structural, and functional aspects on root sections and the root surface.

Key words Histology, Immunolocalization, Immunogold labeling, Light microscopy, Fluorescence microscopy, Electron microscopy

1 Introduction

Ongoing developments in the field of microscopy has resulted in improved resolution and increasingly advanced methodology. The application of these new innovative techniques enable researchers to study in ever-increasing detail the morphology of tissues and cells, the dynamics of biological processes, and the interactions of molecules [1–4]. Nevertheless we should realize that all microscopy studies stand or fall with sample preparation and that with a good experimental design even conventional light microscope systems can shed new light on important scientific questions, be it in the field of cell biology, molecular biology, biotechnology, systems biology, or taxonomy [5–7].

Conventional light microscopy (CLM) offers the description of cell size, cell shape, and also the investigation of the pattern of cells

Wendy A. Harwood (ed.), *Barley: Methods and Protocols*, Methods in Molecular Biology, vol. 1900,
https://doi.org/10.1007/978-1-4939-8944-7_10,

that build up tissues and organs [8]. In combination with basic histological staining techniques additional information, such as number and size of cell organelles or the composition of the cell wall, are achievable [9].

Fluorescence microscopy and particularly confocal laser scanning microscopy (CLSM) allow a spectroscopical analysis of tissues under investigation. The emission spectra collected after illumination with specific wavelengths can be used to establish the distribution patterns of a wide variety of fluorescent molecules. These include both naturally present molecules, e.g., chlorophyll, and molecules added experimentally by staining or genetic transformation. The use of fluorescence enables the localization and visualization of structures and organelles below the level of microscopic resolution. For more precise location studies molecules are labeled with antibodies conjugated with gold particles and examined in a transmission electron microscope (TEM) or a scanning electron microscope (SEM) may be used [10].

In this chapter different protocols for sample preparation of barley roots for CLM, CLSM, TEM, and SEM analysis are described. The specific preparation will allow the analysis of cell differentiation in root tissues on a cellular and subcellular level (Fig. 1).

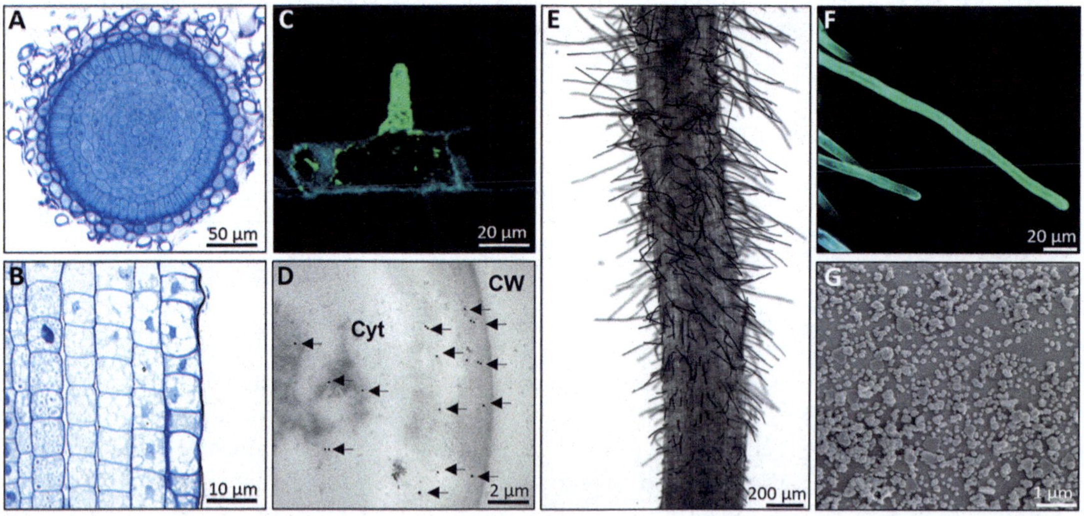

Fig. 1 Investigation of barley root hair development using different microscopy techniques. CLM image of a 2.5 μm cross section of meristematic zone (**a**) and longitudinal section of the early differentiation zone (**b**) after application of histological staining with methylene blue/azur II. Localization of arabinogalactan protein epitope LM2 on sections (**c**, **d**) and the root surface (**e**–**g**) using CLM (**e**), CLSM (**c**, **f**), TEM (**d**), and SEM (**g**). Fluorescent labeling on 0.5 μm sections using Alexa 488, with green indicating the signal of the fluorophore and blue the autofluorescence of cell walls (**c**). 10 nm protein A-gold labeling on 70 nm ultrathin sections (**d**). Whole mount immunolocalization of LM2 epitope on the surface of root hairs using CLM and 10 nm protein A-gold with silver enhancement (**e**), CLSM using Alexa-488 (**f**) and SEM using 10 nm protein A-gold with silver enhancement (**g**). Cyt—cytoplasm, CW—cell wall, arrows indicate single 10 nm protein A-gold particles conjugated to the secondary antibody

2 Materials

2.1 Fixation and Dehydration Components

1. Cacodylate-buffer (0.1 M, 250 mL, pH 7.0): Dissolve 5.35 g of cacodylic acid sodium salt trihydrate ($C_2H_6AsNaO_2 . 3H_2O$) in 200 mL of millipore water (MQ W). Adjust pH to 7.0 with 1 M hydrochloric acid (HCl) and fill up to 250 mL. Store at 4 °C.
2. Fixative solution (20 mL, pH 7.2): Add paraformaldehyde (PFA), to MQ W (all proportions were given in Table 1) in a beaker. Protect from evaporation and place on magnetic stirrer (60 °C, 250 rpm). When the suspension reaches 60 °C, add 1 M sodium hydroxide (NaOH) and wait until the solution becomes transparent. Add 1 M HCl solution, 0.1 M cacodylate-buffer. Finally add 25% glutaraldehyde (GA). Always prepare solution fresh, if stored at 4 °C use within 24 h.
3. Osmium tetroxide (OsO_4) (1% solution, 100 mL): Transfer a glass ampoule containing 1 g of crystalline OsO_4 into liquid nitrogen. Once cooled, use tweezers to carefully tap the ampoule on a solid surface to release the OsO_4 crystals. Open ampoule with a glass cutter and suspend the OsO_4 crystals in 100 mL ultrapure water in an acid bottle with cap. Place the flask on a shaker for 24 h (200 rpm). To avoid possible contamination with glass fragments filter the solution using a Whatman glass microfiber filter. Wrap the acid bottle with aluminum foil and store at RT in a fume hood (*see* **Note 1**).
4. Acetone and ethanol series for dehydration: 20, 30, 40, 50, 60, 70, 80, and 90% (V/V) are prepared in MQ W and stored at RT.

Table 1
Recipe for 20 mL of fixative solutions for immunolocalization (2% of PFA and 0.5% of GA) and ultrastructure (2% of PFA and 2% of GA) studies

Component	Immunolocalization	Ultrastructure
PFA	0.4 g	0.4 g
Distilled water	9.4 mL	8.2 mL
NaOH (1 M)	100 μL	100 μL
Cacodylate-buffer (0.1 M)	10 mL	10 mL
HCl (1 M)	96 μL	96 μL
25% GA	400 μL	1.6 mL

2.2 Resins Infiltration Components

1. Low-viscosity epoxy resin Spurr embedding for ultrastructural analysis: Spur kit components can be mixed in a glass beaker under stirring (250 rpm) with a 6 cm magnetic stir bar. First mix 26 g of nonenyl succinic anhydride (NSA) with 10 g of ERL-4221 (ERL) and stir for 20 min (RT, 250 rpm). Next add 6 g of diglycidyl ether of polypropylene glycol (DER) and stir for additional 30 min. Finally add 0.4 g of dimethylaminoethanol (DMAE). To ensure a good and fast mixing, release the DMAE 2–3 cm under the surface of a vigorously stirred Spurr solution. Split up in 15–20 mL aliquots, the solution can be stored at −20 °C for at least 6 months and longer.

   For resin infiltration 25, 50, and 75% (V/V) Spurr solutions in acetone (*see* **Note 2**) are prepared immediately before use.

2. Low-viscosity acrylic resin LR White embedding for immunolabeling: Complete LR White medium hardness is ready for use. Keep resin at 4 °C. For resin infiltration 25, 50, and 75% (V/V) LR White solutions in ethanol (*see* **Note 2**) are prepared immediately before use.

3. Lowicryl HM20 embedding for immunolabeling: For medium hardness HM20 mix 170.2 g of Monomer E, 29.8 g of Crosslinker D, and 1 g of Initiator C for 20 min in a 250 mL glass brown bottle on a magnetic stirrer (250 rpm) in a fume hood at RT. Wrapped in aluminum foil the solution can be stored at RT in a fume hood.

   For resin infiltration 25, 50, and 75% (V/V) solutions of HM20 in ethanol (*see* **Note 2**) are freshly prepared immediately before use.

   For thermal polymerization at 60 °C, take 100 mL of a ready solution of HM20 and add 0.5 g of benzoyl peroxide ($C_{14}H_{10}O_4$). Stored in a fridge at 4 °C the solution remains functional for several months.

2.3 Contrasting Components

1. Lead citrate ($Pb(NO_3)_2$), (3%, 50 mL): Ready solutions can be ordered or prepared yourself. For the latter dissolve 1.33 g of lead citrate and 1.76 g of sodium citrate dihydrate ($Na_3C_6H_5O_7$. $2H_2O$) in 30 mL MQ W. Stir for 30 min (RT, 200 rpm, solution should remain cloudy). Add 5 mL of 1 M NaOH and wait until the solution becomes clear before adding another 15 mL MQ W. After stirring for an additional 10 min the solution can be stored in a fridge at 4 °C and protected from light.

2. Uranyl acetate ($UO_2(CH_3COO)_2$. $2H_2O$; UAc) (2%, 100 mL): Standard solutions are 1–3% in MQ W. For a 2% solution dissolve 2 g of UAc in 100 mL MQ W and filter through Whatman glass microfiber filters. Since UAc is sensitive to light, especially UV, solutions should be stored in a brown bottle, at RT.

3. Methanolic UAc, (10 mL): Dissolve 6 g of UAc in 10 mL of 70% methanol and shake for 5 min (200 rpm) at RT. Store without filtering in a brown bottle at 4 °C.

2.4 Immuno-localization Components

1. Washing buffer (WB) (pH 7.6, 0.1% of BSA): Dissolve 148 mg sodium phosphate dibasic (Na_2HPO_4), 43 mg potassium dihydrogen phosphate (KH_2PO_4), 720 mg sodium chloride (NaCl), 65 mg sodium azide (NaN_3), 100 μL Tween 20, and 100 mg bovine serum albumin (BSA) in 75 mL MQ W. Stir for 20 min (RT, 250 rpm), fill up with MQ W to 100 mL and store at 4 °C.
2. Antibody buffer (AB) (pH 7.6, 1% of BSA): Dissolve 0.9 g of BSA in 100 mL WB. Aliquot in 2 mL portions in eppendorf tubes and store at −20 °C until use.
3. Blocking buffer (BB) (pH 7.6, 1% of BSA): Dissolve 2.9 g of BSA in 100 mL WB. Aliquot in 2 mL portions in eppendorf tubes and store at −20 °C until use.

2.5 Silver Enhancing Components

1. For silver enhancement a commercially available kit for Light and Electron Microscopy (BBInternational, Cat. No. SEKB-251, Cardiff, UK) might be use according to manufacturers' instructions.

2.6 Components for Histological Staining

1. Methylene blue (1% stock solution): Dissolve 1 g of borax ($Na_2B_4O_7 \cdot 10H_2O$) in 100 mL MQ W. Ad 1 g of methylene blue and store at RT.
2. Azure II (1% stock solution): Dissolve 1 g of Azure II in 100 mL MQ W. Store at RT.
3. Methylene blue/Azure II staining solution. Mix equal amounts of 1% methylene blue and 1% azure II and filter the mixture over a Whatman glass microfiber filter. Store the filtered solution at RT. Since precipitations accumulate over time, take the staining solution up in a 10–20 mL plastic syringe and dispense the stain through a 0.8 μm filter.

3 Methods

Protocols for the preparation of barley tissues for immunolocalization or ultrastructure studies are largely identical. Until explicitly indicated, the following description can be applied for both immunological and ultrastructural research.

3.1 Tissue Fixation, Dehydration, and Infiltration with Resin

Tissue processing for resin embedding consists of a consecutive exchange of fixation, dehydration, and infiltration solutions. The traditional way to facilitate the exchange of solutions is by placing probes on a shaker with protocols running over several days. The

Table 2
Protocol for conventional fixation, dehydration, and infiltration of barley roots

Process	Chemical	Immunolocalization	Ultrastructure
Primary fixation	Fixative solution[a]	4 h on shaker, RT	4 h on shaker, RT
Washing	Cacodylate-buffer (50 mM) Distilled water (×2)	15 min on shaker, RT 15 min on shaker, RT	15 min on shaker, RT 15 min on shaker, RT
Secondary fixation[b]	1% OsO_4	–	1 h on shaker, RT, protect from light
Washing[b]	Distilled water (×2)	–	15 min on shaker, RT
Dehydration	Alcohol series[c]: 20, 30, 40, 50, 60, 70, 80, 90% 100% alcohol[c] (×2)	20 min on shaker, RT after every solution change 20 min on shaker, RT	20 min on shaker, RT after every solution change 20 min on shaker, RT
Resin infiltration	Resin in alcohol[d]: 25, 50, 75% 100% resin[d] (×2)	6 h on shaker, RT after every solution change 6 h on shaker, RT	6 h on shaker, RT after every solution change 6 h on shaker, RT

[a]For immunolocalization studies use 2% of PFA and 0.5% of GA fixative solution, whereas for ultrastructure analysis use 2% of PFA and 2% of GA

[b]Secondary fixation and washing step after this fixation are necessary only for ultrastructure analysis. Treatment with 1% OsO_4 resulted in antigenicity decrease of many epitopes present in tissues

[c]Use acetone for resin infiltration of Spurr, ethanol for LR White or HM20

[d]Use LR White or HM20 for infiltration of specimens for immunolocalization studies and Spurr for ultrastructure studies

introduction of microwave-assisted protocols has significantly reduced processing time while improving preservation. The microwave-assisted protocol described here was carried out with a Pelco 34700 Biowave Laboratory Microwave Processor.

1. Prepare fresh fixative as described in Subheading 2.1, **item 2**: Use 2% of PFA and 0.5% of GA for immunolocalization studies or 2% of PFA and 2% of GA for ultrastructure analysis.
2. Barley roots, grown in aeroponic or hydroponic conditions, are placed in a droplet of fixative and cut into 2 mm fragments (*see* **Note 3**).
3. Transfer specimens into a 2.0 mL eppendorf tube with 2 mL of fixative solution and apply gentle vacuum infiltration of fixative using a desiccator or vacuum pump. For conventional embedding follow the steps shown in Table 2 and for microwave-assisted embedding follow the steps shown in Table 3.

3.2 Polymerization of Spurr

Spurr polymerization may be carried out in polyethylene Beem capsules or in silicone embedding molds. The pre-polymerization of a thin resin layer in silicone embedding molds or Beem capsules is not essential, but facilitates the subsequent handling of sample resin blocks during trimming and sectioning.

Table 3
Protocol for microwave-assisted fixation, dehydration, and infiltration of barley roots

		Microwave tissue processor		
Process	**Chemical**	**Power [V]**	**Time [min]**	**Vaccum [mm Hg]**
Primary Fixation	Fixative solution[a]	150	1	15
		0	1	15
		150	1	15
		0	1	15
		150	1	15
		0	1	15
Washing	Cacodylate-buffer (50 mM)	150	0.45	0
		0	0.45	0
	Distilled water	150	0.45	0
		0	0.45	0
		150	0.45	0
		0	0.45	0
		After each step of washing shake for 5 min in RT		
Secondary fixation	*1% OsO_4*	*0*	*1*	*15*
		80	*2*	*15*
		0	*1*	*15*
		80	*2*	*15*
		After that 4 steps shake for 1 h and protect from light		
Washing	*Distilled water*	*150*	*0.45*	*0*
		0	*0.45*	*0*
Dehydration	Alcohol series[b]: 20, 30, 40, 50, 60, 70, 80, 90%	150	0.45	0
		Use the same settings for each solution and after each step shake for 5 min, RT		
	100% alcohol[b]	150	0.45	0
		Shake for 5 min, RT		
		150	0.45	0
		Shake for 5 min, RT		
Resin infiltration	Resin in alcohol[c]: 25, 50, 75%	250	3	15
		After each step shake for 1 h, RT		
	100% resin[c]	250	3	15
		Shake for 1 h, RT		
		250	3	15
		Shake for 1 h, RT		

Steps italicized for the secondary fixation are obligatory only during preparation of samples for ultrastructure studies; omit these steps for protein immunolocalization approach

[a]For immunolocalization studies use 2% of PFA and 0.5% of GA fixative solution, whereas for ultrastructure analysis use 2% of PFA and 2% of GA

[b]Use acetone for resin infiltration of Spurr, ethanol for LR White or HM20

[c]Use LR White or HM20 for infiltration of specimens for immunolocalization studies and Spurr for ultrastructure studies

1. Fill capsules or embedding molds to about 1/3 with resin (*see* **Note 4**). Place the forms in an oven set at 60 °C for 4–6 h, or until the Spurr is no longer "sticky." Avoid overnight pre-polymerization since it can result in different properties between this and the subsequent Spurr layer, affecting section quality.
2. After the pre-polymerization fill up the capsules or molds with 100% Spurr and place the root fragment on the bottom of the form in the desired orientation. Transfer the forms to an oven set at 60 °C for 24 h. Under these conditions polymerization will become noticeable after 2–3 h. As samples may drift, orientation should be checked before the resin becomes too viscous, i.e., between 2 and 3 h after the start of polymerization.

3.3 Polymerization of LR White

LR White does not polymerize in the presence of oxygen. Polymerization should be carried out in closed polyethylene capsules. A nitrogen atmosphere is helpful but not necessary. Also here, pre-polymerization is not essential, but will facilitate subsequent handling of sample resin blocks during trimming and sectioning.

1. For pre-polymerization add 50 μL of LR White Accelerator to 2 mL of LR White and immediately pipet 200 μL aliquots in empty capsules (*see* **Note 5**). Close the capsules and leave for 5 min to complete polymerization.
2. Once hardened, fill up the capsule with pure LR White and place the root fragment in the desired orientation at the bottom of the capsule.
3. To minimize exposure to oxygen add more LR White to obtain a convex meniscus before closing the capsule.
4. Polymerize at 70 °C for 48 h. As in some cases the lids may open during heat incubation, place a metal block on top of the capsule to keep the lids closed.

3.4 Polymerization of HM20

Polymerization of HM20, whether at −20 °C under UV light or thermic at +60 °C, is best performed in an automated freeze substitution system (ASF) as this enables the introduction of undisrupted nitrogen atmosphere. In the protocols presented here a Leica EM ASF2 was used. Work as much as possible in a fume hood.

1. For pre-polymerization fill gelatine capsules with HM20 (150–200 μL), and place the capsules into an ASF with nitrogen atmosphere. If polymerization is to be carried out under UV light (−20 °C for 72 h), capsules can be left open during the polymerization for 72 h. For thermic polymerization (60 °C for 24 h) fill capsules with 200 μL of the appropriate HM20 solution and put a lid on the capsule to reduce evaporation. Capsules pre-polymerized under UV or thermic should

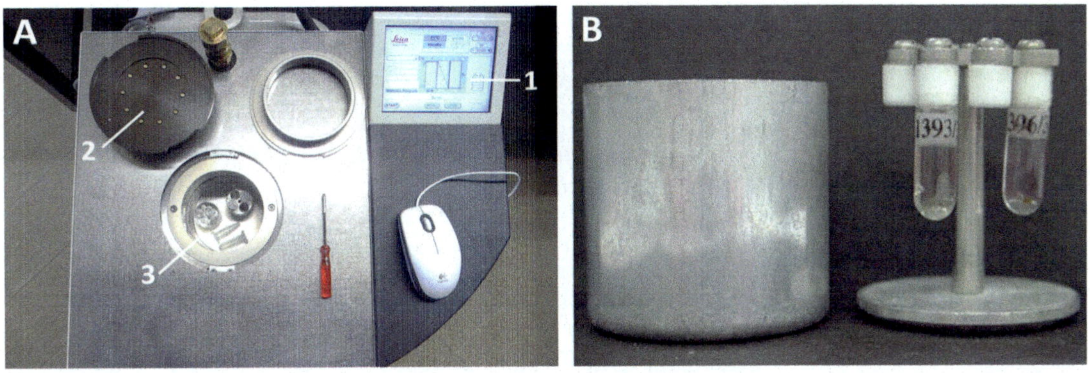

Fig. 2 Polymerization using ASF system. (**a**) Overview of Leica EM ASF2, (**b**) gelatine capsules filled with HM20 and container with 100% ethanol. 1—control panel of a Leica EM ASF2, 2—UV lamp, 3—polymerization chamber

Table 4
Program for HM20 polymerization using a Leica EM AFS2

Step no.	Time (h)	Temperature (°C)	UV light
1.	2	20	No
2.	3	−10	No
3.	72	−20	Yes

only be used for subsequent UV-assisted or thermic embedding, respectively.

2. For polymerization use Leica accessories (Fig. 2). Fill up the capsule after pre-polymerization with 100% HM20 and transfer the root fragments. As the surface of the pre-polymerized HM20 is strongly concave, probes will almost invariably settle in the center of the capsule and do not need reorientation.
3. For polymerization under UV light, close the capsules with ring-attached slightly conical Teflon stubs (Fig. 2), thus ensuring the exclusion of oxygen. Transfer the now ring-attached capsules to an AFS set at −20 °C and place onto an aluminum container filled with ethanol as a cooling liquid. Program for polymerization is listed in Table 4.
4. For thermic polymerization transfer the aluminum cylinder holding the capsules into an AFS set at 60 °C for 24 h. Fill additional HM20 to obtain convex surfaces before placing a lid on the capsules.

3.5 Cutting of Resin Blocks

1. Trim away excessive resin, thus creating a trapezoid surface with 45° angled sides. We have used the Leica EM Trim trimming device though trimming can also be done by hand using razor blades.

2. For histological studies we cut 1–2 μm thick sections on a Leica Ultracut UCT ultramicrotome using a Diatome Histo Knife. Collect sections on 15 μL droplets of MQ W on poly-L-lysine-coated glass slides and dry on a hot plate set at 80 °C.
3. For immunolocalization studies using a fluorescence microscope 0.5 μm sections are cut and collected on Teflon-coated 8-well diagnostic microscopy slides. Slides are dried on a hot plate set at 40 °C.
4. For ultrastructure and immunolocalization studies in a transmission electron microscope 70 nm thick sections are cut with a Diatome Ultra Knife. Collect sections on coated 70 mesh copper grids after which the grids are dried at RT.

3.6 Histological Staining

1. Place slides on warming plate set at 80 °C and apply ample methylene blue/azur II solution to cover all sections. Incubate for 2 min, then remove the staining solution under tap water and wash the slides two times 5 min with distilled water.
2. Dry slides on heating plate and mount cover slips with Entellan or similar media.

3.7 Contrasting Grids for TEM

The contrast in the TEM depends primarily on the electron diffraction pattern elements. Since the lighter elements of biological tissue show only poor contrast, they often need additional contrast prior to the analysis. Standard contrasting with uranyl acetate and lead citrate can be performed using an automated contrasting system or by hand with the solutions described in the Material section.

1. Automated grid-contrasting was performed with a QG-3100 Automated TEM Stainer (RMC Boeckeler, München, Germany) which has the capacity to contrast up to 40 grids at once (*see* Table 5). Stained grids were dried at RT and stored in grid boxes.
2. Contrasting by hand is performed on 6 × 6 cm parafilm in a Petri dish. Prepare two Petri dishes, one for UAc staining and the other for lead citrate staining. In the Petri dish for lead citrate staining pellets of NaOH are added to prevent precipitation of the lead citrate by exposure to CO_2 (Fig. 3).
3. For UAc staining pipet 20 μL droplets of 2% UAc on the parafilm. Place grids on the drop and close the Petri dish. Incubate for 20 min, then wash grids four times by dipping in a series of four 250 mL beakers with MQ W. In case of methanolic UAc stain for no more than 10 min, then wash grids by dipping into a series of 100% methanol, 75% methanol, 50% methanol, and 25% followed by a series of four 250 mL beakers with MQ W.

Table 5
Protocol for grids staining using the QG-3100 Automated TEM Stainer

Step no.	Process	Time (min)
1.	Fill H_2O	1
2.	Wait	10
3.	Fill uranyl acetate	0.4
4.	Stain with uranyl acetate	10
5.	Dispose uranyl acetate	2
6.	Wait	0.1
7.	Fill lead citrate	0.4
8.	Stain with lead citrate	5
9.	Dispose lead citrate	2
10.	Wait	0.1
11.	Fill H_2O	1

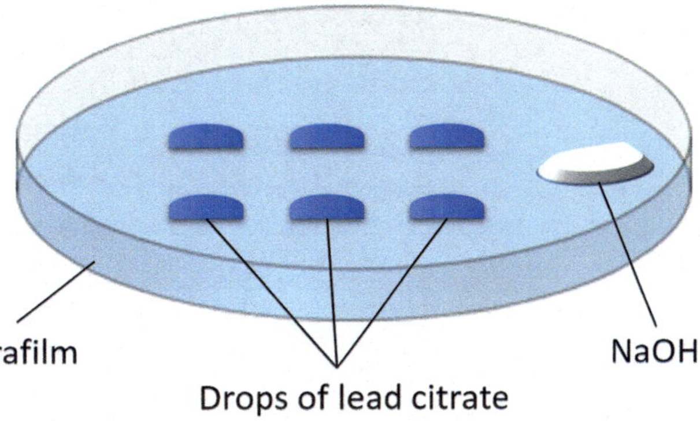

Fig. 3 Scheme of Petri dish prepared for grid staining using lead citrate. Drop number of staining solution corresponds to the grid number

4. For subsequent lead citrate staining place grids on a drop of lead citrate and close the Petri dish. Avoid breathing onto grids. After 10 min wash the grids four times as described in the previous steps, dry at RT, and store in grid boxes (Fig. 4).

3.8 Immuno-localization of Proteins on Sections

1. Teflon-coated 8-well diagnostic microscopy slides are extremely well suited for immunolabeling on semi-thin sections. The hydrophobic Teflon lining prevents the mixing of solutions, thus allowing application of multiple antibodies on one and the same slide. Start with placing the slide in a Petri dish lined with wet tissue paper which serves as a moist chamber. For each well apply 50 μL of appropriate solutions.

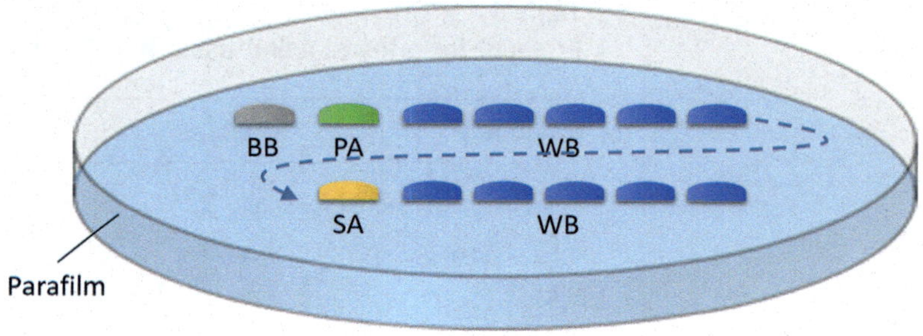

Fig. 4 Scheme of Petri dish prepared for immunolocalization of proteins on grids. *BB* blocking buffer, *PA* primary antibody, *WB* washing buffer, *SA* secondary antibody

Table 6
Protocol for immunolocalization studies

Step no.	Solution	Time (min)
1.	Blocking buffer	10
2.	Primary antibody	120
3.	Washing buffer	5
4.	Washing buffer	5
5.	Washing buffer	5
6.	Washing buffer	5
7.	Washing buffer	5
8.	Secondary antibody	30
9.	Washing buffer	5
10.	Washing buffer	5
11.	Washing buffer	5
12.	Washing buffer	5
13.	Washing buffer	5

Standard immunolabeling protocol is summarized in Table 6. Antibody incubation typically is between 30 min at 37 °C or overnight at 4 °C. The concentration of primary and secondary antibody has to be adjusted experimentally for each individual antibody.

After the final washing step pipet a 10 μL of antifade and cover the slide with a cover slip. Thus treated sections can be stored in a fridge at 4 °C for months without drying out or losing fluorescence.

2. Immunolocalization on grids basically follows the same protocol as used for fluorescence labeling described above. Conventionally, labeling is performed on a paraplast film placed in a Petri dish. A labeling series consisting of small aliquots (20 μL) of all solutions are pipetted onto the paraplast and grids are transferred from one droplet to another. After labeling, grids can be stained with UAc and lead citrate as described before.

3.9 Whole Mount Immunolocalization of Proteins

Whole mount immunolocalization is used to analyze the presence of specific proteins on the root surface. For visualization in a fluorescence microscope secondary antibodies conjugated with fluorophores are used. If observations are to be performed under bright field conditions the secondary antibodies are enzyme or gold conjugated. The latter can also be used for analysis in SEM using back-scatter detection. Since conjugated gold particles themselves are extremely small (1–5 nm) visualization of their distribution is increased by silver enhancing.

1. Primary fixation and washing steps were described in Subheading 3.1.
2. Whole mount labeling of root fragments can be performed in 0.2 mL PCR tubes (use protocol described in Subheading 3.6).
3. After use of secondary antibodies conjugated with a fluorophore samples can be investigated using fluorescence microscopy.
4. For silver enhancing with the adequate kit (e.g., BBInternational, Cat. No. SEKB-251, Cardiff, UK) root fragments are transferred into 2 mL eppendorf tubes and treated according to the manufacturers' instruction. Basically, roots are washed once with MQ W for 5 min. Following, 500 μL of silver enhancing solution is added and the roots incubated for another 5 min. After two more washes with MQ W, roots can be examined in a CLM under bright field conditions. Alternatively, samples are prepared for SEM observation. In this case roots are dehydrated in an ethanol series followed by critical point drying in a Bal-Tec Critical Point Dryer CDP 030 and covered with a gold layer of approximately 10 nm thickness using an Edwards Sputter Coater S150B.

4 Notes

1. OsO_4 is a highly toxic agent even at low concentration. It is essential to work in a fume hood, always wear gloves and avoid inhaling its vapors. Waste of OsO_4 should be neutralized by twice the volume of 20% ascorbic acid.

2. To prepare anhydrous ethanol and acetone, 20 g of molecular sieve 0.4 nm were added per 500 mL of 99.8% ethanol or 99.8% acetone and left overnight at RT before use.
3. Fragments with root meristems should also be cut longitudinally into two parts to facilitate the later stages of resin infiltration.
4. Spurr stored at −20 °C is semi-fluid, so it is recommended to adjust the resin to RT before use.
5. Mixture of LR White and LR White Accelerator will start to polymerize within 2–3 min; this step should be carried out quickly.

References

1. Barr M, Fahy A, Martens J, Jardine AP, Ward DJ, Ellis J, Allison W, Dastoor PC (2016) Unlocking new contrast in a scanning helium microscope. Nat Commun 7:10189. https://doi.org/10.1038/ncomms10189
2. Henning W, Andersen JB, Højgaard L, Greisen G, Law I, Thorseth A, Christensen AN (2015) A new self-made digital slide scanner and microscope for imaging and quantification of fluorescent microspheres. J Biomed Graph Comput 5:33
3. Haller E, Hudson J, Kelly A, Cotta DA, Peaudecerf B, Bruce GD, Kuhr S (2015) Single-atom imaging of fermions in a quantum-gas microscope. Nature Phys 11:738–742
4. Liu CS, Lin YC, Hu PH (2013) Design and characterization of precise laser-based autofocusing microscope with reduced geometrical fluctuations. Microsyst Technol 19:1717–1724
5. Foissner W (2014) An update of 'basic light and scanning electron microscopic methods for taxonomic studies of ciliated protozoa'. Int J Syst Evol Microbiol 64:271–292
6. King GA, Gross P, Bockelmann U, Modesti M, Wuite GJ, Peterman EJ (2013) Revealing the competition between peeled ssDNA, melting bubbles, and S-DNA during DNA overstretching using fluorescence microscopy. Proc Natl Acad Sci U S A 110:3859–3864
7. Li H, Pu Y, Kumar R, Ragauskas AJ, Wyman CE (2014) Investigation of lignin deposition on cellulose during hydrothermal pretreatment, its effect on cellulose hydrolysis, and underlying mechanisms. Biotechnol Bioeng 111:485–492
8. Marzec M, Melzer M, Szarejko I (2013) Asymmetric growth of root epidermal cells is related to the differentiation of root hair cells in *Hordeum vulgare* (L.). J Exp Bot 64: 5145–5155
9. Ruzin SE (1999) Plant microtechnique and microscopy, vol 198. Oxford University Press, New York
10. Marzec M, Szarejko I, Melzer M (2015) Arabinogalactan proteins are involved in root hair development in barley. J Exp Bot 66:1245–1257

Chapter 11

Preparation of Barley Pollen Mother Cells for Confocal and Super Resolution Microscopy

Sybille Mittmann, Mikel Arrieta, Luke Ramsay, Robbie Waugh, and Isabelle Colas

Abstract

Recombination (crossover) drives the release of genetic diversity in plant breeding programs. However, in barley, recombination is skewed toward the telomeric ends of its seven chromosomes, restricting the re-assortment of about 30% of the genes located in the centromeric regions of its large 5.1 Gb genome. A better understanding of meiosis and recombination could provide ways of modulating crossover distribution and frequency in barley as well as in other grasses, including wheat. While most research on recombination has been carried out in the model plant *Arabidopsis thaliana*, recent studies in barley (Hordeum *Vulgare*) have provided new insights into the control of crossing over in large genome species. A major achievement in these studies has been the use of cytological procedures to follow meiotic events. This protocol provides detailed practical steps required to perform immunostaining of barley meiocytes (pollen mother cells) for confocal or structured illumination microscopy.

Key words Barley, Meiosis, Immuno-cytochemistry, Confocal Microscopy, 3D-SIM, Antibodies

1 Introduction

Meiosis is a specialized cell division that underpins sexual reproduction by producing haploid male and female gametes. When the parent plants are heterozygous, each gamete will carry a unique set of alleles. During meiosis, the chromosomes need to pair and recombine in order to segregate properly [1]. Cytological and immunochemical methods have been used to study meiosis in model organisms for decades [2–4]. In crops, *in situ* DNA hybridization has been particularly helpful in both pre-breeding programs and academic research [5–7]. Despite this, the use of immunocytochemistry in large genome crops, such as barley, has been limited and as such remains novel. The application of immuno-

Sybille Mittmann and Mikel Arrieta contributed equally to this work.

Wendy A. Harwood (ed.), *Barley: Methods and Protocols*, Methods in Molecular Biology, vol. 1900,
https://doi.org/10.1007/978-1-4939-8944-7_11,

cytochemical methodologies has recently elucidated the timing of meiotic events and provided insight into the control of crossover formation [8–10]. With the rapid progress in sequencing the barley genome [11], it is now possible to create custom tools such as antibodies and protocols to study meiosis and recombination in this large genome crop [12–14]. Confocal microscopy is a standard method to analyze chromosome behavior during cell division, and its application in barley is straightforward [15, 16]. The emergence of super resolution microscopy such as 3D Structured Illumination Microscopy (3D-SIM) has pushed our understanding of meiosis and has helped illustrate once more the differences between small and large genome species [9, 16–19]. Here we describe a method to prepare barley pollen mother cells (male meiocytes) for investigating the progression of meiosis through analysis with confocal or 3D-SIM. We detail a protocol using custom barley antibodies to follow synapsis and recombination in male meiocytes and/or sections through developing anthers.

2 Materials

All solutions should be sterile and prepared with ultrapure water and microscopy grade reagents. Prepare and store all reagents as indicated in the protocol. Wear the appropriate laboratory protective equipment and follow local waste disposal regulations. Local suppliers or catalog number can be used unless stated otherwise for specific reagent/equipment (e.g., slides).

2.1 Plant Material

Barley *cv. Bowman* (*see* **Note 1**) plants were grown from seeds under controlled environment in growth cabinets (SANYO) under the following conditions: 16 h of light at 18–20 °C and 8 h of dark at 14–16 °C. The humidity was kept between 60% and 80%.

2.1.1 Lab Equipment

- Orbital shaker.
- 4 °C incubator (or fridge).
- Dark wet chamber (*see* **Note 2**).
- Fine forceps and/or insulin needles.
- Glass rod.
- Glass Coplin staining jar.
- Embryo Dish 30 mm Diameter.
- Cavity Slide Single Depression.
- 1.5 mL, 2 mL, and 50 mL centrifuge tubes.
- Cryostat CM1100 (Leica) (*see* **Note 3**).
- Glass Microscope slides (for staging).

- Microscope Polysine® slide or Superfrost® plus.
- Glass coverslip, 22 × 50 mm, thickness No 1.5.
- Glass coverslip, 18 × 18 (or 22 × 22), thickness No 1 or 1.5.
- Dark Dropping bottle (for aceto-carmine solution).
- Aluminum slide tray, 20 slides capacity (Brunel Microscopes Ltd.).
- Razor blade (single edge).
- Tissue paper.
- Nail varnish.
- PAP pen liquid blocker for immunostaining.

2.1.2 Microscopes

- Stereomicroscope with 10× Wide Field eyepiece.
- Light microscope with 10×, 40× Objectives (Optional: 60× objective).
- Confocal microscope LSM 710 (Carl Zeiss), or equivalent, equipped with but not limited to:
  - Objective (for Zeiss LSM 710): 60× APO C-apochromat 63×/1.20 W Korr M27 or equivalent.
  - DAPI fluorescence (blue 405, UV lamp) (Hoechst33342).
  - Fitted 405-30 diode laser 405 nm (for Hoechst33342/DAPI acquisition).
  - Fitted Argon laser 488 nm (for Alexa Fluor® 488 acquisition).
  - Fitted DPSS 561-10 laser 561 nm (for Alexa Fluor® 561 acquisition).
- Deltavision OMX (GE Healthcare group) or equivalent, equipped with 405, 488, and 561 laser lines.

2.2 Solutions

1. **Aceto-carmine:** In a fume hood, bring a solution of 45% acetic acid (45 mL glacial acetic acid/mixed with 55 mL of distilled water) to boiling point and add 0.5 g of Carmine. Stir under heat for 15–20 min, let cool and store at 4 °C overnight. Filter to remove any precipitates and store in a dark bottle (*see* **Note 4**).
2. **PBS (Phosphate Buffered Saline):** For 10× stock solution, dissolve 80 g of NaCl, 2.0 g of KCl, 14.4 g of Na_2HPO_4, and 2.4 g of KH_2PO_4 in 800 mL dH_2O. First, adjust pH to 7.4 and then adjust the volume to 1 L with additional dH_2O. Sterilize by autoclaving. Dilute with dH_2O to make 2× PBS or 1× PBS working solution. Add 5 mL of Triton™ X100 and 5 mL of

Tween® 20 to make 1 L working solution of 1× PBS, 0.5% Triton™ X100, 0.5% Tween® 20.

3. **SSC (Saline Sodium Citrate):** Dissolve 175.3 g of NaCl, 88.2 g of **Trisodium citrate dehydrate** ($Na_3C_6H_5O_7$) in 800 mL of distilled H_2O. Adjust the pH to 7.0 with 1 M HCl, then adjust the volume to 1 L with additional distilled H_2O. Sterilize by autoclaving. Dilute in distilled water to make 1 L of 4× SSC solution and add 5 mL of Tween® 20.
4. **Fixative:** In the fume hood, dissolve 1 g of paraformaldehyde (microscopy grade) in 12.5 mL of water, add 2–3 drops of 5 M NaOH, and bring to 60 °C (do not boil), stirring. Once the solution is clear (it takes about 5–10 min), remove from the heat and let cool for 20 min (or 5 min on ice). Then add 12.5 mL of 2× PBS. The solution is ready to use and can be stored in 50 mL tube in the fridge for 1 week and up to 2 weeks, (*see* **Note 5**).
5. **Blocking solution:** 1× PBS, 0.5% Triton™ X100, 0.5% Tween® 20, containing 5% Goat Serum Albumin and 5% Donkey Serum Albumin. Make it fresh.
6. **Anti-fading agent:** Vectashield anti-fade mounting medium containing DAPI (Vector Laboratories, catalog number: H-1200) (*see* **Note 6**).
7. **Primary antibodies:** Custom antibodies or commercial antibodies with good reactivity for plant tissue raised in different hosts (rabbit, rat, etc.). Dilute the primary antibody in blocking solution. The dilution needs to be worked out empirically for each antibody and generally ranges from 1:200 to 1:1000.
8. **Secondary antibodies:** We use a combination of IgG (H + L) Secondary Antibody, Alexa Fluor® conjugate (Thermo Fisher Scientific) listed in Table 1 (*see* **Note 7**).

Table 1
Secondary antibodies used in this study

Reactivity	Alexa Fluor® dye	Catalog number
Anti-Rabbit	568	A10042
Anti-Rabbit	488	A21206
Anti-Rat	568	A11077
Anti-Rat	488	A21208

9. **Tissue Freezing Medium**: Embedding medium for cryosection. Although we use Jung tissue freezing medium, it is possible to use other alternatives from Leica, Tissue Tek or Electron Microscopy Sciences (e.g., TFM™, catalog number 72592).
10. **Counterstaining solution:** (optional) Hoechst 33342 (10 mg/mL); Dilute in dH_2O to obtain a working solution at 1 μg/mL. Store in the fridge and protect from light.

3 Methods

3.1 Meiocyte Preparation

1. Harvest the barley stems and keep them in water until use. For the Bowman cultivar, meiosis normally starts 6–7 weeks after sowing. You can estimate the size of the spike by gently press the stem with two fingers up and down along the stem and feel the spike (Fig. 1a). Meiosis progresses from the base of the barley spike (Fig. 1b).
2. Open the stem to release the spike (the barley inflorescence) (Fig. 1a, b, e, j). Barley produces three anthers in each developing spikelet (Fig. 1f, h, k, m). Spikelets (individual barley flowers) are arranged either side of the central axis or rachis. Use a mounted insulin needle (or very fine forceps) to dissect out one of the three developing anthers from the center of the developing spike (Fig. 1e, j, arrow) and check under the microscope to determine the overall meiotic stage. To do this, place the anther on a microscope slide, add one drop of acetocarmine (Fig. 1c), place a coverslip (18 × 18 mm), and tap very quickly but very gently to squeeze the meiocytes out of the anther. Look under 10× objective to locate the anther and then to 40× (or 60× Objectives) to locate the meiocytes (Fig. 1d, g, i, l, n). Use a filter paper to remove the excess staining solution if the cells are floating too much on the slide.
3. Keep the other two anthers in an embryo dish (or a cavity slide) in 1× PBS containing 0.5% Triton™ X100 and 0.5% Tween® 20 at room temperature (RT). Repeat for the other florets and pool the anthers at the same stage until there is enough material to fix (about 10–20 anthers).
4. In a fume hood, remove any liquid, then add the fixative (*see* **Note 8**). Fix the samples for 15– 20 min.
5. Wash the sample twice for 5 min with 1× PBS, 0.5% Triton X100, 0.5% Tween® 20. After the last wash, the sample is safe outside of the fume hood. Use it immediately.
6. Prepare a polysine (or superfrost) slide by making a well in the center with the PAP pen and let it dry fully.
7. Remove most of the liquid (leave at least 5–10 μL) and tap out the fixed anthers repeatedly with a glass rod to release the

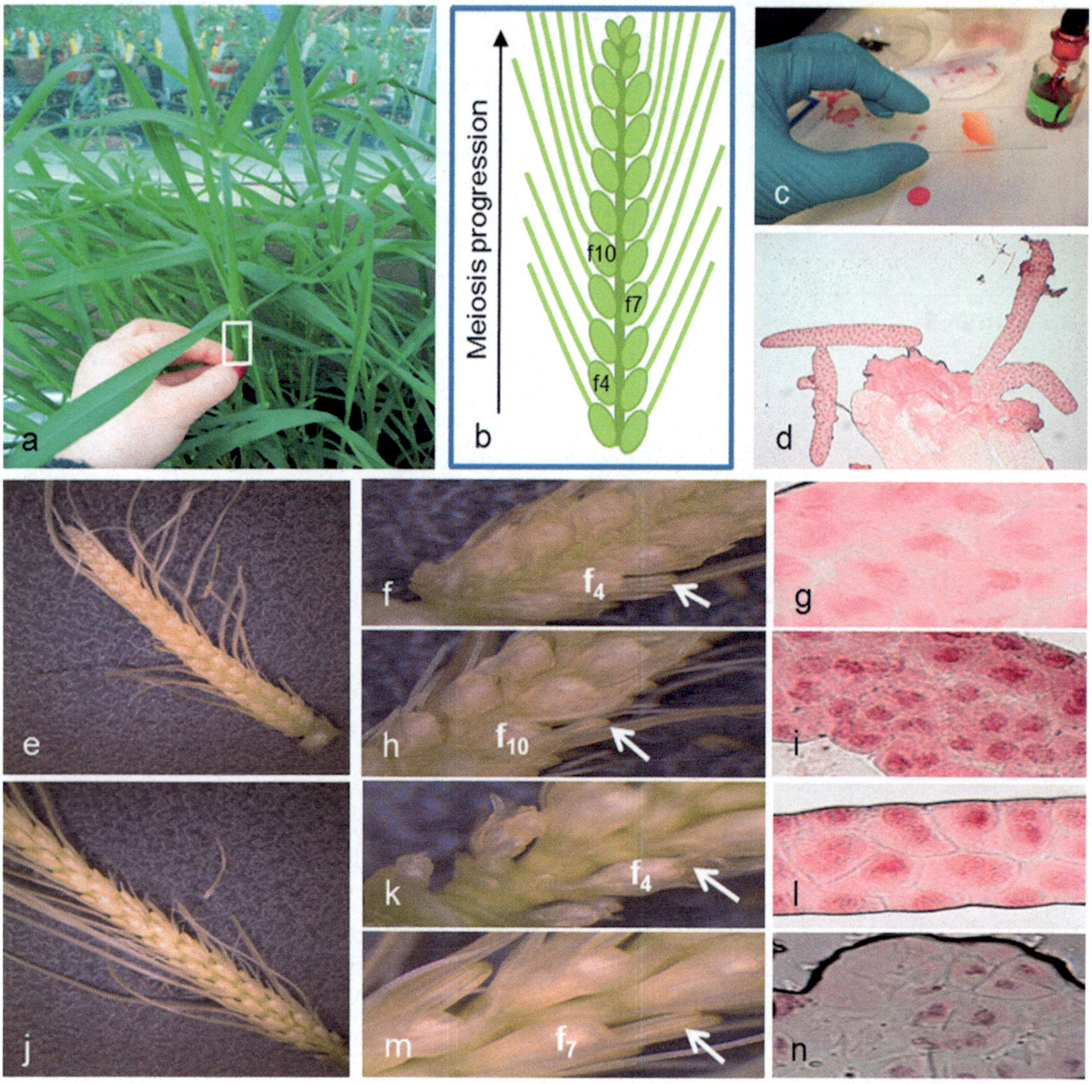

Fig. 1 Sample collection and meiosis staging. Spikes are located inside the barley stem (white box) (**a**). Meiosis progresses from the base of the barley spike toward the top (**b**). One anther is checked under aceto-carmine (**c**) and gently tapped to release the contents of the loculi (**d**). In 2 cm spikes (**e**) it is possible to have meiocytes in early prophase at the base of the spike (**f**, **g**) and later prophase in the middle of the spike (**h**, **i**)—white arrow indicates anthers (**f**, **k**, **h**, **m**). In a 3 cm spike (**j**), it is still possible to identify material in prophase in the fourth flower from the base (**k**, **l**), but most of the anthers will be at tetrad or pollen stages (**m**, **n**)

meiocytes (it takes about 5 min per sample). Add 30 μL of 1× PBS, 0.5% Triton X100, 0.5% Tween-20, mix gently, and transfer 25 μL of the solution onto the prepared slide. Let it evaporate at RT slowly so that the meiocytes fall at the bottom of the slides and stick to the polysine. Use the slide immediately.

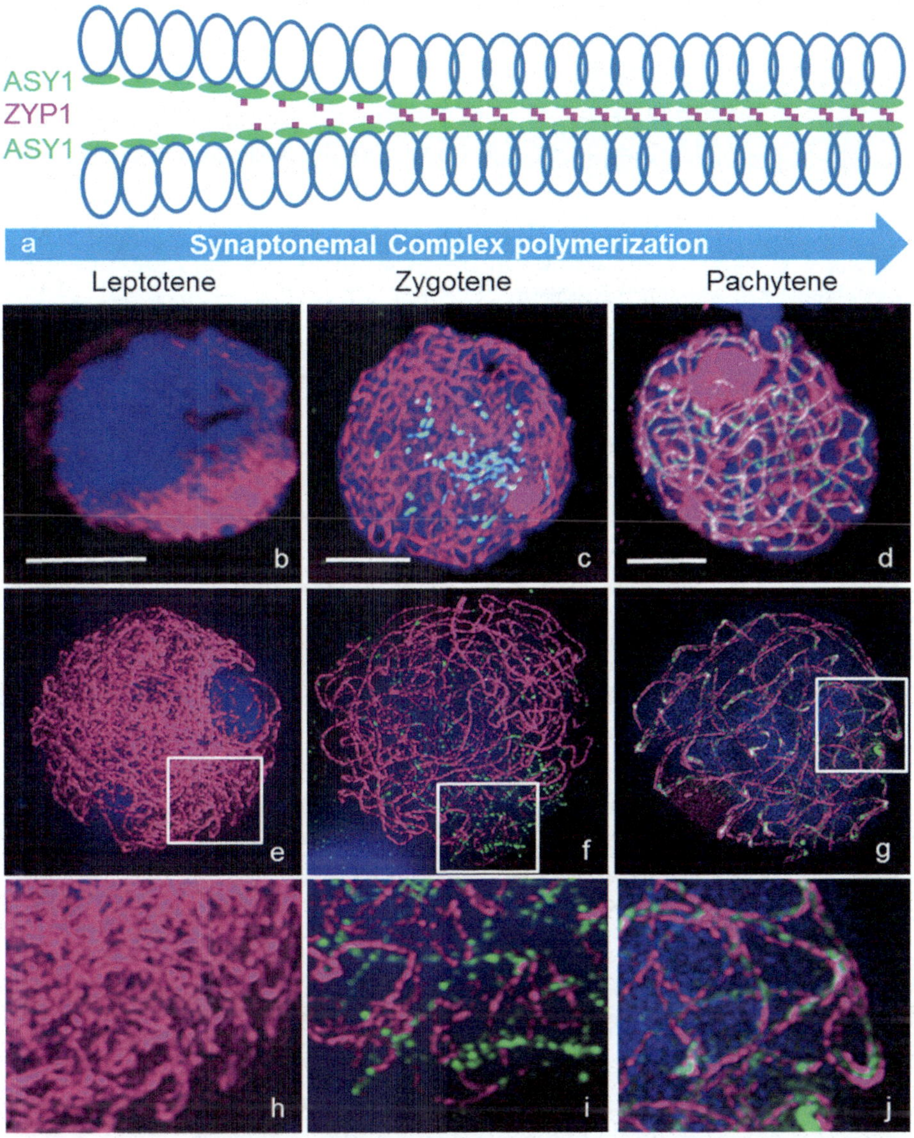

Fig. 2 Following synapsis with confocal and SIM. The synaptonemal complex (**a**) is a tri-partite structure formed between the homologous chromosomes during synapsis. We can follow chromosome axis formation (**a**) and synapsis using antibodies against the protein ASY1 (green) and ZYP1 (magenta) respectively. The same slide preparation can be used for confocal (**b**–**d**) and 3D-SIM (**e**–**j**). At leptotene (**b**, **e**, **h**), chromatin is organized around the ASY1 axes (magenta). During zygotene (**c**, **f**, **i**), chromosomes synapse as the synaptonemal complex is formed between the homologs. At pachytene, (**g**, **j**), chromosomes (magenta) are aligned entirely and linked closely by ZYP1 (green). With 3D-SIM (**e**–**j**), is it possible to see the synaptonemal complex more clearly with ASY1 threads separated by polymerized ZYP1 (**j**). Scale bar 5 μm

3.2 Immunostaining Protocol

1. Freshly dilute the primary antibodies in blocking solution. For most of our antibodies, we use a 1:500 dilution, but this depends on each antibody (commercial or custom) and the tissue. We recommend starting with 1:200 for a new antibody

and work out the best concentration by serial dilution. We used ASY1 at 1:2000 and ZYP1 at 1:500 to produce the images in Fig. 2.

2. (Optional) Add 50–100 μL of blocking solution onto the sample that is bound to the slide and leave to incubate for 30 min at RT. Remove the solution by tipping the slide over.
3. Add 30 μL of the primary antibodies solution onto the slide and add a soft plastic coverslip (parafilm, or cut autoclave bags are suitable).
4. Leave for 30 min (up to 1 h) at RT in the dark wet chamber. This step helps the solution to penetrate into the sample.
5. Move the dark wet chamber into a cold room or fridge (4–6 °C) for 24–48 h. The time of incubation also varies from one antibody to another and also on the type of tissue (e.g., meiocyte spread or section). We found that for meiocytes, 36 h is the optimum time for most of our antibodies.
6. Remove the wet chamber from the fridge and leave to warm up at RT for at least 30 min (up to 1 h).
7. Remove the plastic coverslip gently and wash the slides in 4× SSC, 0.5% Tween® 20 for 10 min and then in 1× PBS for 5 min, shaking gently (*see* **Note 9**).
8. Apply 40 μL of secondary antibody solution consisting of a 1:600 dilution of secondary antibodies (anti-rabbit, anti-rat) in blocking solution. Apply a soft plastic coverslip and leave to incubate at RT in the dark wet chamber for 90 min–2 h (*see* **Note 10**).
9. Wash the slides as previously in **step 7** (*see* **Note 11**).
10. Remove as much PBS without drying the slide and add one drop of Vectashield containing DAPI.
11. Add a full-length glass cover slip (Glass coverslip, 22 × 50 mm), press gently upside down on paper towel, and seal ends with nail varnish.
12. Leave the slide on an aluminum tray at RT to dry, and store in slide box, in the dark at 4 °C for longer storage (*see* **Note 12**).
13. Proceed to either confocal or structured illumination microscopy (Fig. 2).

3.3 Section Preparation

1. Collect stems and check one another at the base, the middle and top of each spike to get an idea of the range of meiotic stages present in the spikes. Follow **items 1** and **2** from Subheading 2.1.
2. Leave the spike 15 min in 1× PBS, 0.5% Triton™ X100, 0.5% Tween®20, in a 2 mL tube on ice.

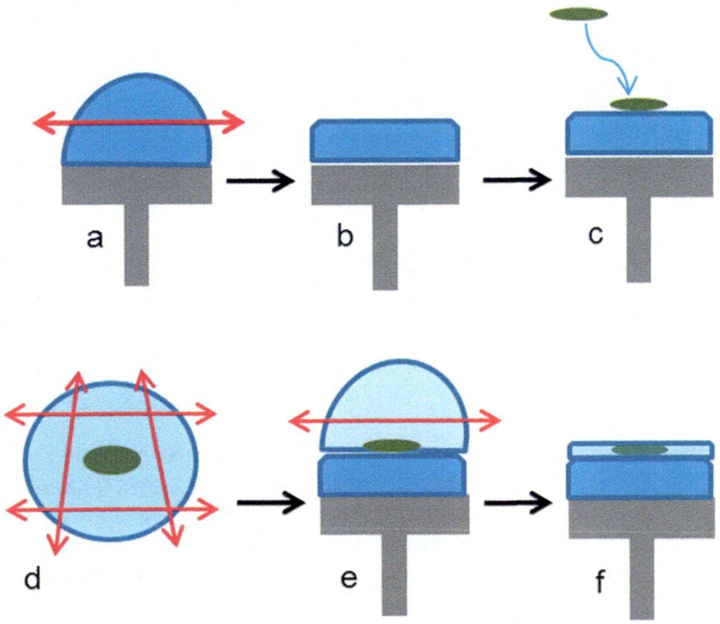

Fig. 3 Mounting the sample for cryo-sectioning. The gray is the Cryostat holder, blue shapes represent the freezing medium, dark green oval is for the sample and the double red arrows are the plane of sectioning. First, prepare the base (gray) with freezing medium (**a**) and slice it flat before placing the sample (**b**). Place the sample on the frozen surface (**c**), and add some freezing medium. Let it freeze and cut excess of medium around the sample in a trapezoid shape (**d**). Discard the first slices (**e**) until you reach your sample (**f**)

3. In the fume hood, substitute the 1× PBS, 0.5% Triton™ X100, 0.5% Tween® solution with fresh fixative. Fix the sample for 35 min (*see* **Note 13**).
4. Wash the sample twice for 10–15 min each with 1× PBS. The best quality results will be achieved when preparing the slide immediately, even though fixed spikes can be kept for 1–2 days in the fridge until use.
5. Start the Cryostat until the temperature reaches −20 °C.
6. Prepare the Specimen disk by spreading a thin layer of Tissue freezing medium on the surface (Fig. 3a), let if freeze, and cut few slices to make the surface horizontal (Fig. 3b, c).
7. If the spike is very small (less than 1 cm), cut the whole spike longitudinally. For longer spikes, use transverse sections of single florets.
8. Mount the tissue on the flat freezing medium surface (Fig. 3d), cover it with more medium, and leave it to freeze fully (Fig. 3e). Cut slices until you reach the sample (Fig. 3f).

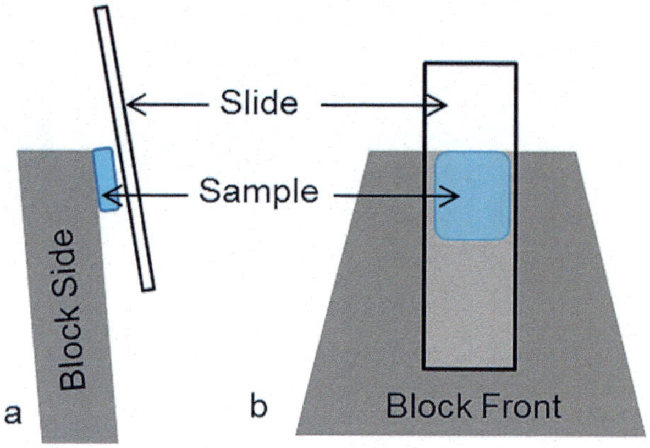

Fig. 4 Section transfer onto slides. Side view (**a**) of the cutting block (gray) that the 20 μm slice (blue) is sticking to. Approach the slide (dark rectangle) carefully toward the sample slice, and once the sample touches the slide (and defrosts), remove it quickly upside down to flatten the sample and let it dry. Front view (**b**) of the cutting block (gray), sample slice (blue), and slide (dark rectangle)

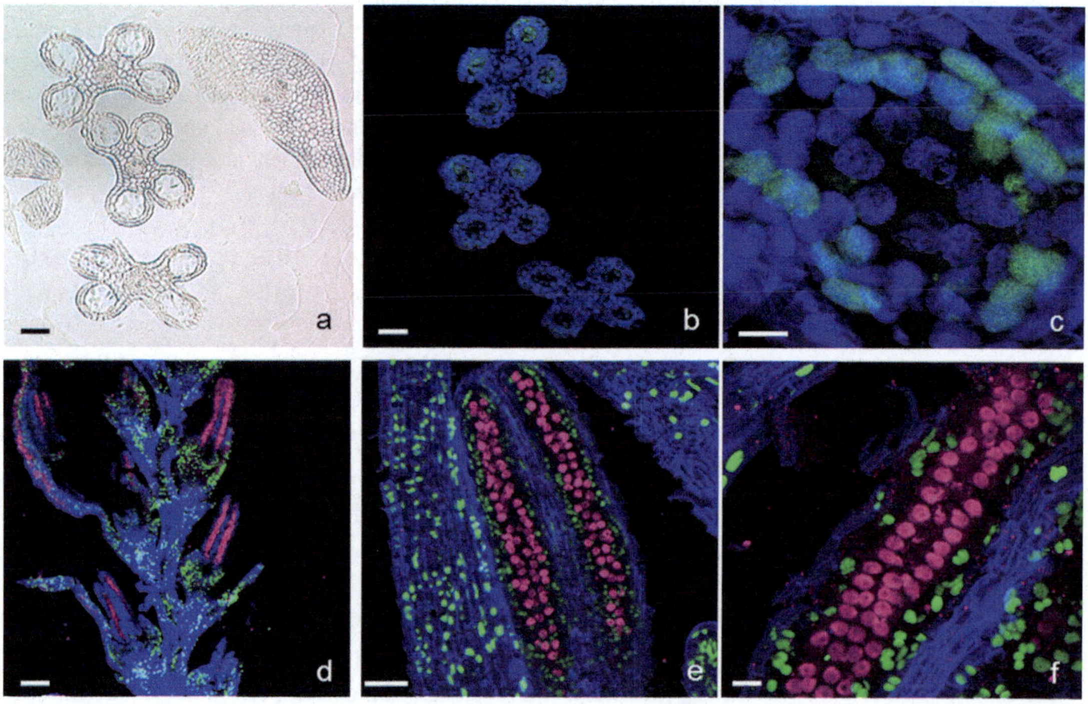

Fig. 5 Immunostaining on sections. Cross sections (**a**–**c**) of barley anthers sectioned with the cryostat (**a**) and DAPI stained (blue) (**b**). Close up of a cross-section with tapetal cells labeled with EDU (green—*see* **Note 11**) around the meiocytes (arrows). Longitudinal section (**d**–**f**) of barley anthers showing cells labeled with EDU (green) and meiocytes labeled with ASY1 (magenta). Samples were collected 43.5 h after EDU injection. Bars: 50 μm (**a**, **b**, **e**); 10 μm (**c**); 200 μm (**d**); 20 μm (**f**)

9. Cut slices (20 μm thick) and stick them on the slide by approaching the RT slides near the sample (*see* **Note 3** and Fig. 4).
10. Leave at RT until almost dry and delimit the borders of the samples with a PAP pen.
11. Proceed to immunostaining as per Subheading 3.2 (Fig. 5).

4 Notes

1. Growing conditions such as temperature or humidity might affect growing time and therefore entry into meiosis. Although the cultivar Bowman was used in this study, other cultivars such as Golden Promise or Morex can be used, but spike sizes and developmental time may vary.
2. Dark wet chambers are commercially available or can be easily made [12].
3. The cryostat CM100 manual is available at this link: https://drp8p5tqcb2p5.cloudfront.net/fileadmin/downloads_lbs/Leica%20CM1100/User%20Manuals/Leica_CM1100_IFU_1v3B_en.pdf. Although this is a different model form the one we use, the approach to transfer sample onto the slide is similar that is described in this video at 2:45 min: https://www.youtube.com/watch?v=z-WoPn2YLRQ&t=178s
4. Chromosomes are darker stained in iron *aceto-carmine* solution and this could be helpful for early meiotic stages. Simply add a "mordant" by soaking a piece of iron (for example, an iron nail) in your solution until the solution gets darker. Alternatively, use the following protocol. Make two solutions of 100 mL aceto-carmine, but in one of the solution add 5 g of Ferric oxide. Add the Ferric oxide solution drop by drop to 50 mL of the first Carmine solution until a precipitate starts to appear. Immediately add the remaining 50 mL of first Carmine solution to the titrate mixture and filter to remove any precipitate.
5. Paraformaldehyde waste needs to be disposed of in appropriate bottles. All other material can be decontaminated overnight in a solution of sodium metabisulfite before rinsing with plenty of water.
6. Other anti-fading agents can be used but we recommend Vectashield for 3D-SIM because of its correct refraction index. If using an anti-fading agent without DAPI, it is possible to add an extra step by adding 50 μL of DAPI or HOECHST33342 (1 μg/mL) for 10 min in the dark.

7. We found that the Alexa dyes are quite robust but can be replaced by other alternatives (Dylight, for example).
8. We use 40 μL fixative for cavity slides, 200 μL for embryo dish, and 1.9 ml for the whole spike (in a 2 mL tube).
9. Washing steps can be conducted using a combination of buffers with different stringencies. We found that for "easy" samples, 15 min in 1X PBS is usually sufficient. Slides are safe in 1× PBS for up to 2 h at RT.
10. Longer incubation in secondary antibody or higher concentration can increase background.
11. This method is compatible with time course studies using EDU (5-ethynyl-2′-deoxyuridine). In this case, EDU is injected into the transpiration stream of the sample below the inflorescence and meiocytes are collected at different times [9, 13]. Samples are then fixed and immunostained as described in this protocol. After the secondary antibody wash, EDU detection can be performed as per supplier's protocol and then slides can be washed and mounted.
12. Slides can be kept in the fridge for a few months (up to 1 year) without any degradation of the chromatin or immunofluorescence signal, even after a few confocal sessions. In general, after 2 or 3 SIM sessions, slides can only be kept up to 3–4 months.
13. Vacuum Infiltration can improve the fixation of large tissue such as spike. In this case, 5–10 min prior to fixation is enough (20).

Acknowledgments

The research leading to these results has received funding from the European Community's Seventh Framework Programme *FP7/ 2007-2013* under grant agreement n° 222883 and ERC-SHUFFLE (Project no. 669182) to R.W. PhD studentship from BBSRC EastBIO and an ESR position within the EU FP7 ITN COMREC supported S.M. and M.A. respectively. Use of the OMX microscope was supported by the Euro-BioImaging PCS to I.C. and through the MRC Next Generation Optical Microscopy Award (Ref: MR/K015869/1). L.R. and R.W. were also funded from the Scottish Government's Rural and Environment Science and Analytical Services work package RD 2.1 and RD 2.2. The authors declare no conflict of interest. Sybille Mittmann and Mikel Arrieta contributed equally to this work.

References

1. Zickler D, Kleckner N (1999) Meiotic chromosomes: integrating structure and function. Ann Rev Gen 33:603–754
2. Baudat F, Imai Y, de Massy B (2013) Meiotic recombination in mammals: localization and regulation. Nat Rev Gen 14(11):794–806
3. Mercier R, Grelon M (2008) Meiosis in plants: ten years of gene discovery. Cytogenet Genome Res 120(3–4):281–290
4. Mercier R, Mézard C, Eric Jenczewski E et al (2014) The molecular biology of meiosis in plants. Annu Rev Plant Biol 66:297–327
5. Bass HW, Birchler J (2011) Plant cytogenetics: genome structure and chromosome function. Springer, New York
6. Colas I, Shaw P, Prieto P et al (2008) Effective chromosome pairing requires chromatin remodeling at the onset of meiosis. PNAS 105 (16):6075–6080
7. Schwarzacher T, Anamthawat-Jónsson K, Harrison GE et al (1992) Genomic in situ hybridization to identify alien chromosomes and chromosome segments in wheat. Theor Appl Genet 84(7–8):778–786
8. Barakate A, Higgins JD, Vivera S et al (2014) The synaptonemal complex protein ZYP1 is required for imposition of meiotic crossovers in barley. Plant Cell 26(2):729–740
9. Colas I, Macaulay M, Higgins JD et al (2016) A spontaneous mutation in MutL-Homolog 3 (HvMLH3) affects synapsis and crossover resolution in the barley desynaptic mutant *des10*. New Phytol 212(3):693–707
10. Phillips D, Jenkins G, Macaulay M et al (2015) The effect of temperature on the male and female recombination landscape of barley. New Phytol 208(2):421–429
11. IBGSC et al (2012) A physical, genetic and functional sequence assembly of the barley genome. Nature 491(7426):711–716
12. Colas I, Baker K, Flavell AJ (2016) Cytology and microscopy immunolocalization of covalently modified histone marks on barley mitotic chromosomes. Bio-Protocol 6(12): e1841
13. Higgins JD (2013) Analyzing meiosis in barley. In: Pawlowski W, Grelon M, Armstrong S (eds) Plant meiosis: methods and protocols, methods in molecular biology, 990. Springer, New York, pp 135–144
14. Phillips D, Wnetrzak J, Nibau C et al (2013) Quantitative high resolution mapping of HvMLH3 foci in barley pachytene nuclei reveals a strong distal bias and weak interference. J Exp Bot 64(8):2139–2154
15. Ramsay L, Colas I, Waugh R (2014) Modulation of meiotic recombination. In: Kumlehn J, Stein N (eds) Biotechnological approaches to barley improvement, biotechnology in agriculture and forestry, vol 69. Springer Verlag, Berlin Heidelberg, pp 311–329
16. Golubovskaya IN, Wang CJ, Timofejeva L et al (2011) Maize meiotic mutants with improper or non-homologous synapsis due to problems in pairing or synaptonemal complex formation. J Exp Bot 62(5):1533–1544
17. Pawlowski WP, Wang CJ, Golubovskaya IN et al (2009) Maize AMEIOTIC1 is essential for multiple early meiotic processes and likely required for the initiation of meiosis. PNAS 106(9):3603–3608
18. Phillips D, Nibau C, Wnetrzak J et al (2012) High resolution analysis of meiotic chromosome structure and behaviour in barley (Hordeum vulgare L.). PLoS One 7(6):e39539
19. Kuo J (2007) Processing plant tissues for ultrastructural study. Methods Mol Biol 369:35–45

Chapter 12

Microarrays for High-Throughput Gene Expression Analysis of Barley

Jenny A. Morris and Pete E. Hedley

Abstract

Microarrays remain an extremely robust and cost-efficient technology for transcriptome analysis. A custom Agilent microarray has been designed from the predicted gene transcripts of the barley genome reference. Here, we describe total RNA extraction, fluorescent labeling, hybridization, and image analysis of barley arrays. Quality control measures are detailed, along with recommendations for downstream data analysis.

Key words Microarrays, Gene expression, RNA, Transcriptomics, Agilent, Barley

1 Introduction

Determining the relative activities of genes at defined growth stages or under certain conditions or treatments can help prioritize candidates for targeted processes, such as tissue specificity or biotic and abiotic stresses. Transcriptomics enables high-throughput gene expression profiling on a whole-genome scale. There are two primary methods for transcriptomic analysis: microarrays and RNA sequencing (RNAseq) [1, 2]. Recently, there has been significant transition to Next Generation Sequencing (NGS) based methods, specifically RNAseq; however, there are still very good reasons to utilize microarray technology, which has been the method of choice for the last 15 years. Firstly, microarray platforms and downstream analysis approaches are very well established, ensuring that sample processing and data handling can be performed accurately and efficiently. In addition, data accessibility strategies are very well established, following strict MIAME (Minimal Information About a Microarray Experiment; http://fged.org/projects/miame/) guidelines. Importantly, microarrays do not require dedicated bioinformatics support, which is often a cost- and rate-limiting factor for NGS projects. Although microarrays are a "closed" system, thereby requiring pre-determination of gene

Wendy A. Harwood (ed.), *Barley: Methods and Protocols*, Methods in Molecular Biology, vol. 1900,
https://doi.org/10.1007/978-1-4939-8944-7_12, © Springer Science+Business Media, LLC, part of Springer Nature 2019

content, many crop species including barley now have a high quality, well annotated genome reference available with the majority of genes defined. Finally, from a cost perspective, microarrays are still up to fourfold less expensive per sample compared to RNAseq. If the primary goal of the experiment is simply to determine relative gene expression changes between sample sets, and not complex isoform analysis or alternative splicing, then microarrays remain a cost-effective and reliable alternative.

Choice of microarray platform is still important, but the commercial supplier Agilent offer free custom-design by user-driven software, thereby enabling any transcript set to be represented in an affordable highly flexible format. We have developed custom microarrays for several crop species, including potato [3], tomato [4], raspberry [5], and blackcurrant [6]. Here we describe the design and utilization of a custom microarray for barley and, based on our experience, provide some recommendations to help ensure trouble-free processing and downstream analysis.

2 Materials

2.1 Total RNA Extraction

1. TRI Reagent solution (Part No. T9424, Sigma-Aldrich).
2. General laboratory solutions: isopropanol; 0.8 M sodium citrate/1.2 M NaCl; 75% (v/v) ethanol; nuclease-free water; saturated phenol/chloroform/IAA; absolute ethanol; 70% (v/v) ethanol.

2.2 Agilent Microarray Processing

1. Custom Agilent barley 60 k microarray (JHI_Hv_60k_v1) (*see* **Note 1**) (Part No. G4102A; Design AMADID 039514; Agilent Technologies).
2. One-Color RNA Spike-In Kit (*see* **Note 2**) (Part No. 5188-5279; Agilent Technologies).
3. Low Input Quick Amp Labeling Kit (*see* **Note 3**) (Part No. 5190-2305; Agilent Technologies).
4. Quality control instrumentation. RNA integrity: Bioanalyzer 2100 (Agilent Technologies), RNA 6000 Nano Kit (Part No. 5067-1511); RNA purity and cRNA labeling: NanoDrop Spectrophotometer (Thermo Fisher Scientific).
5. 25× Fragmentation Buffer, 10 mL (Part No. 5185-5974; Agilent Technologies).
6. 2× Hi-RPM Hybridization Buffer, 25 mL (Part No. 5190-0403; Agilent Technologies).
7. 10× Gene Expression Blocking Agent (Part No. 5188-5281; Agilent Technologies).

8. Microarray gasket slides: 8 per slide (*see* **Note 4**) (Part No. G2534-60014; Agilent Technologies).
9. Hybridization chamber (Part No. G2534A; Agilent Technologies).
10. Hybridization oven (Part No. G2545A; Agilent Technologies).
11. Gene Expression Wash Buffer 1, 4 L (Part No. 5188-5325); Gene Expression Wash Buffer 2, 4 L (Part No. 5188-5326); Gene Expression Wash Buffer Kit (Part No. 5188-5327; Agilent Technologies).
12. High-Throughput Wash Station baths (Part No. HTW; ArrayIt).
13. Slide centrifuge: Microarray High-Speed Centrifuge (*see* **Note 5**) (Part No. MHC; ArrayIt).
14. Microarray scanner: SureScan Scanner (*see* **Note 6**) (Part No. G2505B; Agilent Technologies).

2.3 Microarray Data Analysis

1. Image processing software: Feature Extraction v 10.7.3.1 (Agilent Technologies).
2. Microarray analysis software: Genespring v 7.3 (*see* **Note 7**) (Agilent Technologies).

3 Methods

3.1 Total RNA Extraction

Total RNA is extracted from barley tissue using either the RNeasy Plant Mini Kit (Qiagen), as recommended, or TRI Reagent (Sigma-Aldrich), as detailed below:

1. Grind tissue (100–200 mg fresh frozen material) under liquid nitrogen using a baked mortar and pestle.
2. Add TRI Reagent while tissues are still partially frozen (1 mL per 50–100 mg of tissue) and mix thoroughly (*see* **Note 8**).
3. Once thawed, transfer to 1.5 mL microfuge tubes.
4. Optional additional spin (*see* **Note 9**): centrifuge at 12,000 × *g*, 10 min at 4 °C, transfer supernatant to new 1.5 mL tubes.
5. Leave at room temperature for 5 min.
6. Spin for 10 min at 12,000 × *g* at 4 °C, and transfer supernatant to new 1.5 mL tube.
7. Add 0.2 mL chloroform (per 1 mL TRI Reagent), and vortex for 15 s. Allow to stand for 2–15 min.
8. Centrifuge at 12,000 × *g* for 15 min at 4 °C.
9. Transfer upper aqueous phase to a fresh tube and add 0.5 mL of isopropanol per 1 mL TRI Reagent and mix well.

10. Optional cleanup step (*see* **Note 10**): instead of adding 0.5 mL isopropanol, add 0.25 mL isopropanol and 0.25 mL 0.8 M sodium citrate/1.2 M NaCl.
11. Leave at room temperature for 10 min, then centrifuge at 12,000 × *g* for 10 min at 4 °C. The RNA precipitate forms a pellet on side/bottom of tube.
12. Remove supernatant and wash pellet with 1 mL 75% (v/v) ethanol per mL of TRI Reagent. Vortex and centrifuge at 7500 × *g* for 5 min at 4 °C (if pellet floats, centrifuge at 12,000 × *g*) (*see* **Note 11**).
13. Discard supernatant and leave pellet to air-dry.
14. Add nuclease-free water, 100 μL at a time, to dissolve pellet (up to a maximum of 400 μL).
15. Add an equal volume of saturated phenol/chloroform/IAA, vortex, and centrifuge at 12,000 × *g* for 10 min at 4 °C.
16. Transfer upper phase (up to 300 μL) to a new 1.5 mL tube on ice.
17. Add 100 μL water to the original tubes containing phenol/chloroform/IAA, vortex, and spin at 12,000 × *g* for 5 min at 4 °C.
18. Remove 100 μL of upper phase and combine with the 300 μL previously isolated.
19. Precipitate RNA by adding 40 μL 3 M NaOAc and 1 mL absolute ethanol, mix well, and place at −80 °C for 2 h to overnight.
20. Centrifuge at 12,000 × *g* for 10 min at 4 °C and remove supernatant. RNA will pellet on side/bottom of tube.
21. Wash pellet with 1 mL 70% (v/v) ethanol, vortex, and spin at 12,000 × *g* for 5 min. Remove supernatant and dry at room temperature.
22. Resuspend pellet in 100 μL nuclease-free water and store at −80 °C.

3.2 Quality Control of Total RNA

Quantification of RNA and detection of contaminants is measured by spectrophotometry, and RNA integrity is determined by gel electrophoresis using a Bioanalyzer assay.

1. Measure absorbances at 280 nm, 260 nm, and 230 nm of 1.5 μL of undiluted stock total RNA on a Nanodrop (*see* **Note 12**).
2. Run 1 μL of RNA at a concentration of 250 ng/μL (*see* Fig. 1; also *see* **Note 13**).

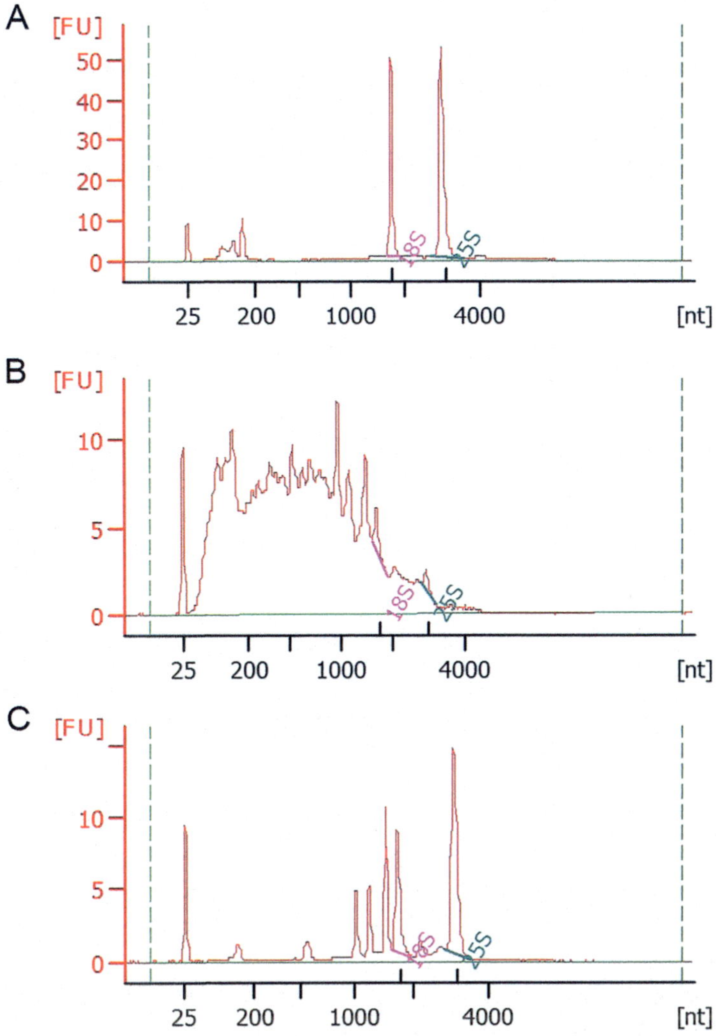

Fig. 1 Examples of total RNA extracted from plant tissue run on the Bioanalyzer RNA Nano Assay: (**a**) high-quality RNA (RIN 9.5); (**b**) degraded RNA (RIN 2.5) and; (**c**) RNA from photosynthetic tissue (RIN 7.8), note additional small rRNA banding present. X-axis represents size (nt) and y-axis represents fluorescent units (FU)

3.3 Fluorescent Labeling of RNA

This is carried out using a modified version of the recommended manufacturer's protocol (One-Color Microarray-Based Gene Expression Analysis v 6.9.1; Agilent Technologies). Also *see* **Note 14**. The protocol is summarized below:

1. Prepare the RNA spike mix (*see* **Note 15**). Pre-warm the stock solution to 37 °C prior to making serial dilutions with nuclease-free water: for labeling 100 ng total RNA, this comprises 1:20, then 1:25, and finally 1:20.
2. Set water baths to 65 °C, 80 °C, and 40 °C.

3. Aliquot 1.5 μL RNA (67 ng/uL, 100 ng in total) into a 1.5 mL microfuge tube. To this, add 2 μL spike-in dilution, 0.8 μL T7 Promoter Primer, and 1 μL nuclease-free water. This can be made up as a Primer Spike-In Master Mix to cover all RNA samples.
4. Denature by incubating at 65 °C for 10 min.
5. Place the reactions on ice for 5 min. Briefly spin each sample tube and return to ice.
6. Immediately prior to use, gently mix components (and multiples thereof) for the cDNA Master Mix (*see* **Note 16**) in this order: 2 μL 5× First Strand Buffer, 1 μL 0.1 M DTT, 0.5 μL 10 mM dNTP Mix, and 1.2 μL AffinityScript RNase Block, per sample.
7. Add 4.7 μL cDNA Master Mix to each sample tube and mix by pipetting. Each tube now contains a volume of 10 μL.
8. Incubate samples at 40 °C for 2 h.
9. Move samples to a 70 °C and incubate for 15 min.
10. Move samples to ice and incubate for 5 min. Spin briefly.
11. Immediately prior to use, gently mix components by pipetting (and multiples thereof) for the Transcription Master Mix (*see* **Note 17**): 0.75 μL nuclease-free water, 3.2 μL 5× Transcription Buffer, 0.6 μL 0.1 M DTT, 1 μL NTP Mix, 0.21 μL T7 RNA Polymerase, and 0.24 μL Cy3 Dye, per sample.
12. Add 6 μL Transcription Master Mix to each sample and mix by pipetting.
13. Incubate samples at 40 °C for 2 h (*see* **Note 18**).

3.4 Purification of Labeled cRNA

Remove unincorporated dye and purify the labeled cRNA using an RNeasy Mini Column (Qiagen).

1. To each sample, add 84 μL nuclease-free water followed by 350 μL RLT Buffer. Mix well by pipetting.
2. Add 250 μL absolute ethanol and mix thoroughly by pipetting. Do not centrifuge.
3. Transfer 700 μL of the cRNA sample to an RNeasy Mini column in a 2 mL collection tube. Centrifuge the sample at 12,000 × *g* for 30 s at 4 °C. Discard the flow-through and collection tube.
4. Transfer the RNeasy column to a new collection tube and add 500 μL RPE Buffer (pre-diluted with ethanol) to the column. Centrifuge at 12,000 × *g* for 30 s at 4 °C. Discard the flow-through. Reuse the collection tube.

5. Add another 500 µL RPE Buffer to the column. Centrifuge at 12,000 × *g* for 60 s at 4 °C. Discard the flow-through and the collection tube.
6. Elute the cRNA sample by transferring the RNeasy column to a new 1.5 mL collection tube. Add 30 µL nuclease-free water directly onto the RNeasy filter membrane, incubate 60 s, then centrifuge at 12,000 × *g* for 30 s at 4 °C.
7. Maintain the cRNA on ice.

3.5 Quantification of Labeled cRNA

1. Start the NanoDrop software.
2. Click the "Microarray Measurement" tab.
3. Before initializing the instrument, clean the sample-loading area with nuclease-free water.
4. Once the instrument has initialized, select "RNA-40" as the "Sample Type".
5. Blank the instrument by pipetting 1.5 µL nuclease-free water and click "Blank".
6. Clean the sample-loading area with a laboratory wipe. Pipette 1.5 µL sample onto the instrument sample-loading area and click "Measure".
7. Record the following values:
   (a) Cyanine 3 (or cyanine 5) dye concentration (pmol/µL).
   (b) RNA absorbance ratio (260 nm/280 nm).
   (c) cRNA concentration (ng/µL).
8. Determine the yield and specific activity of each reaction:
   (a) Use concentration of cRNA (ng/µL) to determine the cRNA yield: (concentration of cRNA × 30 µL (elution volume))/1000 = µg of cRNA.
   (b) Use concentration of cRNA (ng/µL) and cyanine 3 (pmol/µL) to determine the specific activity: (concentration of Cy3 × 1000)/concentration of cRNA = pmol Cy3 per µg cRNA.
9. Examine the yield and specific activity results (*see* **Note 19**).

3.6 Hybridization of Labeled cRNA to Microarrays

1. Add 1.25 mL nuclease-free water to the lyophilized 10× Blocking Agent supplied with the Gene Expression Hybridization Kit.
2. Mix by gently vortexing. If the pellet does not go into solution completely, heat the mix for 5 min at 37 °C. Centrifuge briefly (*see* **Note 20**).
3. Equilibrate a water bath to 60 °C.
4. For each microarray, add the following components to a 1.5 mL microfuge tube: 600 ng Cy3-labeled cRNA, 5 µL

10× Blocking Agent, 1 μL 25× Fragmentation Buffer and nuclease-free water to 25 μL total volume.

5. Mix gently by vortexing and incubate at 60 °C for exactly 30 min to fragment cRNA.
6. Immediately cool on ice for 60 s.
7. Add 25 μL 2× Hi-RPM Hybridization Buffer to stop the reaction. Mix by pipetting (*see* **Note 21**).
8. Spin for 60 s at 12,000 × *g* at room temperature. Place sample on ice and load onto the array as soon as possible.
9. Load a clean gasket slide into the Agilent SureHyb chamber base with the label facing up and aligned with the rectangular section of the chamber base (*see* **Note 22**).
10. Slowly dispense 40 μL hybridization sample onto the gasket well in a "drag and dispense" manner.
11. Slowly place an array "active side" down onto the SureHyb gasket slide, so that the "Agilent" barcode is facing down and the numeric barcode is facing up (*see* **Note 23**).
12. Place the SureHyb chamber cover onto the sandwiched slides and slide the clamp assembly onto both pieces. Hand-tighten the clamp onto the chamber.
13. Vertically rotate the assembled chamber to wet the gasket and assess the mobility of the bubbles (*see* **Note 24**).
14. Clamp assembled slide chamber onto rotisserie in the hybridization oven set to 65 °C. Set the oven to rotate at 10 rpm.
15. Hybridize at 65 °C for 17 h.

3.7 Washing and Scanning Microarrays

1. Pre-warm Gene Expression Wash Buffer 2 at 37 °C overnight (*see* **Note 25**).
2. Completely fill Dish 1 with Gene Expression Wash Buffer 1 at room temperature.
3. Fill another Dish 2 with sufficient Gene Expression Wash Buffer 1 at room temperature to cover a slide rack.
4. Place an empty Dish 3 on the stir plate. Do not add the pre-warmed Gene Expression Wash Buffer 2 until the first wash step has begun.
5. Remove one hybridization chamber from the oven and disassemble. Place the hybridization chamber assembly on a flat surface and loosen the thumbscrew, turning anticlockwise. Slide off the clamp assembly and remove the chamber cover.
6. Remove the array-gasket sandwich from the chamber base by grasping the slides from their ends. Keep the microarray slide numeric barcode facing up and transfer to Dish 1.

7. Submerge the array-gasket sandwich into slide-staining Dish 1 containing Gene Expression Wash Buffer 1.
8. While completely submerged in Gene Expression Wash Buffer 1, pry the slides apart from the barcode end only (*see* **Note 26**).
9. Let the gasket slide drop to the bottom of the staining dish.
10. Remove the microarray slide and place into slide rack of Dish 2 containing Gene Expression Wash Buffer 1 at room temperature (*see* **Note 27**).
11. Repeat wash steps for additional slides (*see* **Note 28**).
12. When all slides in the group are placed into the slide rack of Dish 2, stir for 1 min.
13. During this wash step, pour pre-warmed Gene Expression Wash Buffer 2 into Dish 3.
14. Transfer slide rack to Dish 3 and stir for 1 min (*see* **Note 29**).
15. Spin-dry each slide in the slide centrifuge for 10 s, ensuring this is done with the active side facing upward.
16. Put the slides into a slide holder for scanning. Scan slides immediately using recommended settings (*see* **Notes 30** and **31**).
17. Check visually images for hybridization and washing artifacts (*see* **Note 32**; *see* Fig. 2).

3.8 Feature Extraction

Microarray images are aligned with the design grid to enable spot (feature) identification and extraction of signal intensities for both foreground and local background of every feature. The resulting data files can then be imported into analysis software.

1. Open Feature Extraction (FE) v 10.7.3.1 (Agilent Technologies).
2. Import microarray images (tif files) from the scanning process into the FE software.
3. Download the Grid Template from eArray and import into FE software. For the barley 60 k microarray, 039514_D_F_20120314 is required.
4. The FE Protocol to be used is GE1_107_Sep09 and the QC Metric Set GE1_QCMT_Sep09.
5. Use default settings, save the project, and run. This can take up to 20 min per slide.
6. Check for any error messages and visualize the QC Report generated for each array (*see* **Note 33**).

3.9 Data Preprocessing and Analysis

There are many different options for analyzing microarray data, including free academic and commercial software (*see* **Note 34**). Detailed description of these analysis tools is outwith the scope of

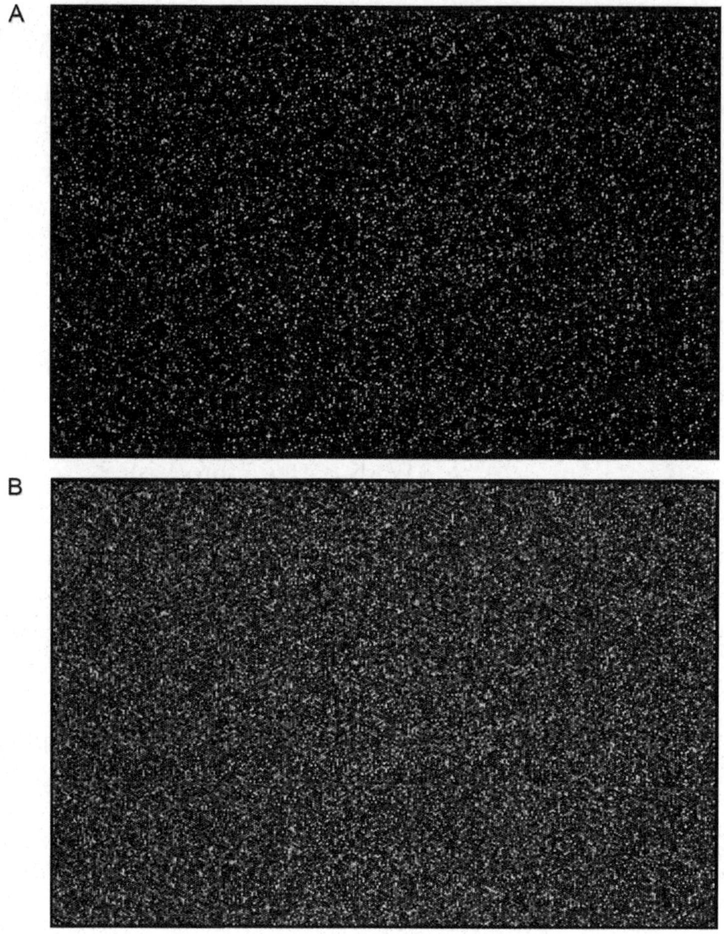

Fig. 2 Examples of images obtained from scanning barley Agilent microarrays: (**a**) high-quality hybridization; (**b**) graining artifact with high background (*see* **Note 32**)

this chapter; however, we would recommend GeneSpring GX (Agilent Technologies) software as a comprehensive package for preprocessing and statistical analysis, and it provides good visualization options.

4 Notes

1. The basis of the design for the custom barley microarray was the predicted transcript set of the 2012 barley reference genome sequence generated by the International Barley Sequencing Consortium [7]. A representative transcript from each high-confidence annotated gene was put into a single fasta formatted file and uploaded to Agilent eArray design software. eArray is a simple free-to-use web interface which enables

design of 60 mer probe sequences with high transcript specificity, targeted to the 3′-end of the genes of interest. Probes are then grouped, along with probes for quality control measures, and subsequently configured into the most appropriate microarray format, in this case 8 × 60 k (that is 8 identical microarrays per slide which can be processed with independent samples). The Agilent design identifier for the barley 60 k microarray (JHI_Hv_60k_v1) is AMADID 039514 and the design can be found at ArrayExpress data repository (https://www.ebi.ac.uk/arrayexpress/; accession A-MEXP-2357). Flexibility of the Agilent custom format allows simple design changes, which is very useful as new iterations of genome references become available, potentially leading to versioned designs.

2. Agilent's RNA Spike-In Kit allows efficient monitoring of the microarray workflow for linearity, sensitivity, and accuracy. Each spike mix contains ten in vitro synthesized, polyadenylated transcripts in predetermined ratios. When hybridized to Agilent microarray control probes (included in the custom design), the data is utilized for quality control in Feature Extraction software. Two-Color RNA Spike-In Kits and One-Color RNA Spike-In Kits are available.
3. Labeling of sample RNA can be performed by conversion to cDNA or cRNA, and as one-color or two-color formats, which are both compatible with Agilent microarrays.
4. Type of gasket slide depends on the format of arrays: there are gasket slides for 1, 2, 4, and 8 arrays per slide and they can be purchased in quantities of 5, 20, or 100.
5. Microarray High-Speed Centrifuge provides rapid desktop centrifugation of microarray slides, with drying complete in 10 s.
6. Other commercial microarray scanners are available, but for Agilent microarray compatibility and resolution requirements this scanner is recommended.
7. Other commercial and academic analysis software are available.
8. This volume can be increased to 2 mL if samples are very starchy and split between two tubes then combined on completion. Using TRI Reagent, especially with the optional extra steps, will increase both the yield and quality of RNA from starchy material. The starch can very easily clog column-based extraction methods, such as Qiagen, and drastically reduces the resulting RNA yield.
9. This additional step aids separation from insoluble material.
10. Optional step to help remove polysaccharides.
11. Pellet can be stored in ethanol at −20 °C for up to 1 year.

12. In order for RNA to be labeled efficiently, ratios of absorbance for 260 nm/280 nm and 260 nm/230 nm should each be >1.8. Low 260 nm/230 nm ratio can indicate phenol contamination from TRI Reagent buffer carry-over. This can lead to drastic reduction in labeling efficiency.
13. Unlike standard gel electrophoresis, the Bioanalyzer requires only a very small volume of RNA, which may be in short supply. It gives a RIN value (RNA Integrity Number) which can be used to compare similar samples within an experiment. The highest RIN value is 10; however, this needs to be assessed depending on the origin of plant material: in general, samples such as root and potato tuber will give a higher RIN value (usually 9–10) than leaves or stems (6-9). This is due to RNA from green material containing additional chloroplast rRNA bands. It is more important that your sample is not degraded and that samples of similar origin are of consistent values, rather than having a very high RIN value.
14. One- or two-color processing of Agilent microarrays can be performed depending upon the experimental design chosen. Due to differing labeling dye qualities, we would recommend using Cy3 only wherever possible to reduce dye-bias. Fluorescent dyes are subject to fading through light and ozone exposure; however, Cy3 dye in combination with Agilent microarrays appears to suffer less in this respect compared to other systems. We have found that the maximum number of samples which can be comfortably handled in a day by one person is 16. This allows sufficient time for the washing steps to be completed without columns drying out and allows the setting up of hybridizations to be completed in a short time frame.
15. Store the Agilent RNA Spike-In Kit at −80 °C for up to 1 year. The first dilution of the Agilent Spike Mix positive controls can be stored up to 2 months at −80 °C and freeze/thawed up to eight times. After use, discard the second, third, and fourth dilution tubes.
16. Pre-warm the 5× First Strand Buffer at 80 °C for 4 min to ensure adequate resuspension of the components, vortex, and spin briefly. Keep at room temperature until needed. AffinityScript RNase Block Mix is a blend of enzymes which needs to be kept on ice and added to the cDNA Master Mix just before use. Use only the 10 mM dNTP mix from the kit. It is recommended to use Safe Lock microfuge tubes to prevent them from opening during heating.
17. T7 RNA polymerase blend is a blend of enzymes, which needs to be kept on ice and should be added to the Transcription Master Mix just before use.

18. If you do not plan to proceed to the next step immediately, samples can be stored at −80 °C.
19. For 8× format microarrays, a minimum yield of 825 ng and a specific activity of 6 pmol Cy3 per μg cRNA are required. If the yield and the specific activity is less than the appropriate values, do not proceed to the hybridization step: repeat cRNA preparation.
20. 10× Blocking Agent can be prepared in advance and stored at −20 °C for up to 2 months. Aliquot 50 μL into tubes, which is sufficient for processing eight arrays. After thawing, repeat vortexing and centrifugation procedures before use. Fragmentation buffer, hybridization buffer, and blocking agent can be bought in a kit together: Hi-RPM Gene Expression Hybridization Kit, Large Volume, Part No. 5190-0404 (Agilent Technologies).
21. Do not vortex and avoid introducing air bubbles.
22. Ensure that the gasket slide is sitting flush with the chamber base.
23. Verify that the sandwiched slides are properly aligned.
24. If necessary, tap the assembly gently on a hard surface to move stationary bubbles.
25. Make sure that you prepare enough of Wash Buffer 2 the night before you plan to do the microarray wash. Add Triton X-102 to Gene Expression wash buffers, this step is optional but highly recommended as it reduces the possibility of array wash artifacts. Add the Triton X-102 to both Gene Expression Wash Buffer 1 and 2 when the vessel is first opened. Ensure you pre-warm Wash Buffer 2 at 37 °C overnight.
26. Slip one of the blunt ends of the forceps between the slides and gently turn the forceps to separate them.
27. Minimize exposure of the slide to air. Touch only the barcode portion of the microarray slide or its edges.
28. For uniform washing, separate a maximum of eight microarray slides.
29. The elevated temperature of the second wash step is usually around 31 °C due to cooling.
30. Minimize the impact of environmental oxidants on signal intensities. If necessary, store slides in a 50 mL tube wrapped in foil with 200 μL 0.1 M DTT in the bottom.
31. Custom barley 60 k microarray slides are scanned at 3 μm resolution and, with the G2505B scanner (Agilent Technologies), the array format is automatically detected.
32. There is a known "graining" artifact with Agilent arrays which can sometimes occur, where the microarray images have a very

high background. This is usually not an RNA sample issue, but due to manufacturing defects of the gasket slides or the arrays themselves. Should this occur, store the labeled cRNA overnight at −20 °C and re-hybridize to fresh arrays. More often than not the problem disappears.

33. The FE QC reports are useful to assess accurate grid alignment, any scanning artifacts and determine if the labeling has been successful: actual spike-in RNA signals should have good correlation with predicted signals. Look for an R^2 value of >0.98.

34. Raw data can be analyzed with standard statistical packages; however, dedicated microarray analysis software enables accurate normalization of data to account for differences in labeling efficiency and signal levels. A good academic open-source option is Bioconductor (http://www.bioconductor.org/).

Acknowledgments

This work was funded by the Scottish Government.

References

1. Kogenaru S, Qing Y, Guo Y et al (2012) RNA-seq and microarray complement each other in transcriptome profiling. BMC Genomics 15:629–645
2. Zhao S, Fung-Leung W-P, Bittner A et al (2014) Comparison of RNA-Seq and Microarray in Transcriptome Profiling of Activated T Cells. PLoS One 9(1):e78644
3. Morris WL, Hancock RD, Ducreux LJM et al (2014) Day length dependent restructuring of the leaf transcriptome and metabolome in potato genotypes with contrasting tuberisation phenotypes. Plant Cell Environ 37: 1351–1363
4. Jupe J, Stam R, Howden AJM et al (2013) Phytophthora capsici-tomato interaction features dramatic shifts in gene expression associated with a hemi-biotrophic lifestyle. Genome Biol 14:R63
5. Gotame TP, Cullen DW, Graham J et al (2014) Effect of short-term exposure to high-temperature on total gene expression in the leaves of four raspberry (Rubus idaeus L.) cultivars. J Hortic Sci Biotechnol 89:532–541
6. Cerekovic N, Jarret D, Pagter M et al (2015) The effects of drought stress on leaf gene expression during flowering in blackcurrant (Ribes nigrum L.). Eur J Hortic Sci 80:39–46
7. Mayer KFX, Waugh R, Langridge P et al (2012) A physical, genetic and functional sequence assembly of the barley genome. Nature 491:711–716

Chapter 13

Genome Engineering Using TALENs

Goetz Hensel and Jochen Kumlehn

Abstract

Genome engineering involves methods of genetic modification of cells at predefined genomic sites. Here, we used transcription activator-like effector nucleases (TALENs) for the site-directed mutagenesis in barley. Target gene-specific TALEN-encoding expression units were designed and delivered to totipotent cells of either cultivated embryogenic pollen or immature embryos. The analysis of resulting transgenic plants revealed that the described approach allows for the generation of site-specific, heritable mutations at reasonable efficiency.

Key words Customized endonuclease, Nonhomologous end-joining, Site-directed mutagenesis, *Triticeae*

1 Introduction

In contrast to conventional genetic engineering where the integration site of recombinant DNA within the host genome is unpredictable, genome engineering allows for the genetic modification of genomic target sequences of choice. This sequence-specific modification can be highly effectively triggered by the generation of DNA strand breaks using customizable endonucleases. In recent years, such endonucleases, which can be precisely tailored to digest DNA at just about any target sequence of choice, have revolutionized the field of genetic engineering. Several platforms of customizable endonucleases have been established and are meanwhile being widely used in plant research and breeding. These include meganucleases [1], zinc-finger nucleases [2], transcription activator-like effector nucleases [3], and RNA-guided endonucleases (*see* Chapter 14). In its simplest form, genome engineering is used for site-directed mutagenesis. In this approach, the cellular double strand repair via the error-prone process of nonhomologous end-joining entails the formation of largely unpredictable sequence alterations at the predefined target site. By contrast, when customized endonucleases are used in combination with a DNA repair

Wendy A. Harwood (ed.), *Barley: Methods and Protocols*, Methods in Molecular Biology, vol. 1900,
https://doi.org/10.1007/978-1-4939-8944-7_13,

template that itself shares homology with the target sequence and additionally carries a modification of choice, genomic DNA can be edited in homology-directed fashion so that the outcome is a precisely designed sequence. This principle was recently exemplified in barley [4], yet its implementation in this species is still confined to the cell level, because useful genome edits do not commonly provide a selective advantage that could be used during cell proliferation and plant regeneration.

The present protocol facilitates site-directed mutagenesis in barley using transcription activator-like effector nucleases (TALENs). These chimeric proteins consist of a target-specifically customized DNA-binding domain and the nucleolytic domain of the DNA endonuclease FokI from *Flavobacterium okeanokoites* [5]. The principle of DNA binding derives from transcription activator-like (TAL) effectors secreted into plant cells by phytopathogenic bacteria of the genus *Xanthomonas* which thereby manipulate host gene expression to their own benefit. The sequence specificity of such domains is brought about by near-identical tandem repeats involving basically four repeat variants with binding preference to one of the four nucleobases that represent the genetic alphabet [6]. Coding sequences for binding domains specific for target motifs of choice can be assembled using sophisticated molecular cloning strategies (e.g., [7]). Since FokI is nucleolytically active only in a dimerized state, two TALEN units are required to bind a DNA target in inverse orientation and appropriate distance (ca. 16 bp) to each other. Targeted mutagenesis using TALENs involves four major steps; selection of target motifs within the gene-of-interest, generation of target-specific TALENs and construction of transformation vectors, generation of TALEN-expressing plants, and identification of mutant plants to be used for the production of genetically homogeneous progeny. Here, expression units of *Gfp* or *Mlo* gene-specific TALEN pairs were delivered by hypervirulent *Agrobacterium tumefaciens* to totipotent cells of *Gfp*-transgenic embryogenic pollen [8] or wild-type immature embryos [9], respectively. Out of the primary transgenic plants, twenty two and between six and fifty per cent, respectively, carried detectable mutations in the target gene when embryogenic pollen or immature embryos were used for gene transfer.

2 Materials

Prepare all solutions using ultrapure water (prepared by purifying deionized water to attain a sensitivity of 18 M Ω cm at 25 °C) and analytical grade reagents. Prepare and store all reagents at room temperature (unless indicated otherwise).

2.1 Reagents for Molecular Cloning

1. pGEM®-T Easy cloning vector (Promega, Mannheim, Germany).
2. pUbi-AB-M intermediate vector (DNA-Cloning-Service, Hamburg, Germany).
3. p6i-d35S-TE9 binary vector (DNA-Cloning-Service, Hamburg, Germany).
4. *Sfi*I and corresponding 10× buffer.
5. *Bcu*I and corresponding 10× buffer.
6. *Hin*dIII and corresponding 10× buffer.
7. *Asc*I and corresponding 10× buffer.
8. *Eco*47III and corresponding 10× buffer.
9. T4 DNA ligase and 10× T4 DNA ligase buffer.
10. Chemically competent *E. coli* XL1.
11. Electro-competent *A. tumefaciens* AGL1 [10].
12. LB medium and plates (*see* recipes) containing 100 μg/mL ampicillin (for pGEM and pUbi-AB-M) or 100 μg/mL spectinomycin (for p6i-d35S-TE9).
13. MG/L medium and plates (*see* recipes) containing 100 μg/mL spectinomycin, 100 μg/mL carbenicillin, and 50 μg/mL rifampicin.
14. 15% and 100% (v/v) glycerol.
15. Plasmid miniprep kit.
16. QIAquick PCR Purification Kit.
17. Primers (Table 1).
18. Gene Pulser II (Bio-Rad, Munich, Germany) and corresponding Gene Pulser Electroporation Cuvettes, 0.2 cm gap.
19. PCR thermal cycler.
20. Temperature-controlled incubators equipped with a rotary shaker.
21. Spectrophotometer.

2.2 Materials for the Generation of Transgenic Plants

1. Forceps, scalpel, spatula, preparation needles.
2. Preparative microscope.
3. 6-well culture plates with 3-cm diameter wells.
4. 3.5-, 5.5-, and 9.0-cm-diameter Petri dishes with lid.
5. Plastic boxes with a lid (107 × 94 × 96 cm).
6. Pipettes and disposable autoclaved filter tips (200–1000 μL and 1000–5000 μL).
7. Filter paper (7 cm diameter, ash-free, autoclaved).

Table 1
Primers used for cloning and validation

Name	Sequence	Purpose
GH-Ubi-F1	TTCCGCAGACGGGATCGATC TAGG	Confirmation of constructs and transgenic plants
TALEN-R2	TGGCGGCTTGGCGCGTGACAG	Confirmation of constructs and transgenic plants
GH-GFP-R2	TACGGCAAGCTGACCCTGAA	Target PCR and sequencing
GH-GFP-F1	GGTCACGAACTCCAGCAGGA	Target PCR and sequencing
GH-MLO-F3	ACATCGGTCGTTTCTCACTC	Target PCR and sequencing
GH-MLO-R1	ACGCGCACGCCGATTACAAC	Target PCR and sequencing
35S-F2-Catrin	CATGGTGGAGCACGACACTCTC	Confirmation of transgenic plants
GH-HYG-R5	GATTCCTTGCGGTCCGAATG	Confirmation of transgenic plants

8. Desiccator and vacuum pump.
9. Magnetic stirrer.
10. Refrigerated centrifuge equipped with swing-out baskets.
11. Waring blender (heat-sterilizable).
12. Sterile screw-cap polypropylene centrifuge tubes (round bottom 12 mL and conical 50 mL).
13. Magenta boxes (autoclaved).
14. Nylon mesh (100 μm grid, autoclaved).
15. Hemocytometer (type Rosenthal).
16. Sterile filter-stopped tips (1 mL) for standard pipettes.
17. Erlenmeyer flasks (100 mL) with chicane.
18. Sterile screw-cap polypropylene cryotubes (1.5 mL).
19. Sterile screw-cap polycarbonate round-bottomed tubes (12 mL).
20. Flow cytometer, e.g., PloidyAnalyser I (Partec, Münster, Germany).
21. Nylon-mesh filters, e.g., CellTrics (30 μm mesh diameter, Partec).
22. Staining buffer, e.g., CyStain UV Ploidy Staining Solution (Partec).
23. Wire brush.
24. 0.1% w/v colchicine, supplemented with 3 drops of Tween 20 per 100 mL.

2.3 Stock Solutions and Nutrient Media

2.3.1 Mineral Salts

1. K4N macro minerals (20×): 6.4 g/L NH_4NO_3, 72.8 g/L KNO_3, 6.8 g/L KH_2PO_4, 8.82 g/L $CaCl_2{\cdot}2H_2O$, 4.92 g/L $MgSO_4{\cdot}7H_2O$, filter-sterilized and stored at room temperature (RT).
2. K macro minerals (20×): 1.6 g/L NH_4NO_3, 40.4 g/L KNO_3, 6.8 g/L KH_2PO_4, 8.8 g/L $CaCl_2{\cdot}2H_2O$, 4.9 g/L $MgSO_4{\cdot}7H_2O$, filter-sterilized and stored at room temperature (RT).
3. K micro minerals (1000×): 8.4 g/L $MnSO_4{\cdot}H_2O$, 3.1 g/L H_3BO_3, 7.2 g/L $ZnSO_4{\cdot}7H_2O$, 120 mg/L $Na_2MoO_4{\cdot}2H_2O$, 25 mg/L $CuSO_4{\cdot}5H_2O$, 24 mg/L $CoCl_2{\cdot}6H_2O$, 170 mg/L KI, filter-sterilized and stored at room temperature (RT).
4. Copper sulfate (10 mM): 2.5 g/L $CuSO_4{\cdot}5H_2O$, filter-sterilized and stored at 4 °C.
5. Ethylenediamine-tetraacetic acid, ferric-sodium salt (NaFe-EDTA, 75 mM): 27.53 g/L NaFe-EDTA, filter-sterilized and stored at 4 °C.
6. Calcium chloride (1 M): 147 g/L $CaCl_2{\cdot}2H_2O$, filter-sterilized and stored at 4 °C.
7. Potassium dihydrogen phosphate (1 M): 136 g/L KH_2PO_4, filter-sterilized and stored at 4 °C.
8. Dipotassium hydrogen phosphate (1 M): 174 g/L K_2HPO_4, filter-sterilized and stored at 4 °C.
9. Phosphate buffer 1 (1 M, $pH = 5.0$): 97.5% 1 M KH_2PO_4, 2.5% 1 M K_2HPO_4, filter-sterilized and stored at 4 °C.
10. Phosphate buffer 2 (1 M, $pH = 5.9$): 90% 1 M KH_2PO_4, 10% 1 M K_2HPO_4, filter-sterilized and stored at 4 °C.

2.3.2 Carbohydrates

1. Mannitol (0.4 M): 72.9 g/L mannitol, autoclaved and stored at 4 °C.
2. Maltose (1 M, Sigma Grade I): 360 g/L maltose·H_2O, filter-sterilized and stored at RT.
3. Maltose (1 M, standard quality): 360 g/L maltose·H_2O, filter-sterilized and stored at RT.
4. Maltose (0.55 M, standard quality): 198 g/L maltose·H_2O, filter-sterilized and stored at RT.

2.3.3 Vitamins

1. Gamborg B5 vitamins (1000×): 112 g/L ready-to-use product (Duchefa, no. G 0415.0100), filter-sterilized and stored at −20 °C.
2. KM vitamins (100×): 100 mL ready-to-use product (Sigma, no. K-3129) filter-sterilized and stored at −20 °C.
3. Biotin (0.1 g/L): filter-sterilized and stored at 4 °C.
4. Thiamine·HCl (1 g/L): filter-sterilized and stored at 4 °C.

2.3.4 Growth Regulators

1. 6-Benzylaminopurine (6-BAP, 1 mM): 0.225 g/L, dissolved in a few drops of 1 M NaOH, then made up to the final volume in H_2O, filter-sterilized, and stored at 4 °C.
2. Dicamba (2.5 g/L): dissolved in a few drops of 96% heated ethanol, then made up to the final volume in warmed water, filter-sterilized, and stored at 4 °C.
3. 2,4-Dichlorophenoxyacetic acid (2,4-D, 1 mM): 221 mg/L dissolved in few drops of 50% ethanol by heating gently, made up to the final volume with hot H_2O, filter-sterilized, and stored at 4 °C.

2.3.5 Amino Acids

1. L-Glutamine (0.25 M): 36.6 g/L, dissolved in a few drops 0.1 M KOH, then made up to the final volume in warmed H_2O, filter-sterilized, and stored at −20 °C.

2.3.6 Selective Agents

1. Bialaphos (41 mg/mL): filter-sterilized and stored at −20 °C.
2. Carbenicillin (100 mg/L): filter-sterilized and stored at −20 °C.
3. Hygromycin (50 mg/L): ready-to-use product (Roche, no. 10843555001), stored at 4 °C.
4. Rifampicin (50 mg/L): filter-sterilized and stored at −20 °C.
5. Spectinomycin (100 mg/L): filter-sterilized and stored at −20 °C.
6. Timentin (150 mg/L): filter-sterilized and stored at −20 °C.

2.3.7 Gelling Agent

1. Phytagel (0.4% w/v): 4 g/L dissolved in cold water, autoclaved, the temperature of this stock as well as the remaining components of the medium are set to ca. 50 °C before mixing.
2. Phytagel (2×): 6 g/L suspended in cold water (at best 1.5 g Phytagel per 250 mL unit), autoclaved, and stored at room temperature.

2.3.8 Other Additives and Solutions

1. Acetosyringone (1 M): 196.2 g/L dissolved in dimethylsulfoxide (DMSO) and stored at −20 °C.
2. Ethanol (70%): 729.2 mL/L of 96% ethanol, stored at RT.
3. Glycerol (15%): 150 g/L glycerol, autoclaved and stored at RT.
4. Sodium hypochlorite (2.4% w/v): 200 mL/L of 12% (w/v) NaOCl, to which is added 0.1% (v/v) Tween 20. This solution should be freshly prepared.
5. Morpholinoethanesulfonic acid (MES, $pH = 5.0$, $pH = 5.5$, and $pH = 5.9$): 212.2 g/L dissolved in 40 mL H_2O, the pH adjusted with KOH (few pellets), stored at room temperature overnight, and the *pH* readjusted with either 1 M KOH or 1 M HCl, made up to the required final volume, filter-sterilized, and stored at 4 °C (*see* **Note 1**).

2.3.9 Medium for A. tumefaciens

1. MG/L medium: 250 mg/L KH_2PO_4, 100 mg/L NaCl, 100 mg/L $MgSO_4 \cdot 7H_2O$, 1 g/L L-glutamic acid, 5 g/L mannitol, 5 g/L tryptone, 2.5 g/L yeast extract. The pH is adjusted to 7.0, the solution is autoclaved and cooled, and then 10 μL/L biotin stock is added. To make the solidified MG/L medium, 1.2% (w/v) agar is added.

2.3.10 Media for Plant Transformation Using Immature Embryos

1. Liquid cocultivation medium (BCCM): The medium contains 4.3 g/L MS mineral salts (ready-to-use product, Sigma-Aldrich No. M 5524), 1 g/L casein hydrolysate, 0.69 g/L proline, 0.25 g/L myo-inositol, 30 g/L maltose·H_2O, 1 mL/L Dicamba stock, 1 mL/L thiamine HCl stock, 800 mg/L L-cysteine (*see* **Note 2**). The pH is adjusted to 5.8, the solution is filter-sterilized, and then 500 μL/L acetosyringone stock (*see* **Note 3**) is added. A 100 mL volume of BCCM is sufficient for culturing up to 300 immature embryos processed within a single 6-well plate.
2. Solid callus induction medium (BCIM): This medium is identical to BCCM, but lacks L-cysteine and acetosyringone, while instead it includes 0.5 mL/L $CuSO_4 \cdot 5H_2O$ stock added prior to pH adjustment (*see* **Note 4**). Depending on the construct used, either 750 μL/L bialaphos or 1 mL/L hygromycin stock as well as 1 mL/L Timentin stock is added. The solution is filter-sterilized, and one volume is mixed with three volumes of Phytagel stock.
3. Solid regeneration medium (BRM): The medium contains 50 mL/L K4N macro minerals, 1 mL/L each of the NaFeEDTA, K micro minerals, vitamin B5 and 6-BAP stocks, 4 mL/L L-glutamine stock, 100 mL/L maltose stock, 490 μL/L $CuSO_4 \cdot 5H_2O$ stock and the pH is adjusted to 5.8. Depending on the construct used, either 100 μL/L bialaphos or 0.5 mL/L hygromycin stock as well as 1 mL/L Timentin stock is added, the solution is filter-sterilized, and one volume is mixed with three volumes of Phytagel stock.

2.3.11 Media for Plant Transformation using Embryogenic Pollen

1. Barley pollen culture (KBP) medium: 50 mL/L K macro minerals, 1 mL/L K micro minerals, 1 mL/L NaFeEDTA, 10 mL/L KM vitamins, 12 mL/L L-glutamine, 4 mL/L BAP, and 250 mL/L maltose (1 M, Sigma Grade I) stocks, pH adjusted to 5.9 and stored at 4 °C.
2. Coculture (CK) medium: 50 mL/L K macro minerals, 1 mL/L K micro minerals, 1 mL/L NaFeEDTA, 10 mL/L KM vitamins, 250 mL/L maltose (1 M, Sigma Grade I), 2 mL/L BAP, 0.5 mL/L acetosyringone, 10 mL/L MES ($pH = 5.9$), and 50 mL/L phosphate buffer 2 ($pH = 5.9$) stocks, stored at 4 °C.

3. AgroStop (ASt) medium: 50 mL/L K macro minerals, 1 mL/L K micro minerals, 1 mL/L NaFeEDTA, 10 mL/L KM vitamins, 250 mL/L maltose (1 M, Sigma Grade I), 10 mL/L MES ($pH = 5.0$), 25 mL/L phosphate buffer 1 ($pH = 5.0$), 7 mL/L $CaCl_2$, 4 mL/L L-glutamine, 10 mL/L 2,4-D, 2 mL/L BAP, 0.5 mL/L acetosyringone, 750 μL/L bialaphos, and 1.3 mL/L Timentin stocks, stored at 4 °C.
4. Selection medium (S3): 50 mL/L K macro minerals, 1 mL/L K micro minerals, 1 mL/L NaFeEDTA, 10 mL/L KM vitamins, 250 mL/L maltose (1 M, standard quality), 10 mL/L MES ($pH = 5.5$), 12 mL/L L-glutamine, 1 mL/L BAP, 750 μL/L bialaphos, and 1.3 mL/L Timentin stocks, stored at 4 °C.
5. Solid barley pollen culture (KBP4PT) medium: 50 mL/L K macro minerals, 1 mL/L K micro minerals, 1 mL/L NaFeEDTA, 10 mL/L KM vitamins, 250 mL/L maltose (1 M, standard quality), 1 mL/L BAP, 12 mL/L L-glutamine, 750 μL/L bialaphos, and 1.3 mL/L Timentin stocks; appropriate amounts for a final volume of 0.5 L KBP4PT made up to an intermediate volume of 250 mL (doubled concentration), adjusted to $pH = 5.9$, heated to about 40 °C, then mixed (1:1) with 250 mL Phytagel stock (melted by heating).
6. Regeneration (K4NBT) medium: 50 mL/L K4N macro minerals, 1 mL/L K micro minerals, 0.75 mL/L NaFeEDTA, 1 mL/L B5 vitamins, 4 mL/L L-glutamine, 100 mL/L maltose (1 M, standard quality), 1 mL/L BAP, 0.49 mL/L $CuSO_4$, 100 μL/L bialaphos, and 1.3 mL/L Timentin stocks; appropriate amounts for a final volume of 0.5 L K4NBT made up to an intermediate volume of 250 mL (doubled concentration), adjusted to $pH = 5.9$, heated to about 40 °C, then mixed (1:1) with 250 mL Phytagel stock (melted by heating).

3 Methods

3.1 Selection of Target Sites

The DNA-binding modules typically consist of 15–20 nucleotide-specific repeat units [7, 11]. Since the cleavage domain of FokI is only active as a dimer, two TALEN expression units have to be designed for target motifs residing in opposite orientation on both DNA strands and in a distance between 12 and 31 nucleotides to one another [12, 13].

To ensure effective knockout mutations, conserved domains and/or motifs within the 5′-region of the gene-of-interest's coding sequence are to be selected as targets. The presence of a restriction enzyme-binding site within the spacer region (between the binding motifs of the left and right TALEN units) is not essential, yet can facilitate the screening of potentially mutated target sites (*see* **Note 5**).

The selection of target sequences can be preliminarily carried out using the barley reference genome of cultivar Morex [14]. To avoid any deleterious nucleotide polymorphisms, the selected target motifs are to be confirmed in the genotype used for site-directed mutagenesis, i.e., cultivars Igri and Golden Promise in the present protocol.

1. Design primers using the available genomic data of the gene-of-interest and amplify the target gene by PCR [15] using a high-fidelity polymerase.
2. Clone the PCR product into pGEM®-T Easy vector and sequence several independent clones (at least 5–10).
3. Design the binding modules using a web-based tool such as "TALEN Targeter" (Cornell University). Choose your favorite TALEN pair depending on length of TALEN unit, spacer length, availability of a restriction site, and frequency of potential off-target sites within the genome.
4. The synthesis of customized TALEN-coding sequences is offered by commercial service providers such as GeneArt/ Thermo Fisher Scientific (*see* **Notes 6** and 7).

3.2 Generation of Plasmids

3.2.1 Assembly of the TALEN Expression Units

To assemble a single (left or right) TALEN expression unit, the customized TALEN-coding sequence is inserted into the intermediate vector pUbi-AB-M between the maize *UBIQUITIN1* promoter with first intron and the *A. tumefaciens NOPALINE SYNTHASE* termination sequence (*see* **Note 8**).

1. Double-digest both donor and acceptor plasmids independently with *Bcu*I/*Hin*dIII Fastdigest enzymes (*see* **Note 9**).
   (a) For pUbi-AB-M, prepare the following reaction mix:
      2 μg of pUbi-AB-M
      5 μL of 10× Fastdigest Green buffer (supplied with restriction enzymes)
      1 μL of Fastdigest *Bcu*I
      1 μL of Fastdigest *Hin*dIII
      Distilled, deionized H_2O to 50 μL.
      Incubate the digest reaction at 37 °C for 1 h.
   (b) For pTALEN left and right, prepare the following reaction mixes:
      2 μg of pTALEN left or right
      5 μL of 10× Fastdigest Green buffer (supplied with restriction enzymes)
      1 μL of Fastdigest *Bcu*I
      1 μL of Fastdigest *Hin*dIII

Distilled, deionized H_2O to 50 μL.

Incubate the digest reaction at 37 °C for 1 h.

2. After digest, remove restriction enzymes with QIAquick PCR Purification Kit according to the manufacturer's recommendations.
3. Run a 5 μL aliquot of each digest reaction for size confirmation of the expected fragments on an 0.8% agarose gel in 0.5× TBE buffer at 200 V for 45 min. One to two bands will show up in the gel (4.7 kb for pUbi-AB-M, 3.2 and 2.1 kb for pTALEN left or right).
4. Ligate pUbi-AB-M with pTALEN left or right using the following ligation reaction:

   1 μL purified pUbi-AB-M

   3 μL purified pTALEN left or right

   2 μL 10× T4 DNA ligase buffer

   1 μL 5 U/μL T4 DNA ligase

   Distilled, deionized H_2O to 20 μL.

   Incubate the ligation reaction for 1 h at room temperature or 4 °C overnight.
5. Transform 3 μL ligation product into competent *E. coli* cells (*see* **Note 10**) using standard protocols and plate 100 μL on LB plates containing 1 mL/L ampicillin stock, incubate at 37 °C overnight.
6. Pick up to 10 colonies to start 2 mL LB medium with 1 mL/L ampicillin stock for plasmid verification. Shake overnight at 37 °C and 180 rpm.
7. Isolate plasmid DNA using a miniprep kit and perform control digest using *Bcu*I/*Hin*dIII double-digest as described for **step 5**. Run a 0.8% agarose gel of digest reaction as described for **step 3**.
8. Sequence 2–5 plasmids to confirm correct integration of the expression cassettes using primers given in Table 1 (*see* **Note 11**). Align sequences obtained with plasmid maps generated *in silico*.
9. Start another liquid overnight culture at 37 °C and 180 rpm and mix 900 μL culture with 100 μL 100% (v/v) glycerol, incubate at room temperature for 3 h and store at −80 °C for further use.

3.2.2 Generation of Binary Vectors and Their Transfer into Agrobacterium

After assembly, each TALEN expression unit is transferred into the generic binary vector p6i-d35S-TE9 using the *Sfi*I restriction site variants A and B that are present in both the donor and acceptor plasmids.

1. Digest all three vectors independently with *Sfi*I enzyme.
   (a) For p6i-d35S-TE9, prepare the following reaction mix:

      1 μg of p6i-d35S-TE9

      2 μL of 10× Buffer G (supplied with restriction enzyme)

      1 μL of *Sfi*I

      Distilled, deionized H_2O to 20 μL.

      Incubate the digest reaction at 50 °C for 3 h.

   (b) For pUbi-TALEN left and right, prepare the following reaction mixes:

      2 μg of pTALEN left or right

      2 μL of 10× Buffer G (supplied with restriction enzymes)

      1 μL of *Sfi*I

      Distilled, deionized H_2O to 20 μL.

      Incubate the digest reaction at 50 °C for 3 h.

2. Run a 5 μL aliquot of each digest reaction for size confirmation of the expected fragments on an 0.8% agarose gel in 0.5× TBE buffer at 200 V for 45 min. One to two bands will show up in the gel (10 kb for p6i-d35S-TE9, 2.9 and 5 kb for pUbi-TALEN left or right).
3. Ligate p6i-d35S-TE9 with pUbi-TALEN left or right using the following ligation reaction (*see* **Note 12**):

   1 μL p6i-d35S-TE9

   3 μL pUbi-TALEN left or right

   2 μL 10× T4 DNA ligase buffer

   1 μL 5 U/μL T4 DNA ligase

   Distilled, deionized H_2O to 20 μL.

   Incubate the ligation reaction at 4 °C overnight (*see* **Note 13**).
4. Transform 3 μL ligation product into competent *E. coli* cells (*see* **Note 14**) using standard protocols and plate 100 μL on LB plates containing 1 mL/L spectinomycin stock, incubate at 37 °C overnight.
5. Pick up to 10 colonies to start 2 mL LB medium with 1 mL/L spectinomycin stock for plasmid verification. Shake overnight at 37 °C and 180 rpm.
6. Isolate plasmid DNA using a miniprep kit (*see* **Note 15**) and perform control digest using SfiI as described for **step 1**. Run a 0.8% agarose gel of digest reaction as described for **step 2** (*see* **Note 16**).
7. Transform 1 μL of 1/10-diluted binary vector into electro-competent *Agrobacterium* cells using Gene Pulser II device and Gene Pulser Electroporation Cuvettes with 0.2 cm gap

according to the manufacturer's protocol. Incubate for 3 h at 28 °C and spread 30 μL on MG/L medium plates containing 1 mL/L spectinomycin stock, 1 mL/L carbenicillin stock, and 1 mL/L rifampicin stock. Incubate at 28 °C for two days.

8. Pick two colonies (*see* **Note 17**) to start 3 mL MG/L medium containing 1 mL/L spectinomycin, 1 mL/L carbenicillin, and 1 mL/L rifampicin. Shake overnight at 28 °C and 180 rpm.
9. Isolate plasmid DNA using a miniprep kit (*see* **Note 18**) and perform control digest (*see* **Note 19**) using *Sfi*I digest as described for **step 6**. Run a 0.8% agarose gel of digest reaction as described for **step 1**.
10. Start another liquid overnight culture at 28 °C and 180 rpm and mix several tubes with 200 μL overnight culture with 200 μL 15% (v/v) glycerol each, incubate at room temperature for 3 h and store at −80 °C for further use.

3.3 Generation of Transgenic Plants and Mutants

3.3.1 Gene Transfer to Immature Embryos

The production of transgenic barley via gene transfer to immature embryos comprises six major steps: growth of donor material, dissection of immature embryos, coculture of immature embryos with *Agrobacterium*, callus induction, plantlet regeneration, and establishment of plants in soil.

1. Germinate grains of spring-type cultivar Golden Promise (*see* **Note 20**) in trays filled with soil substrate (*see* **Note 21**) in a growth chamber set to provide a 12-hour photoperiod (136 μmoL $m^{-2}s^{-1}$ photon flux density) and a light/dark temperature regime of 14/12 °C. Ten to twenty grains are germinated at fortnightly intervals to ensure a continuous supply of explant material.
2. To induce reproductive growth in winter-type cultivars, seedlings are to be vernalized for eight weeks at 4 °C under an 8 h photoperiod.
3. Transfer seedlings after three weeks into 2 L pots. At the tillering stage (BBCH code 29/30), 15 g Osmocote (Scotts, Netherlands) is given to each pot (*see* **Note 22**). Transfer the plants at the tiller elongation stage (BBCH code 39) to a glasshouse maintained at 18/16 °C with a 16 h photoperiod (170 μmoL $m^{-2}s^{-1}$ photon flux density).
4. Thaw tubes of *Agrobacterium* glycerol stock of TALEN left and right strains and add the contents to 10 mL antibiotic-free MG/L medium in a 100 mL Erlenmeyer flask with chicane, and incubate overnight at 28 °C on a rotary shaker running at 180 rpm (*see* **Note 23**).
5. Harvest immature caryopses ca. 12–14 days after anthesis, when the embryos turn from glassy to whitish appearance

owing to cellular starch accumulation. Remove the awns and place the caryopses in a 500 mL bottle standing on ice.

6. In a laminar airflow hood, immerse the caryopses for 3 min in 70% ethanol, then in NaOCl solution for 15 min on a shaker. Wash five times in sterile, distilled water.
7. Excise the embryos from the caryopses, and remove the embryo axes using forceps and preparation needle. This is achieved by slitting the caryopsis down the middle, right above the embryo; the embryo axis is removed while the embryo is still in its original position. Then, the embryo can be readily released from the caryopsis using the preparation needle.
8. Immerse 30–50 embryos in 2.5 mL freshly prepared liquid BCCM (*see* **Note 24**) per well of a 6-well plate.
9. Remove the BCCM from the well containing the embryos and replace with 600 μL *Agrobacterium* culture ($OD_{600} = 2.0$).
10. Place the plate in a desiccator and apply a 500 mbar vacuum for 1 min.
11. Keep the plate inside a laminar flow hood for 10 min without any agitation, and then wash the embryos in 2.5 mL BCCM.
12. Add a fresh 2.5 mL aliquot of BCCM and incubate at 21 °C in the dark for 48–72 h without agitation (*see* **Note 25**).
13. Place the embryos with their scutellum side facing down on solid BCIM; use ten embryos per 10-cm Petri dish (*see* **Note 26**).
14. After sealing, incubate the plates at 24 °C in the dark for two weeks, and then transfer the material to fresh solid BCIM for a another two weeks.
15. Transfer the callusing embryos onto BRM and incubate at 24 °C under 136 μmoL m^{-2} s^{-1} photon flux density for 16 h per day.
16. Transfer the material to fresh BRM every two weeks until shoots emerge.
17. Once their leaf length has reached 2–3 cm, transfer the shoots into a plastic box (107 × 94 × 96 cm) containing 80 mL BRM; use up to 12 plantlets per box (*see* **Note 27**).
18. When the plantlets have developed roots, transfer them into soil, maintaining them at high humidity for 7–10 days by covering using a plastic hood. Subsequently, treat primary transgenic plants as described for donor material (**steps 2** and **3**) and grow the resulting plants to maturity (*see* **Note 28**).

3.3.2 Gene Transfer to Embryogenic Pollen

As alternative target tissue for *A. tumefaciens*-mediated gene transfer, barley pollen of winter-type cultivar Igri can be used. Due to the haploid nature of the material, this allows the immediate generation of homozygous mutant plants [3].

1. Germinate grains of winter-type cultivar Igri in trays containing soil substrate (*see* **Note 21**) in a growth chamber providing 12 h light (136 μmoL $m^{-2}s^{-1}$ photon flux density) per day and a light/dark temperature regime of 14/12 °C. Ten to twenty grains are germinated at two-week intervals to ensure a continuous supply of explant material.
2. Vernalize the seedlings for eight weeks at 4 °C under an 8 h photoperiod.
3. Transfer and treat plants as described for **step 3** for cultivar Golden Promise.
4. Harvest spikes within the boots when the awns have just emerged from the sheath of the flag leaf, surface-sterilize boots by spraying with 70% ethanol, remove the flag leaf sheath under aseptic conditions and place dissected spikes on a 10-cm Petri dish containing two moistened, 7 cm filter paper disks (use five spikes per plate). After sealing, the plates are held at 4 °C for 4–5 weeks in the dark.
5. Chop 10 to 15 pretreated spikes (*see* **Note 29**) into ca. 1-cm fragments and macerate in a Waring blender filled with 20 mL 0.4 M mannitol. Two bursts of 15 s each are to be performed with the blender being set on "low" speed.
6. Filter the macerated material through a 100-μm mesh, which is placed on top of a Magenta box. The blender is flushed with 10 mL of 0.4 M mannitol, which is also passed through the mesh.
7. Squeeze remaining debris with tweezers, and then return to the blender for re-maceration (twice at 10 s) in another 10 mL of 0.4 M mannitol. The macerate passed through the mesh, which is followed by flushing the blender once again with 10 mL of 0.4 M mannitol. The latter is also passed through the mesh.
8. Transfer the filtrate to a 50-mL centrifuge tube and centrifuge at 100 × *g* at 4 °C for 10 min.
9. Resuspend the pellet in 3 mL of 0.55 M maltose in a round-bottomed 12-mL tube and flush the centrifuge tube with 2 mL 0.4 M mannitol, which is poured carefully over the top of the 0.55 M maltose suspension, thereby forming two distinct liquid layers.
10. Perform density gradient centrifugation in a swing-out rotor (100 × *g*, 10 min, and 4 °C) with the centrifuge set to give slow

acceleration and deceleration so as to avoid the two layers becoming mixed.

11. Transfer the interphase containing immature pollen to a fresh 50-mL tube by pipetting and add 10 mL 0.4 M mannitol; resuspend by gentle shaking. Place a 100-μL aliquot into a hemocytometer cell to estimate the population density. Spin down the remaining microspores by centrifugation (100 × *g*, 10 min, and 4 °C).
12. Resuspend the pellet in an appropriate volume of KBP medium adjusting the density to 400,000 microspores per mL.
13. Transfer 1 mL pollen suspension aliquots into 3.5-cm Petri dishes. After sealing the plates using Para- or Nescofilm, keep them in the dark for 7–8 days at 25 °C prior to cocultivation with *Agrobacterium* (*see* **Note 30**).
14. Start a 10 mL culture of the appropriate *Agrobacterium* strain as described for **step 4** of the immature embryo protocol. Split each culture into two 12-mL round-bottomed tubes, which are centrifuged at 4000 rpm and room temperature for 10 min. Vortex each pellet in 2.5 mL CK medium, transfer the suspension into a 100-mL Erlenmeyer flask, and shake at 100 rpm 1–3 h at 28 °C.
15. Carefully withdraw the KBP medium used to pre-cultivate the pollen and add 1 mL of CK medium. Keep one out of every ten dishes untouched to later provide feeder cells which support embryogenic development of the pollen after coculture with *Agrobacterium*.
16. Measure OD_{550} to determine the concentration of *Agrobacterium* and challenge each 1 mL pollen culture with 2.5×10^7 cfu (*see* **Note 31**). Place the dishes within a plastic box which is kept in the dark at 21 °C for 48 h with shaking at 65 rpm (*see* **Note 32**).
17. Carefully withdraw the CK medium and rinse the pollen using 0.4 mL ASt medium. Remove the bacterial slime from the wall and bottom of the dish using a pipette tip. This material requires to be left inside the dish, as it contains much of the embryogenic pollen. Replace the medium with a fresh aliquot of 1.1 mL ASt medium, add a 100-μL aliquot of non-cocultivated embryogenic pollen serving as feeder tissue. Incubate the sealed dishes in the dark at 25 °C for one week with shaking at 75 rpm.
18. Replace ASt medium by 1.1 mL S3 medium, using the same pipetting technique described for the previous step; no washing step is required here. Keep the sealed dishes in the dark at 25 °C for one week with shaking at 65 rpm; refresh the S3 medium and incubated for a further week.

19. Transfer the formed microcalli to an ash-free filter disk placed over solidified KBP4PT medium in a 10-cm Petri dish; incubate the sealed dishes at 25 °C in the dark for two weeks.
20. Move calli grown to a diameter > 1 mm into 10-cm Petri dishes containing K4NBT regeneration medium, expose those calli that have not yet reached this size to a further week's culture on fresh KBP4PT medium. Hold the sealed K4NBT plates in the dark at 25 °C for one week, and then transfer them into the light.
21. Calli and any emerging regenerants are transferred after three weeks to tissue culture boxes containing K4NBT, and subcultured twice at three-week intervals.
22. Transfer rooted regenerants to soil as described for **step 18** of the immature embryos protocol. Plants of winter-type cultivars need to be vernalized as described above.

3.4 Screening of Transgenic Plants for Mutations

After successful generation of primary transgenic plants, mutation analysis has to be performed. This includes the following steps: ploidy analysis for regenerants derived from embryogenic pollen cultures, confirmation of presence of T-DNA, amplification of target region, PCR/RE assay if restriction site is present, initial sequencing of PCR product, subcloning and sequencing of individual amplicons.

1. Determine the ploidy level of the regenerants at best during the vernalization period. Isolate nuclei from leaf sample by supplying ice-cold staining buffer; disintegrate leaf tissue gently with a wire brush.
2. Filter the resulting suspension through a Nylon-mesh filter and use a PloidyAnalyser PA I device following the manufacturer's instructions. Adjust the device by controls from wild-type plantlets.
3. Expose the identified haploids to a colchicine treatment directly after vernalization. Wash roots from soil and trim tillers and roots to a length of 5 cm and 3 cm, respectively. Place each plant into a 50-mL polypropylene conical tube containing 0.1% (w/v) colchicine solution. The level of the liquid is adjusted to reach the base of the shoots. The tubes are capped, and then held in the light for 6 h at 21 °C.
4. Remove plants from colchicine and rinse roots carefully in water. Transfer plantlets in 9-cm pots filled with *Petuniensubstrat* and cover the tray with a plastic hood (*see* **Note 33**).
5. Snap-freeze a 200–400 mg sample of fresh leaf tissue in liquid nitrogen. Isolate genomic DNA by kit or as previously described [16].

6. Amplify the target region using primers given in Table 1. Purify the PCR product using a kit.
7. Perform PCR/RE assay; prepare the following mix:

   5 μL of purified target amplicon

   2 μL of 10× Fastdigest Green buffer (supplied with restriction enzymes)

   0.5 μL of Fastdigest enzyme

   Distilled, deionized H_2O to 20 μL.

   Incubate the digest reaction at 37 °C for 1 h.
8. Run each digest reaction for size confirmation of the expected fragments on a 1.5% agarose gel in 0.5× TBE buffer at 200 V for 45 min. If a mutation occurs, the target amplicon will not be digested and will not produce two fragments as expected for wild-type sequence.
9. Sequence and analyze chromatograms for the presence of multiple peaks downstream of the TALEN-binding site.
10. Subclone amplicons of plants with promising chromatograms into pGEM®-T easy vector by setting the following ligation reaction:

    3 μL purified amplicon

    1 μL linearized pGEM®-T easy

    2 μL 10× T4 DNA ligase buffer

    1 μL 5 U/μL T4 DNA ligase

    Distilled, deionized H_2O to 10 μL.

    Incubate the ligation reaction 1 h at room temperature.
11. Transform 3 μL ligation product into competent *E. coli* cells (*see* **Note 10**) and plate 100 μL on LB plates containing 1 mL/L ampicillin stock, incubate at 37 °C overnight.
12. Pick up to 10 colonies to start 2 mL LB medium with 1 mL/L ampicillin stock for plasmid isolation. Shake overnight at 37 °C and 180 rpm.
13. Isolate plasmid DNA using a miniprep kit and sequence the inserts. Align with wild-type sequence to determine differences (indels/mutations) in the targeted region.
14. Confirm the presence of T-DNA elements using primers given in Table 1. Run 0.8% agarose gel in 0.5× TBE buffer at 200 V for 45 min to confirm the expected fragment sizes.

4 Notes

1. A separate stock for each pH value ($pH = 5.0$, $pH = 5.5$, and $pH = 5.9$) is required.
2. L-Cysteine supplementation of the cocultivation medium has been reported to prevent tissue browning in soybean induced by the presence of *Agrobacterium*, and also to increase transformation efficiency. A comparison of concentrations in barley cocultivation experiments suggests that the optimum L-cysteine concentration is 800 mg/L. Adjustment of the pH to 5.8 before the addition of L-cysteine results in a final pH of ca. 5.5.
3. While contradictory results have been published regarding the effect of acetosyringone on the agroinoculation of immature barley embryos, the given concentration is optimal with respect to transformation efficiency under the conditions described here.
4. An increased concentration of $CuSO_4 \cdot 5H_2O$ causes improved plant regeneration.
5. Alternatively, the T7E1 mismatch cleavage assay can be used [17].
6. A synthetic DNA-binding domain consists of a series of nucleotide-specific repeats assembled according to the target sequence within the context of the scaffold of a truncated variant of the *Xanthomonas campestris* pv. *vesicatoria AvrB3* gene [11].
7. Make sure that each synthetic binding domain includes a nuclear targeting sequence encoding the Simian Virus 40 (SV40) Nuclear Localization Signal (NLS) at its N-terminus. Alternatively, such sequence can be inserted by an additional cloning step. Along with the NLS signal, a hemagglutinin (HA) tag can be optionally inserted.
8. Pairs of TALEN expression units can be functionally validated, e.g., using the transient expression test system [18].
9. The reaction can also be performed with regular restriction enzymes.
10. Use strain DH5α or XL1.
11. Due to the high similarity of the RVDs, sequencing of the whole binding module is not recommended.
12. Heat-inactivation of *Sfi*I enzyme is impossible, purification not necessary. Use aliquots of digest reaction directly for ligation.
13. Longer incubation increases the number of positive clones. Usually, 50–90% positive clones are obtained after 16 h of ligation at 4 °C.

14. In order to provide stable amplification of the large binary vectors, the use of DH10B or XL1 strains is highly recommended.
15. Follow recommendations for low-copy plasmids in order to increase yield.
16. Sequencing is not necessary at this step, because fragments were just released by restriction enzyme.
17. According to our experience, nearly 100% of the colonies contain the desired plasmid. Therefore, analysis of two colonies is sufficient.
18. Follow recommendations for low-copy plasmids in order to increase yield.
19. Due to the low content of plasmid DNA in *Agrobacterium*, use 10 μL pDNA to ensure a DNA content allowing for visualization by gel electrophoresis.
20. The same protocol has proven to be effective for a range of cultivars including some winter types; minor modifications are required to optimize transformation in some of these cultivars.
21. The substrate is a 3:1:2 mix of garden mold/sand/white and black peat (Klasmann Substrate 2).
22. Osmocote is a commercially available fertilizer formulated to contain 19% N, 6% P, and 12% K.
23. There is a documented risk that the *Agrobacterium* cells delete nonessential plasmids. However, in the present protocol, binary vectors carrying the pVS1 origin of replication appear to be very stable even in the absence of selective conditions. The advantage of growing *Agrobacterium* without antibiotics prior to inoculation is that the cell suspension can be used directly, and the recipient plant cells are not exposed to any antibiotic challenge.
24. Cocultivation in liquid medium allows many embryos to be processed simultaneously, which is beneficial for throughput efficiency.
25. Pretreated immature embryos are cocultivated on moistened (300 μL BCCM) filter paper in 5.5 cm diameter Petri dishes, each holding 25 embryos.
26. The number of embryos can be increased up to 25 per Petri dish on cost of the development of the individual embryo.
27. Alternatively, up to 16 plantlets can be grown per box on BRM. If desired, individual plantlets can also be raised in a glass tube, which minimizes the risk of cross-contamination by persisting *Agrobacterium*.
28. The regeneration *in vitro* via callus formation causes a significantly reduced fitness of the plants as compared to regular

germination from mature grains. Nonetheless, primary transgenic plants commonly produce more than 50 grains.

29. All materials coming into contact with the explant need to be sterile and precooled to 4 °C, and all solutions should be kept on ice during the pollen isolation and purification procedures. Liquid transfer is most effectively carried out using a filter-stopped 5–10 mL pipette tips attached to a battery-operated pipette. We prefer the Waring blender for spike maceration, while other devices may be useful as well.

30. The standard protocol was optimized for the *Agrobacterium* strain LBA4404/pSB1. Alternatively, the strain AGL-1 can be used, albeit with a reduced transformation efficiency.

31. For example: suppose the OD_{550} of the 1:10 dilution of *Agrobacterium* cells is 0.9, then the required aliquot volume is $[(45/0.9) \times 0.25] = 12.5$ μL. Thus, in this case, 12.5 μL aliquots of the non-diluted inoculum are added to the dishes containing 1 mL embryogenic pollen culture. Dilution of the grown *Agrobacterium* cells is needed to ensure that the spectrophotometer readings are sufficiently precise.

32. The dishes used for cocultivation must not be sealed, because the transformation efficiency of *Agrobacterium* is lowered by limited supply of oxygen. The use of a rotary shaker is essential to prevent the pollen from being immobilized in aggregated bacterial cells at the bottom of the dish. Optimal conditions (quantity of inoculum, medium composition, shaking, and temperature) allow small aggregates to form, each comprising no more than a few dozen pollen.

33. More than 80% of plantlets typically survive the colchicine treatment, and out of the survivors over 90% produce grain, which indicates that diploidization of germline cells has taken place at least in some plant sectors.

References

1. Vu GTH, Cao HX, Watanabe K, Hensel G, Blattner F, Kumlehn J, Schubert I (2014) Repair of site-specific DNA double-strand breaks in barley occurs via diverse pathways, mostly involving the sister chromatid. Plant Cell 26:2156–2167
2. Shukla VK, Doyon Y, Miller JC, DeKelver RC, Moehle EA, Worden SE, Mitchell JC, Arnold NL, Gopalan S, Meng X, Choi VM, Rock JM, Wu Y-Y, Katibah GE, Zhifang G, McCaskill D, Simpson MA, Blakeslee B, Greenwalt SA, Butler HJ, Hinkley SJ, Zhang L, Rebar EJ, Gregory PD, Urnov FD (2009) Precise genome modification in the crop species Zea mays using zinc-finger nucleases. Nature 459:437–441
3. Gurushidze M, Hensel G, Hiekel S, Schedel S, Valkov V, Kumlehn J (2014) True-breeding targeted gene knock-out in barley using designer TALE-nuclease in haploid cells. PLoS One 9:e92046
4. Budhagatapalli N, Rutten T, Gurushidze M, Kumlehn J, Hensel G (2015) Targeted modification of gene function exploiting homology-directed repair of TALEN-mediated double strand breaks in barley. G3 (Bethesda) 5:1857–1863

5. Christian M, Cermak T, Doyle EL, Schmidt C, Zhang F, Hummel A, Bogdanove AJ, Voytas DF (2010) Targeting DNA double-strand breaks with TAL effector 78 nucleases. Genetics 186:757–761
6. Boch J, Scholze H, Schornack S, Landgraf A, Hahn S et al (2009) Breaking the code of DNA binding specificity of TAL-Type III effectors. Science 326:1509–1512
7. Cermak T, Doyle EL, Christian M, Wang L, Zhang Y et al (2011) Efficient design and assembly of custom TALEN and other TAL effector-based constructs for DNA targeting. Nucleic Acids Res 39:e82
8. Kumlehn J, Serazetdinova L, Hensel G, Becker D, Lörz H (2006) Genetic transformation of barley (Hordeum vulgare L.) via infection of androgenetic pollen cultures with Agrobacterium tumefaciens. Plant Biotechnol J 4:251–261
9. Hensel G, Valkov V, Middlefell-Williams J, Kumlehn J (2008) Efficient generation of transgenic barley: the way forward to modulate plant-microbe interactions. J Plant Physiol 165:71–82
10. Lazo GR, Stein PA, Ludwig RA (1991) A DNA transformation competent Arabidopsis genomic library in Agrobacterium. Biotechnology (NY) 9:963–967
11. Miller JC, Tan S, Qiao G, Barlow KA, Wang J et al (2011) A TALE nuclease architecture for efficient genome editing. Nature Biotechnol 29:143–148
12. Mussolino C, Morbitzer R, Lütge F, Dannemann N, Lahaye T et al (2011) A novel TALE nuclease scaffold enables high genome editing activity in combination with low toxicity. Nucleic Acids Res 39: 9283–9293
13. Li T, Liu B, Spalding MH, Weeks DP, Yang B (2012) High efficiency TALEN-based gene editing produces disease-resistant rice. Nat Biotechnol 30:390–392
14. The International Barley Genome Sequencing Consortium (2012) A physical genetic and functional sequence assembly of the barley genome. Nature 491:711–716
15. Kramer MF, Coen DM (2001) Enzymatic amplification of DNA by PCR: standard procedures and optimization. Curr Protoc Mol Biol 56:15.1.1–15.1.14
16. Palotta M, Graham R, Langridge P, Sparrow D, Barker S (2000) RFLP mapping of manganese efficiency in barley. Theor Appl Genet 101:1100–1108
17. Vouillot L, Thélie A, Pollet N (2015) Comparison of T7E1 and surveyor mismatch cleavage assays to detect mutations triggered by engineered nucleases. G3 (Bethesda) 5:407–415
18. Budhagatapalli N, Schedel S, Gurushidze M, Pencs S, Hiekel S, Rutten T, Kusch S, Morbitzer R, Lahaye T, Panstruga R, Kumlehn J, Hensel G (2016) A simple test for the cleavage activity of customized endonucleases in plants. Plant Methods 12:18

Chapter 14

Creating Targeted Gene Knockouts in Barley Using CRISPR/Cas9

Tom Lawrenson and Wendy A. Harwood

Abstract

Knockout mutants are an invaluable reverse genetics tool which have not been well developed in crop species compared to models like Arabidopsis. However, the emergence of CRISPR/Cas9 has changed this situation making the generation of such mutants accessible to many crops including barley. A single T-DNA construct can be transformed into barley immature embryos and stable transgenic lines regenerated through tissue culture which contain targeted mutations. Mutations are detected in T0 plants and go on in subsequent T1 and T2 generations to segregate from T-DNA, leaving lines which are non-transgenic and carrying a variety of mutations at the target locus. These mutations can be targeted to a particular gene of interest in order to bring about a loss of function creating a knockout mutant.

Key words Barley, Vector, CRISPR/Cas9, Knockout, Mutant, Gene edit, Transgene free

1 Introduction

Since the emergence of CRISPR/Cas9 and its first application in plants [1–4] it has proven to be functional in virtually all species where it has been tried. Most often in plants it has been used to introduce double strand breaks (DSB) at a particular locus which are then repaired by the host cells own error prone non-homologous end joining mechanisms (NHEJ). This typically results in the loss or gain of a few nucleotides and these indels can then result in a loss of function if, for example, they are located in the coding region of a particular gene of interest [5].

Cas9 from *Streptococcus pyogenes* is an RNA-guided endonuclease associated with the CRISPR (Clustered Regularly Interspersed Short Palindromic Repeats) type II adaptive immune system in *S. pyogenes* [6]. The bacterium uses the Cas9/guide RNA complex to interrogate and cleave foreign DNA such as from invading bacteriophage. The native guide RNA is composed of two separate RNA molecules which hybridize before complexing with Cas9. As part of repurposing for genome editing the two RNA molecules

Wendy A. Harwood (ed.), *Barley: Methods and Protocols*, Methods in Molecular Biology, vol. 1900,
https://doi.org/10.1007/978-1-4939-8944-7_14,

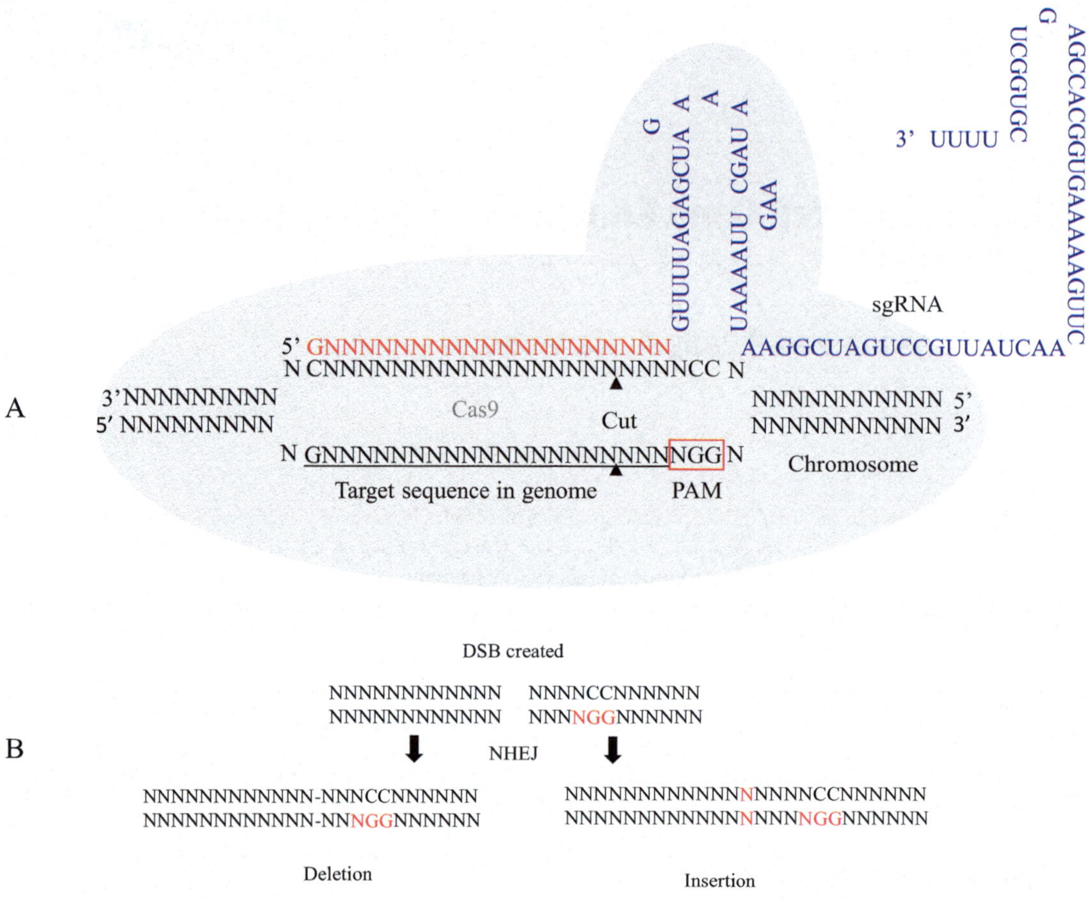

Fig. 1 (**a**) Schematic of Cas9/sgRNA complex having located a target sequence in chromosomal DNA. The 5′ 20 nucleotides of the sgRNA (red) (protospacer) are complementary to the top strand of the chromosome and are directly followed by a PAM which allows Cas9 to make a DSB. The non-variable section of sgRNA (blue) remains identical regardless of the intended target. (**b**) DSBs are repaired by error prone NHEJ resulting in small insertions and deletions (indels)

have been successfully provided as a single guide RNA (sgRNA) [7]. Essentially by tailoring the 5′ end (protospacer) of this sgRNA to be complementary in sequence to a target locus and adding a nuclear-localization signal to the Cas9 it has been possible to introduce DSBs at specific loci in plant and other Eukaryotic genomes.

The target specific 5′ protospacer is just 20 nucleotides in length while the remaining sgRNA remains otherwise identical regardless of the intended target (Fig. 1). The Cas9/sgRNA complex is able to interrogate the genome sequence and introduces a DSB where a suitable match is found. A key determinant of sgRNA functionality is whether the homologous genomic sequence is directly followed by NGG which is commonly referred to as the PAM (protospacer adjacent motif). If the entire 23 base sequence is identified in the genome (20 from the sgRNA plus PAM), then the endonuclease will cut. In higher Eukaryotic plants the predominant repair mechanism is NHEJ and although many repairs will restore a perfect wild-type sequence some will result in indels.

Because the length of the sgRNA responsible for targeting is relatively short, it is sometimes difficult, especially in larger genomes, to ensure this sequence does not exist in other "off-target" locations. There is a chance of off-target mutations arising even if mismatches occur between the off-target locus and the protospacer [5]. BLAST searches against the barley genome using the full 23 nucleotide sequence (protospacer plus PAM) as a query and other CRISPR/Cas9 specific online tools can allow selection of guides with minimal predicted off targets.

The stable transformation of barley using a T-DNA delivered by *Agrobacterium tumefaciens* is well established and an efficient process [8]. Using four transcriptional cassettes and inoculating immature embryos with *Agrobacterium* containing them in a T-DNA binary vector it is possible to introduce CRISPR/Cas9 targeted mutations. Figure 2 shows the structure of a dual guide vector which we use for such mutagenesis in barley. Adjacent to the left border is the first cassette which provides resistance to hygromycin and allows regeneration of barley plants which contain the T-DNA integrated into the genome. The second cassette expresses nuclear-localized Cas9 in a ubiquitous fashion while the third and fourth cassette utilize the wheat U6 promoter to drive transcription of sgRNA specific to the relevant target locus. We have found some sgRNAs work, some do not, and efficiency of mutagenesis can vary greatly depending on which sgRNA is used. While a significant amount of effort has been applied to enable selection of efficient sgRNAs [9, 10], this largely relates to mammalian genomes and we have found the best predictor to be actual usage in transgenic barley. Therefore, in order to minimize the number of transgenic lines required, we prefer to include a pair of sgRNAs in each binary construct.

For each target gene we make two dual sgRNA vectors, each with a pair of unique sgRNAs, making four in total per target gene (Fig. 3). This strategy has enabled us to target a total of 15 genes to date with 100% success as even in the least efficient case, one sgRNA out of four was functional. In many cases two, three, or four sgRNAs were active. Another potential benefit of using a dual sgRNA vector is that if both sgRNAs are active then simultaneous DSBs at the two genomic loci can often result in the entire sequence between being deleted [11], which may, for example, facilitate the removal of an entire exon.

For each binary construct made we routinely make twenty independent transgenic lines (Fig. 3). We then screen the T0 lines for the presence of targeted mutations and take forward into T1 those which are active (contain targeted mutations). After sowing

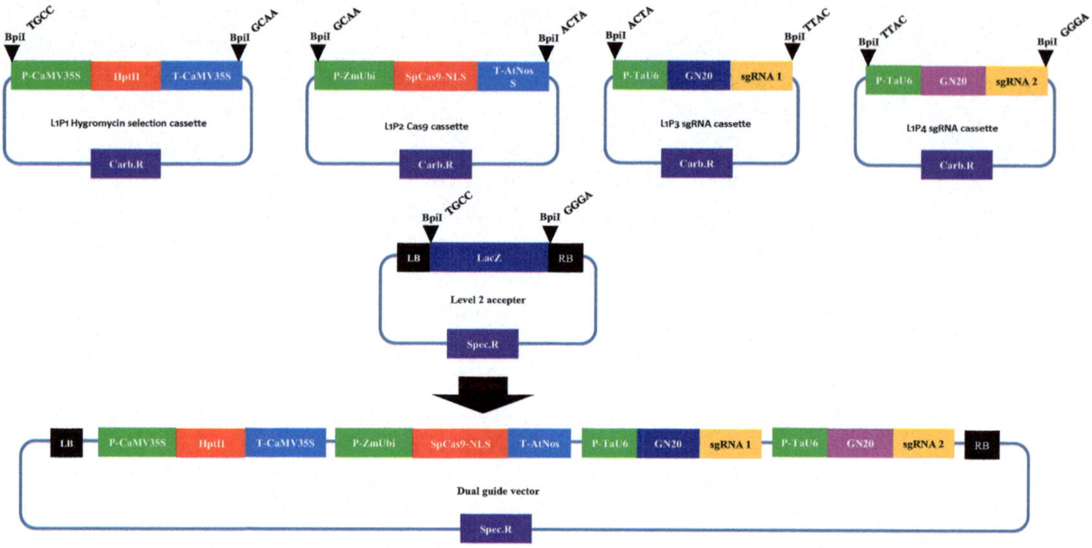

Fig. 2 Level 1 cassettes are assembled into the level 2 accepter via *BpiI* cloning. Level 1 position 1 (L1P1) Hygromycin resistance cassette consisting of the *hygromycin phosphotransferase* coding sequence (*hptII*) driven and terminated by the 35 s promoter (P-CaMV35s). Level 1 position 2 (L1P2) Cas9 expression cassette consisting of sequence encoding Cas9 from *Streptococcus pyogenes* with a carboxy-terminal nuclear-localization signal from Simian vacuolating virus 40 (SpCas9:NLS) driven by a ubiquitin promoter from *Zea mays* (*P-ZmUbi*) and terminated by a nopaline synthase terminator from *Agrobacterium tumefaciens* (*T-AtNos*). Level 1 position 3 and 4 (L1P3/L1P4) single guide RNA (sgRNA1 and sgRNA2) driven by a *Triticum aestivum* U6 promoter *(P-TaU6)*. Selection of these plasmids in bacteria is carbenicillin (Level 1)) and spectinomycin (level 2). Left border (LB) and right border (RB) T-DNA sequences are indicated

out T1 seed we expect to find progenies which have lost their T-DNA via segregation but retain targeted mutations. The T1 progenies are also likely to have segregated for the various mutations seen in the T0 parent and so there is every likelihood of obtaining homozygous mutants which are transgene free in the T1 generation.

There are various ways in which plant material may be screened for the presence of mutations including restriction enzyme/PCR-based methods [12]. However, we have found that direct sequencing of PCR amplicons which cover the target sites is relatively quick and simple to perform and gives a detailed output of events occurring at that locus (Fig. 3). Cas9 cuts within the 20 base target sequence between base 3 and 4 counting from directly 5′ to the PAM (Fig. 1). This means that where indels are present, sequencing chromatograms typically become double or triple peaked from this cut point whereas the preceding sequence consists of clean single peaks (Fig. 3). This indicates a mixture of alleles, many of which will be potentially useful in terms of causing loss of function. Another benefit of using a dual sgRNA strategy is that when both sgRNAs are active, and simultaneous cutting allows the entire fragment between two target sites to be deleted, a PCR amplicon that is

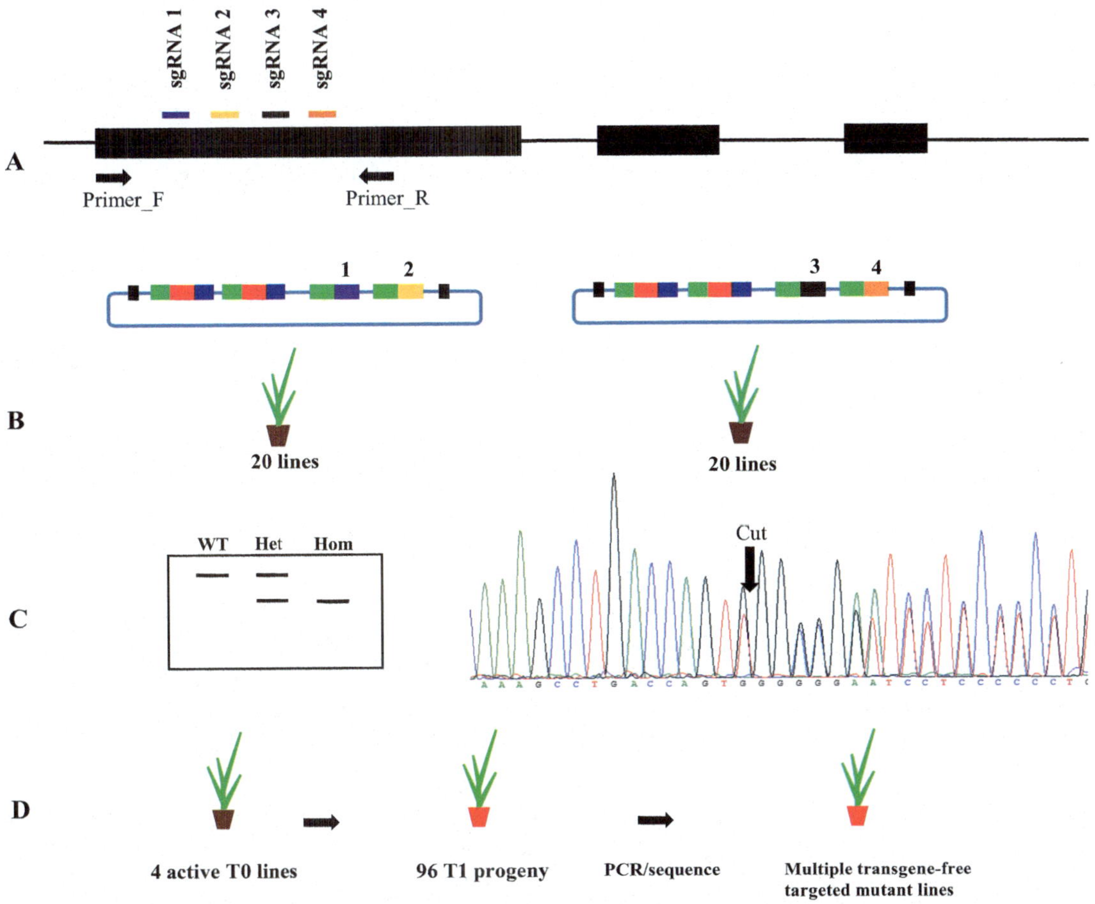

Fig. 3 (**a**) Four sgRNAs are designed to target the first exon of a gene. (**b**) Two dual guided binary constructs are assembled, each containing a pair of sgRNA transcriptional cassettes. Twenty independent transgenic barley lines are made for each of the constructs and (**c**) screened via PCR/sequencing using forward and reverse primers. Some lines may display significant deletions leading to band shift on agarose gels which can be heterozygous or homozygous in appearance. By sequencing the bands which appear as wild type a mixture of indel-based alleles can be revealed by observing sequencing chromatogram files. Double peaks arising from the Cas9 cut point or very close to it are indicative of this. (**d**) By screening 24 T1 progeny from 4 active T0 lines, multiple transgene-free targeted mutants can be recovered

considerably shorter (by approximately the distance between the two target loci) is obtained. This significant difference in size relative to a wild-type amplicon is easily seen on agarose gels as a band shift (Fig. 3).

Active T0 lines in which indels or larger deletions have been identified are kept and grown to seed. Indels should segregate from T-DNA and by taking four active T0 lines per target gene into T1, sowing out 24 of each (96 in total), plenty of T-DNA-free mutants should be identified. T1 plants which contain no T-DNA can be identified by PCR and mutations can be screened using the same PCR/sequencing procedure as in T0. Now the aim is to identify lines containing no T-DNA and ideally homozygous mutations. While the T-DNA should segregate in Mendelian fashion in T1, often the targeted mutations do not. Where editing occurred early in T0, for example, the founder cell of regenerated plants, T1 mutagenesis can reach 100% (all alleles in all siblings of the original T0 plant). However, where editing has occurred later in T0 lines the plants may be chimeric with fewer T1 siblings likely to be mutated and the resulting range of alleles at a locus is often more than two.

2 Materials

2.1 Selection of Target Sequences

Online tools are constantly evolving for this purpose but currently one such facility which links to the barley genome is hosted by Deskgen.com. BLAST searches against the barley genome to check off targets can also be done at Ensembl Plants http://plants.ensembl.org/index.html and at morexGenes hosted by The James Hutton Institute https://ics.hutton.ac.uk/morexGenes/blast_page.html.

2.2 Construct Assembly

1. All oligos and primers can be ordered as PCR grade from suppliers such as Sigma.
2. Level 1 and 2 plasmids are available from AddGene (*see* **Note 1**).
3. *Bsa1* may be purchased from New England Biolabs and stored at −20 °C.
4. T4 ligase is best purchased from New England Biolabs and stored at −20 °C.
5. Water used should be deionized and sterile.
6. All modern thermocyclers should be suitable.
7. Electrocompetent cells are commercially available from ThermoFisher.
8. 2 mm electroporation cuvettes are available from Geneflow Ltd.
9. BioRad Gene pulser 2 or equivalent.
10. Carbenicillin 1000×: 10 mL water, 1 g carbenicillin disodium. Mix to dissolve, then filter sterilize and store in 1 mL aliquots at −20 °C.
11. IPTG 1000×: 238 mg IPTG, 10 mL water. Mix to dissolve, then filter sterilize and store in 1 mL aliquots at −20 °C.
12. X-gal 100×: 200 mg X-gal, 10 mL of DMSO. Mix to dissolve and store in 1 mL aliquots at −20 °C for up to 6 months.
13. Plasmid miniprep kit from Qiagen.

14. BigDye® cycle sequencing kit (Terminator v3.1 ready reaction mix plus 5× buffer).
15. *Esp3I* and *BpiI* are best obtained from ThermoFisher.
16. 0.2 mL PCR tubes.
17. Spectinomycin 1000×: 1 g spectinomycin dihydrochloride pentahydrate, 10 mL water. Mix to dissolve, then filter sterilize and store in 1 mL aliquots at −20 °C.
18. LB agar made from commercially available powder formulations following manufacturer's instructions.
19. Laboratory heat block with metal insert fitting Eppendorf tubes.
20. Hybridization buffer: 10 mM Tris HCl pH 7.5, 50 mM NaCl, 1 mM EDTA.

2.3 Genomic DNA Extraction from Barley

1. 1.5 mL Eppendorf tubes.
2. Buffer1: 200 mM Tris HCl pH 7.5, 250 mM NaCl, 25 mM EDTA, 0.5% SDS.
3. Micropestles to fit 1.5 mL Eppendorf tubes.
4. Propan-2-ol.
5. 70% ethanol.
6. Microfuge.
7. (TE) Tris-EDTA buffer pH 8. Commercially available.

2.4 PCR and Sequencing of Target Loci

1. Commercially available 2× PCR master mix.
2. Modern thermocycler.
3. PCR grade primers.
4. Molecular biology grade agarose.
5. 10× TBE gel running buffer available commercially.
6. Deionized water.
7. Ethidium bromide stock: 10 mg ethidium bromide, 1 mL water. Mix to dissolve and store at 4 °C.
8. Gel loading buffer commercially available.
9. DNA size marker appropriate to expected band sizes.
10. Alkaline phosphatase (1 unit/μL) which is heat inactivated at 65 °C.
11. Exonuclease 1 (10 units/μL) which is heat inactivated at 80 °C.
12. BigDye® cycle sequencing kit (Terminator v3.1 ready reaction mix plus 5× buffer).

2.5 Identification of Transgene-Free Mutant Lines

1. 9 cm diameter Whatman paper disks.
2. 9 cm Petri dishes.
3. Micropore tape.
4. Levington M3 compost.
5. Perlite.
6. Grit.
7. Osmocote slow release fertilizer.
8. 4 cm wide × 6 cm deep pots in 8 × 12 (96) cell format.
9. Commercially available 2× PCR master mix.
10. Modern thermocycler.
11. PCR grade primers.
12. Molecular biology grade agarose.
13. 10× TBE gel running buffer.
14. Deionized water.
15. Ethidium bromide stock: 10 mg ethidium bromide, 1 mL water. Mix to dissolve and store at 4 °C.
16. Gel loading buffer commercially available.
17. DNA size marker appropriate to expected band sizes. Use 100 bp ladder for the *HptII* PCR detecting transgene presence/absence.

3 Methods

3.1 Selection of Target Sequences in Barley Genome

1. 23 Nucleotide sequences which are unique should be identified in the first exon of the target gene(s). The sequence should conform to the template GN20GG (Fig. 1) and can be present on either the sense or antisense strand. Potential off-target sites should be checked to ensure that mutagenesis is as specific as possible (*see* **Note 2**).
2. When four target loci have been identified for the target gene, a PCR amplicon(s) covering the target sites should be designed and tested using template genomic DNA extracted from Golden Promise (*see* **Note 3**). The PCR should be capable of amplifying clean single bands reliably which can be directly sequenced after a SAPX clean up to yield chromatograms with single peaks covering the target regions. *See* Subheading 3.3 below for genomic DNA extraction and SAPX clean up. *See* Subheading 3.4 below for PCR/sequencing of target loci. This sequence can also be used to check whether polymorphisms exist between the database accession used in target selection and the Golden Promise to be used here in transformation.

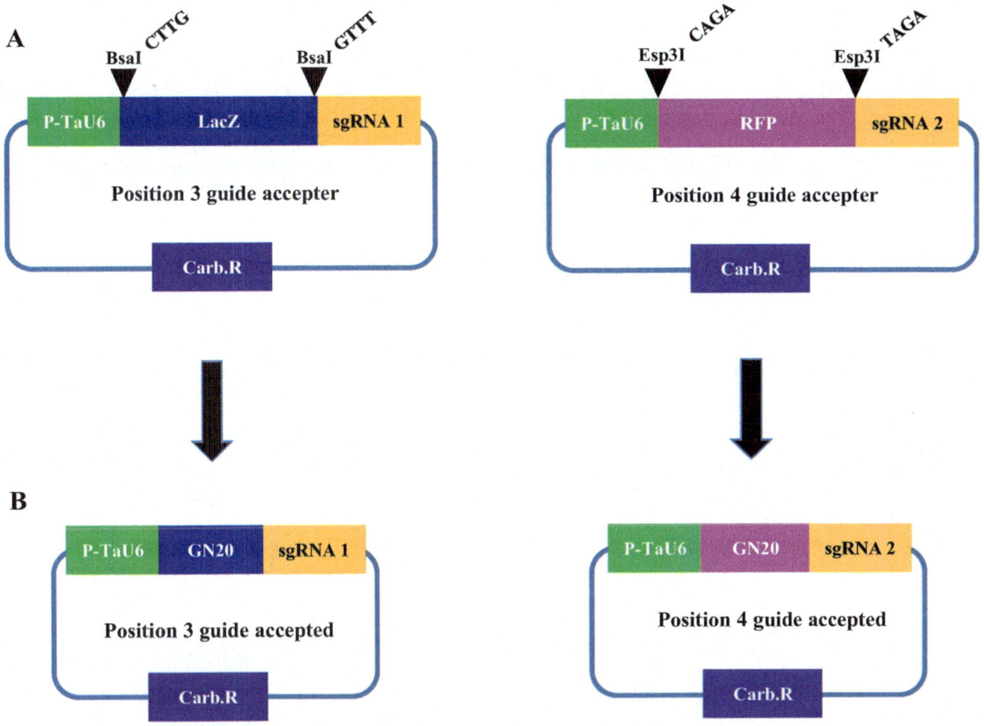

Fig. 4 (**a**) Position 3 and 4 level 1 guide accepters ready to receive oligos via BsaI and Esp3I cloning respectively. Four base GoldenGate overhangs are shown which are compatible to the hybridized oligo pair sticky ends (Table 1). (**b**) Position 3 and 4 complete sgRNA transcriptional cassettes now containing cloned protospacer which has replaced the LacZ and RFP color markers respectively. Single guide RNA (sgRNA1 and sgRNA2) driven by a *Triticum aestivum* U6 promoter (P-TaU6)

3. If **step 2** is achieved, construct assembly can commence; however, if **step 2** has not been achieved, then different target sequences should be selected, in a different region or exon where PCR may be more achievable (*see* **Note 4**).

3.2 Construct Assembly

1. The position 3 guide accepter has an appropriate protospacer added by cloning two complementary oligos using the restriction enzyme *Bsa1* (Fig. 4). The oligos should be designed to include the overhanging ends shown in Table 1 which allow insertion into the *Bsa1* cut accepter vector. Table 1 also shows an example target sequence and how this is represented in the complementary oligo pair. Note that the PAM is not included in the oligo sequence.
2. Complementary oligos are hybridized by preparing the pair at 2 μM in hybridization buffer within an Eppendorf tube and heating to 95 °C for 3 min using a metal heat block. The block is then switched off and allowed to slowly return to room temperature, which should take about 45 min.

Table 1
Template oligos for making sgRNA sequences target specific

	Position 3 sgRNA cloning	Position 4 sgRNA cloning
Enzyme	*BsaI*	*Esp3I*
Generic template	5′ CTTGNNNNNNNNNNNNNNNNNNN 3′ 3′ NNNNNNNNNNNNNNNNNNNCAAA 5′	5′ CAGACTTGNNNNNNNNNNNNNNNNNNNGTTT 3′ 3′ GAACNNNNNNNNNNNNNNNNNNNCAAAATCT 5′
Example target	GTGACCATGGAGGACGTGGT<u>CGG</u>	GACGGCGGCCACGACCTCCA<u>TGG</u>
Example incorporated	5′ CTTGTGACCATGGAGGACGTGGT 3′ 3′ ACTGGTACCTCCTGCACCACAAA 5′	5′ CAGACTTGACGGCGGCCACGACCTCCAGTTT 3′ 3′ GAACTGCCGCCGGTGCTGGAGGTCAAAATCT 5′

Compatible restriction enzyme site overhangs are shown in red. PAM's in example target sequences are underlined. These are not included in the cloned sequences which are shown incorporated in the bottom row

3. The following should be added to a 0.2 mL PCR tube: 100 ng position 3 guide accepter, 1 μL of hybridized oligo pair, 1 μL of 10× T4 ligase buffer, water to 8.5 μL, 0.5 μL (10 units) *Bsa1*, 1 μL (400 units) T4 ligase. The mixture should be immediately cycled as follows: 1 × 20 s at 37 °C, 26 × 37 °C for 3 min/16 °C for 4 min, 1 × 50 °C for 5 min, 1 × 80 °C for 5 min.
4. 1 μL of the cycled mixture should be transformed into electrocompetent *E.coli* cells using 2 mm electroporation cuvettes and conditions specified by the manufacturer, then plated onto LB agar containing 100 mg/L carbenicillin (1 mL of 1000× per liter), 0.1 mM IPTG (1 mL of 1000× per liter), 200 μg/mL X-gal (10 mL of 100× per liter) before overnight incubation at 37 °C. Around three white colonies (blue colonies are not useful) should be inoculated into separate 10 mL liquid LB vessels containing 100 mg/L carbenicillin (1 mL of 1000× per liter) and grown overnight at 37 °C (*see* **Note 5**). The resulting turbid cultures should be used to extract plasmid DNA. The plasmid should be Sanger sequenced using the primer 5'TAGGAGGGAATCGAACTAGGAATATTG3' which will reveal whether the oligo pair has been incorporated correctly. Reactions should contain 200 ng plasmid, water to 6.5 μL, 1.5 μL BigDye 3.1 buffer, 1 μL of 10 μM primer, 1 μL of BigDye version 3.1. Be sure to add the BigDye version 3.1 last. Cycle reactions as follows: 1 × 96 °C for 1 min, 25 × 96 °C for 10 s/50 °C for 5 s/60 °C for 4 min. Completed reactions should be sent to one of the specialist companies dealing with capillary electrophoresis of such material. The correct sequence around the incorporation site is shown below where the sense strand is depicted and the first base of transcription is the **G** directly 5′ to the NX19. Ns indicate protospacer sequence and the **G** directly 3′ to this is the beginning of the non-variable section of

sgRNA. The entire sequence of the position 3 guide accepter vector is available via AddGene (*see* **Note 1**).

5' GCTTGCTGCATCAGACTT**G**(NX19)**G**TTTTAGAG 3'

5. The position 4 guide accepter has an appropriate protospacer added by cloning two complementary oligos using the restriction enzyme *Esp3I* (Fig. 4). The oligos should be designed to include the overhanging ends shown in Table 1 which allow insertion into the *Esp3I* cut accepter vector. Table 1 also shows an example target sequence and how this is represented in the complementary oligo pair. Note that the PAM is not included in the oligo sequence. Hybridize the pair as in Subheading 3.2, **step 2** above.
6. The following should be added to a 0.2 mL PCR tube: 100 ng position 4 guide accepter, 1 μL of hybridized oligo pair, 1 μL of 10× T4 ligase buffer, water to 8.5 μL, 0.5 μL (10 units) *Esp3I*, 1 μL (400 units) T4 ligase. The mixture should be immediately cycled as follows: 1 × 20 s at 37 °C, 26 × 37 °C for 3 min/ 16 °C for 4 min, 1 × 50 °C for 5 min, 1 × 80 °C for 5 min.
7. 1 μL of the cycled mixture should be transformed into electrocompetent *E.coli* cells using 2 mm electroporation cuvettes and conditions specified by the manufacturer, then plated onto LB agar containing 100 mg/L carbenicillin (1 mL of 1000× per liter) before overnight incubation at 37 °C. Around three white colonies (purple colonies are not useful) should be inoculated into separate 10 mL liquid LB vessels containing 100 mg/L carbenicillin (1 mL of 1000× per liter) and grown overnight at 37 °C (*see* **Note 5**). The resulting turbid cultures should be used to extract plasmid DNA. The plasmid should be Sanger sequenced and verified as above in Subheading 3.2, **step 4**.
8. Level 1 components: L1P1 barley hygromycin selection, L1P2 Cas9, L1P3 sgRNA cassette 1 and L1P4 sgRNA cassette 2 can now be assembled into the level 2 accepter (*see* **Note 1**) between the T-DNA left and right border sequences (Fig. 2). The following should be added to a 0.2 mL PCR tube: 100 ng level 2 accepter plasmid, 300 ng L1P1 hygromycin selection cassette, 300 ng L1P2 Cas9 cassette, 300 ng L1P3 sgRNA cassette 1, 300 ng sgRNA cassette 2, 1 μL of 10× T4 ligase buffer, water to 8.5 μL, 0.5 μL (10 units) *BpiI*, 1 μL (400 units) T4 ligase. The mixture should be immediately cycled as follows: 1 × 20 s at 37 °C, 26 × 37 °C for 3 min/16 °C for 4 min, 1 × 50 °C for 5 min, 1 × 80 °C for 5 min.
9. 1 μL of the cycled mixture should be transformed into electrocompetent *E.coli* cells using 2 mm electroporation cuvettes and conditions specified by the manufacturer, then plated onto LB agar containing 100 mg/L spectinomycin (1 mL of 1000× per liter), 0.1 mM IPTG (1 mL of 1000× per liter), 200 μg/mL

X-gal (10 mL of 100× per liter) before overnight incubation at 37 °C. Around three white colonies (blue colonies are not useful) should be inoculated into separate 10 mL liquid LB vessels containing 100 mg/L spectinomycin (1 mL of 1000× per liter) and grown overnight at 37 °C. The resulting turbid cultures should be used to extract plasmid DNA.

10. The plasmid clones should be verified by restriction digest and sequencing. The full sequence for all level 1 and 2 component plasmids is available via AddGene (*see* **Note 1**).
11. The complete dual sgRNA vector can now be transformed into *Agrobacterium tumefaciens* strain AGL1 in preparation for barley transformation—*see* Chapter 8.
12. Once T0 barley plants are established on soil at around 14 weeks post inoculation, leaf material can be sampled for DNA extraction and then used in PCR/sequencing.

3.3 Genomic DNA Extraction from Barley

1. Two pieces of leaf around 1 cm long are taken from different parts of T0 plants and pooled to make a single DNA prep. Samples are collected in 1.5 mL Eppendorf tubes and kept cool on ice during leaf collection.
2. 600 μL of buffer 1 is added to each tube and the leaves are ground using micropestles until all large particles are fragmented and the liquid becomes dark green.
3. Tubes are spun in a benchtop microfuge at full speed for ten minutes and 500 μL of the supernatant moved to a fresh 1.5 mL Eppendorf tube, taking care to leave behind solid matter.
4. An equal volume of propan-2-ol is added to each tube before vortexing and then spinning at full speed in a microfuge for twenty minutes. The liquid is then carefully discarded and the pellet washed with 0.5 mL of 70% ethanol.
5. Tubes are spun again at full speed for ten minutes before carefully removing the liquid and allowing the pellet to air-dry for 20 min. Pellets are resuspended in 100 μL of 1× TE.

3.4 PCR and Sequencing of T0 Target Loci

1. Primers which will produce amplicons covering genomic target loci should be designed using primer 3 http://bioinfo.ut.ee/primer3-0.4.0/primer3/ and tested/optimized using wild-type DNA extracted using the method in Subheading 3.3 to achieve specific clean amplification.
2. PCR can be done in 20 μL volumes using commercially available 2× master mixes adding only primers to a final concentration of 150 nM and using 1 μL of extracted genomic DNA.
3. Following cycling, 5 μL of each reaction should be run in agarose gel made with an appropriate percentage of agarose

relative to the expected band sizes. Gels should be made with 1× TBE supplemented with ethidium bromide to a concentration of 0.5 μg/mL (5 μL of 10 mg/mL stock per 100 mL gel). Running buffer should be 1× TBE and a suitable DNA size marker should be included. Large deletions resulting from targeted mutagenesis may be visible by a band shift to a lower position relative to a wild-type control (Fig. 3c).

4. Prepare the remaining 15 μL of PCR for sequencing by adding 1 unit of alkaline phosphatase (heat inactivatable) and 10 units of exonuclease 1 (heat inactivatable) before incubation at 37 °C for 30 min and then 80 °C for 20 min.
5. Sequence PCR amplicons over target loci on both strands by setting up reactions containing the following: Approximately 20 ng SAPX clean PCR product (up to around 1 kb in length), water to 6.5 μL, 1.5 μL BigDye 3.1 buffer, 1 μL of 10 μM primer, 1 μL of BigDye version 3.1. Be sure to add the BigDye version 3.1 last. Cycle reactions as follows: 1 × 96 °C for 1 min, 25 × 96 °C for 10 s/50 °C for 5 s/60 °C for 4 min (*see* **Note 6**).
6. Completed reactions should be sent to one of the specialist companies dealing with capillary electrophoresis of such material who will return ABI files for chromatogram viewing.
7. Indels at target loci may be visible as double peaks starting from or near to the expected Cas9 cut points (Fig. 3c).
8. T0 lines where mutagenesis has been detected should be grown to seed and examined again in T1 for mutation inheritance.

3.5 Identification of Transgene-Free Mutant Lines

1. Select four active T0 lines and from each collect 40 harvested seeds, placing them on four sheets of wet Whatman paper within 9 cm Petri dishes, sealed with micropore tape. Store the plates at 4 °C for two days before moving to 24 °C in the presence of light. Seeds should germinate uniformly in about 4 days when they can be moved to soil. 24 T1 plants per T0 active line should be established in barley growth mix (2:2:1 mix of Levington M3 compost: Perlite: Grit. Osmocote slow release fertilizer is also included at the manufacturers recommended concentration) in 4 cm pots which are available in 96 cell format for ease of organization. Plants are grown in a controlled environment room at 15 °C day and 12 °C night temperatures, 80% relative humidity and with light levels of 500 μmoL/m^2/s^1 at the mature plant canopy level provided by metal halide lamps supplemented with tungsten bulbs. After around 1 week the plants should be well established and leaf material may be sampled and used for DNA extraction as described previously in Subheading 3.3.

2. DNA should be used for PCR/sequencing as previously described with the aim of identifying indels or larger deletions as before. This time due to chromosome segregation there is a much higher likelihood of identifying homozygous mutants.
3. To test for the presence or absence of the T-DNA, conduct a PCR with primers specific to the *HptII* coding sequence within. Each PCR reaction should contain 10 μL of commercially available 2× PCR master mix, 7 μL water, 1 μL of template DNA and 1 μL each of the following primer pair at 10 μM stock concentration: F_GGATTTCGGCTCCAACAATG, R_TATTGGGAATCCCCGAACATC. Reactions should be cycled as follows: 1 × 94 °C 3 min, 35 × 94 °C 30 s/58 °C for 45 s/72 °C for 30 s. Following cycling, 5 μL of each reaction should be run in 2% agarose gel made with 1× TBE supplemented with ethidium bromide to a concentration of 0.5 μg/mL (5 μL of 10 mg/mL stock per 100 mL gel). Where the T-DNA is present a single band of 107 bp should amplify and where it does not indicates T1 lines which are transgene free. A no template negative control must be included as well as using a positive control such as a genomic DNA sample known to contain *HptII* (*see* **Note 7**). T1 plants which contain no T-DNA but have been observed to contain targeted mutations can now be classified as transgene-free mutants as the mutations must have been inherited and are not merely somatic sectors because Cas9 is no longer present. Homozygous mutants which are transgene free are often found at this stage; however, it may be necessary to go into T2 via another round of segregation in some cases where only heterozygous or biallelic mutant lines have been identified.

4 Notes

1. Plasmids used here are available at AddGene along with the corresponding sequences and maps:

   L1P1 hygromycin selection cassette: pICSL11059*(AddGene #68263).

   L1P2 Cas9 expression cassette: pICSL11056 (AddGene #68258).

   L1P3 guide accepter: pHvL1P3GA (AddGene # 112027).

   L1P4 guide accepter: pHvL1P4GA (AddGene # 112028).

   L2 accepter: pAGM8031 (AddGene #48037).
2. Various online tools are available to aid in guide selection and also give a score for predicted off-target activity. Many are linked to various genomes and one which currently does this for barley can be found at deskgen.com. Mismatches to

off-target sites at the PAM distal end of the sgRNA will be tolerated more so than at the proximal end and will still potentially allow off-target mutagenesis to occur. If mismatches to off-target sites occur in the 8 nucleotides directly adjacent to the PAM within the protospacer, these are less tolerated and much more likely to prevent off-target activity.

3. Shorter amplicons generally work better than longer ones. We prefer to amplify between 300 and 900 base pair fragments where possible. Primers should be designed no closer than 100 nucleotides from the closest target sequence within to ensure that the output from capillary sequencing has settled and is giving clean peaks running into the target.
4. Where target sequences conforming to GN20GG are limiting then it is possible to simplify this to N21GG. Because the preferred start base of transcription from the wheat U6 promoter is G, then this should still be included in the cloned oligo sequence. The only impact this has to the oligo design (Table 1) is that it now contains NX20 instead of NX19. The PAM is still not included in the oligo sequence.
5. Prior to cloning into the L1P3 guide accepter a LacZ cassette is located between the two *Bsa1* sites used. Successful oligo insertion is indicated by a loss of blue coloration when grown on X-gal/IPTG. The L1P4 accepter contains a cassette giving purple coloration between two *Esp3I* sites. Where oligo insertion has not occurred, colonies will develop a faint purple color. Where the oligos have inserted colonies will be white.
6. We have found the amount of SAPX clean PCR product used as Sanger sequencing template quite flexible and generally find that for amplicons up to around 1 kb in length 1 μL is sufficient when the 5 μL tested on agarose gave a bright band. However, the BigDye 3.1 manufacturers recommendations are to use 3–10 ng from 200 to 500 bp, 5–20 ng from 500 to 1000 bp and 10–40 ng from 1000 to 2000 bp.
7. When genotyping T1 material for the presence/absence of T-DNA it is of utmost importance that false positives or negatives are not allowed to occur. Assuming the suggested positive control results in the 107 bp band as it should, the greatest risk is of lines which actually contain no T-DNA giving a spurious 107 bp band. To avoid this problem keep plasmids well away, use fresh solutions and always use filter tips.

Acknowledgments

We acknowledge support from the Biotechnology and Biological Sciences Research Council (BBSRC) via grant [BB/N019466/1] and grant [BB/P013511/1] to the John Innes Centre.

References

1. Feng Z, Zhang B, Ding W, Liu X, Yang DL, Wei P, Zhu J-K (2013) Efficient genome editing in plants using a CRISPR/Cas system. Cell Res 23(10):1229–1232
2. Li JF, Aach J, Norville JE, McCormack M, Zhang D, Bush J, Sheen J (2013) Multiplex and homologous recombination-mediated plant genome editing via guide RNA/Cas9. Nat Biotechnol 31(8):688–691
3. Belhaj K, Chaparro-Garcia A, Kamoun S, Nekrasov V (2013) Plant genome editing made easy: targeted mutagenesis in model and crop plants using the CRISPR/Cas system. Plant Methods 9:39
4. Shan Q, Wang Y, Li J, Zhang Y (2013) Targeted genome modification of crop plants using a CRISPR-Cas system. Nat Biotechnol 31(8):686–688
5. Lawrenson T, Shorinola O, Stacey N, Li C, Østergaard L, Patron N, Harwood W (2015) Induction of targeted, heritable mutations in barley and *Brassica oleracea* using RNA-guided Cas9 nuclease. Genome Biol 16:258
6. Jinek M, Chylinski K, Fonfara I, Hauer M, Doudna JA, Charpentier E (2012) A programmable dual-RNA guided DNA endonuclease in adaptive bacterial immunity. Science 337:816–821
7. Cong L, Ran FA, Cox D, Lin S, Barretto R, Habib N, Hsu PD, Wu X, Jiang W, Marraffini LA, Zhang F (2013) Multiplex genome engineering using CRISPR/Cas systems. Science 339:819–823
8. Harwood WA (2014) A protocol for high-throughput Agrobacterium-mediated barley transformation. Methods Mol Biol 1099:251–260
9. Haeussler M, Schonig K, Eckert H, Eschstruth A, Mianne J, Renaud JB, Schneider-Maunoury S, Shkumatava A, Teboul L, Kent J, Joly JS, Concordet JP (2016) Evaluation of off-target and on-target scoring algorithms and integration into the guide RNA selection tool CRISPOR. Genome Biol 17:148
10. Doench JG, Fusi N, Sullender M, Hedge M, Vaimberg EW, Donovan KF, Smith I, Tothova Z, Wilen C, Orchard R, Virgin HW, Listgarten J, Root DE (2016) Optimized sgRNA design to maximise activity and minimise off-target effects of CRISPR-Cas9. Nat Biotechnol 34:184–191
11. Brooks C, Nekrasov V, Lippman ZB, Van Eck J (2014) Efficient gene editing in tomato in the first generation using clustered regularly interspersed short palindromic repeats/CRISPR-associated9 system. Plant Physiol 166:1292–1297
12. Belhaj K, Chaparro-Garcia A, Kamoun S, Patron NJ, Nekrasov V (2014) Editing plant genomes with CRISPR/Cas9. Curr Opin Biotechnol 32:76–84

Chapter 15

Genotyping-by-Sequencing on the Ion Torrent Platform in Barley

Amina Abed, Gaétan Légaré, Sonia Pomerleau, Jérôme St-Cyr, Brian Boyle, and François J. Belzile

Abstract

The characterization of genetic polymorphism is a crucial step in both genetic studies and breeding programs. Genotyping-by-sequencing (GBS) constitutes one of the most attractive approaches for this purpose, especially in a genome as large as that of barley. The genome sequencing project undertaken by the International Barley Sequencing Consortium (IBSC) has produced a structured reference genome for the cultivar Morex [1] that can serve as an excellent resource for the analysis of GBS data. The genome assembly for this species [2] is thought to adequately capture the gene-rich portion of the genome (~80% of the entire genome). In this chapter, we describe the entire GBS process, from library preparation to the analysis of read data to produce a high-quality catalog of single nucleotide polymorphism (SNP) markers using the barley reference genome.

Key words Barley, High-throughput sequencing, Genotyping, Variant calling, SNP marker, Polymorphism, Reference genome, Next-generation sequencing

1 Introduction

For large and complex genomes, such as barley, and when using high-throughput sequencing to perform genotyping on a large scale, approaches targeting a reduced representation of the entire genome are required. This makes it possible to (i) sequence reproducibly the same regions in several individuals and (ii) significantly reduce the per-sample cost of sequencing [3].

Genotyping-by-sequencing (GBS) constitutes an attractive approach for a genome as large and complex as barley (5.1 Gb) [3, 4]. It has proven a robust, efficient, and simple approach. The main objective is to simultaneously identify informative SNP loci and call the genotype at each of these marker loci for all individuals analyzed. In addition, the same sequence data generated by sequencing the GBS libraries can be reanalyzed over time to

Wendy A. Harwood (ed.), *Barley: Methods and Protocols*, Methods in Molecular Biology, vol. 1900,
https://doi.org/10.1007/978-1-4939-8944-7_15,

discover more useful genetic information as bioinformatics methods are refined [5, 6].

The GBS approach described in this chapter employs two advantageous strategies: (i) a complexity reduction step targeting a subset of the genome, which is ensured by the use of appropriate restriction enzymes and (ii) the multiplexing of several individuals in a single GBS library, thanks to the use of barcodes to trace all sequence reads back to the individual sample from which they originated. These strategies reduce the cost and increase the SNP-calling accuracy without compromising quality [5, 7, 8].

Alongside technical advances in DNA sequencing, new algorithms and computational pipelines for bioinformatics data analysis and detection of SNPs have also been developed [3]. Software commonly used for calling SNP data from GBS include two types of pipelines: (i) ones that do not require a reference genome, such as UNEAK [8] or Stacks [9] and (ii) pipelines based on a reference genome for SNP detection, such as TASSEL-GBS [10], IGST [11], and Fast-GBS [12].

A comparison of these various tools has recently shown that both types can successfully be used to extract SNP data from GBS-derived read data, but that the second type of pipeline generally produces more SNP data [13]. In addition, the resulting information is more useful as the detected SNPs are precisely positioned on the reference genome. No matter which bioinformatics pipeline is used, it is always necessary to perform additional data curation and filtration, steps that need to take into account the final purpose of the SNP data set.

The GBS approach will be mainly subdivided into five major procedures: (i) DNA extraction, (ii) GBS library preparation, (iii) Sequencing, (iv) GBS data analysis, and (v) Further SNP Filtering. We will conclude this chapter by an example in order to illustrate the results we can obtain.

2 Materials

2.1 DNA Extraction

1. Reagents.
   (a) DNeasy 96 plant kit, Qiagen or equivalent.
   (b) Liquid nitrogen.
2. Required lab equipment.
   (a) Equipment for tissue grinding: TissueLyser.
   (b) Water bath or heating block (65 °C).
   (c) Vortex.
   (d) Centrifuge with Plate Rotor 2 × 96 with maximum speed of 6,000 rpm (5,796 × *g*).

Table 1
Oligonucleotide sequences

Oligonucleotide	Sequence
Top barcoded oligo	5'-CCCTGCGTGTCTCCGACTCAG-**[Barcode]**-GATTGCA
Bottom barcoded oligo	5'-ATC-**[Barcode Reverse Complement]**-CTGAGTCGGAGACACGCA GGG
Top common adapter	5'-CGAGATCGGAAGAGCGGGGAGCTTAAGC
Bottom common adapter	5'-CCTCTCTATGGGCAGTCGGTGATCCCGCTCTTCCGATCT
Ion Forward PCR primer	5'-CCATCTCATCCCTGCGTGTCTCCGACTCAG
Ion Reverse PCR primer	5'-CCACTACGCCTCCGCTTTCCTCTCTATGGGCAGTCGGTGAT

2.2 GBS Library Preparation and Sequencing

1. Oligonucleotides.
   (a) The oligonucleotides used to prepare barcoded adapters are ordered as normal oligonucleotides at the 25-nM scale with standard desalting (to be shipped dried). Order oligonucleotides to prepare barcoded oligonucleotides in complementary plates, one for the top and one for the bottom strand. Having corresponding wells in two different plates makes the production of doubled-stranded adapters much easier (Tables 1 and 2).
   (b) The oligonucleotides used to prepare the common adapter are ordered as normal oligonucleotides at the 1-μmoL scale with standard desalting. For each adapter, two oligonucleotides are ordered in complementary pairs and must be annealed, to form the double-stranded adapter (Table 1).
2. Enzymes.
   (a) *Msp*I (R0106L), Hi-fidelity *Pst*I (R3140L), T4 DNA ligase (M0202 L), and Q5 High-fidelity polymerase (M0491L) from New England Biolabs.
3. Solutions.
   (a) Elution buffer (EB): 10 mM Tris-Cl pH 8.0.
   (b) 10× Annealing buffer (10× AB): 500 mM NaCl, 100 mM Tris-Cl pH 8.0.
   (c) 80% ethanol freshly prepared.
4. Reagents.
   (a) Qiaquick PCR Purification Kit or equivalent.
   (b) Axygen PCR Clean Up kit or equivalent.

Table 2
Barcode sequences

CTAAGGTAAC	TCTATTCGTC	TCGCATCGTTC	CGGACAATGGC	TTCCTACCAGTC	CTAGGACATTC
TAAGGAGAAC	AGGCAATTGC	TAAGCCATTGTC	TTGAGCCTATTC	TCAAGAAGTTC	CTTCCATAAC
AAGAGGATTC	TTAGTCGGAC	AAGGAATCGTC	CCGCATGGAAC	TTCAATTGGC	CCAGCCTCAAC
TACCAAGATC	CAGATCCATC	CTTGAGAATGTC	CTGGCAATCCTC	CCTACTGGTC	CTTGGTTATTC
CAGAAGGAAC	TCGCAATTAC	TGGAGGACGGAC	TCCACCTCCTC	TGAGGCTCCGAC	TTGGCTGGAC
CTGCAAGTTC	TTCGAGACGC	TAACAATCGGC	CAGCATTAATTC	CGAAGGCCACAC	CCGAACACTTC
TTCGTGATTC	TGCCACGAAC	CTGACATAATC	TCTGGCAACGGC	TCTGCCTGTC	TCCTGAATCTC
TTCCGATAAC	AACCTCATTC	TTCCACTTCGC	TCCTAGAACAC	CGATCGGTTC	CTAACCACGGC
TGAGCGGAAC	CCTGAGATAC	AGCACGAATC	TCCTTGATGTTC	TCAGGAATAC	CGGAAGGATGC
CTGACCGAAC	TTACAACCTC	CTTGACACCGC	TCTAGCTCTTC	CGGAAGAACCTC	CTAGGAACCGC
TCCTCGAATC	AACCATCCGC	TTGGAGGCCAGC	TCACTCGGATC	CGAAGCGATTC	CTTGTCCAATC
TAGGTGGTTC	TCGACCACTC	TGGAGCTTCCTC	TTCCTGCTTCAC	CAGCCAATTCTC	TCCGACAAGC
TCTAACGGAC	CGAGGTTATC	TCAGTCCGAAC	CCTTAGAGTTC	CCTGGTTGTC	CGGACAGATC
TTGGAGTGTC	TCCAAGCTGC	TAAGGCAACCAC	CTGAGTTCCGAC	TCGAAGGCAGGC	TTAAGCGGTC
TCTAGAGGTC	TCTTACACAC	TTCTAAGAGAC	TCCTGGCACATC	CCTGCCATTCGC	TTCGCAATGAAC
TCTGGATGAC	TTCTCATTGAAC	TCCTAACATAAC	CCGCAATCATC	TTGGCATCTC	TTCCGCACGGC

   (c) Quant-iT Picogreen dsDNA assay kit or equivalent.
   (d) 2% Agarose Gel Cassettes (BEF2010) (Sage Science, Beverley, MA, USA).
5. Required lab equipment.
   (a) Thermocycler.
   (b) Magnet for magnetic bead purification.
   (c) Blue Pippin or Pippin prep; a workflow station for DNA size selection and library preparation (Sage Science, Beverley, MA, USA).
   (d) Bioanalyzer 2100 or equivalent; a workflow station for sizing, quantitation, and purity assessments for DNA (Agilent Technologies, Santa Clara, CA, USA).
   (e) Ion Proton sequencer (Thermo Fisher Scientific, Waltham, MA, USA).
6. Recommended lab equipment.
7. Plate mixer (Mix mate, Eppendorf, Hamburg, Germany).
8. Repeater stream with advanced combitips (Eppendorf, Hamburg, Germany).
9. Ion CHEF; a workflow station for template preparation and chip loading (Thermo Fisher Scientific, Waltham, MA, USA).

3 Methods

3.1 DNA Extraction

1. High molecular weight genomic DNA is extracted from 75 mg (fresh weight) of young leaves. If used fresh, tissues need to be frozen (with liquid nitrogen) just prior to sample grinding. More conveniently, fresh leaf samples (cuttings, punches) are air-dried directly in wells/1.5 mL tubes in the presence of silica gel. Grinding is performed either with small disposable plastic pestles in 1.5 tubes or using a mixer mill for 96-well plates, in which case one tungsten bead is included in each well containing leaf tissue. DNA of the highest purity can be obtained using a commercial kit, but CTAB-based protocols can also be used successfully.
2. The DNA concentration (ng/μL) of each sample is measured with a spectrophotometer (Nanodrop 1000, Fisher Scientific) for samples devoid of RNA contamination (prepared with a commercial kit). For samples obtained with CTAB-based protocol, we may have some residual RNA. In the latter case, using a fluorometric quantification method (e.g., PicoGreen) may prove more precise (*see* **Note1**). A total of 200 ng per sample is used for the preparation of the GBS libraries.

3.2 GBS Library Preparation

This part will consist of (i) common and barcoded adapter preparation, (ii) complexity reduction using enzymes, and (iii) multiplexing using barcoded adapters. The described protocol is largely inspired from the original procedure developed in the Poland Lab [3, 7]. We have mainly optimized and improved the procedure over time. In what follows, we will describe a "standard" procedure based on 96-plex library preparation and sequencing (*see* **Note 2**).

1. Double-stranded barcoded adapter preparation (Stock BC adapter plate—0.1 μM final):
   (a) Resuspend dried single-stranded oligonucleotides to 100 μM in EB.
   (b) In a PCR plate, make 100 μL of 10 μM double-stranded barcoded adapters by mixing:
      - 10 μL of top single-stranded oligo at 100 μM.
      - 10 μL of bottom single-stranded oligo at 100 μM.
      - 10 μL of 10× AB.
      - 70 μL of H20.

   (c) Seal the plate, mix using a mixmate, then spin down.
   (d) In a thermocycler, heat to 95 °C for 1 min, then cool down to 30 °C at the rate of 1 °C per minute, then hold at 4 °C (*see* **Note 3**).
   (e) Dilute 1/10 using 1× AB (*see* **Note4**).
   (f) Repeat **step 4** once to bring barcoded adapters to 0.1 μM.
2. *Msp*I common adapter preparation (10 μM final):
   (a) Resuspend dried single-stranded oligonucleotides to 100 μM in EB.
   (b) In a PCR plate, make 100 μL of 10 μM double-stranded common adapter by mixing:
      - 10 μL of top single-stranded oligo at 100 μM.
      - 10 μL of bottom single-stranded oligo at 100 μM.
      - 10 μL of 10× AB.
      - 70 μL of H_2O.

   (c) Seal the plate, mix with mixmate, then spin down.
   (d) In a thermocycler, heat to 95 °C for 1 min and then cool at the rate of 1 °C per minute, then hold at 4 °C.
3. Make Working Adapter Plates:
   (a) Each well in the working adapter plates will have 0.02 μM of a unique barcoded adapter and 1 μM of the common adapter.
   (b) In a 96-well plate add:

- 20 μL Barcoded Adapters at 0.1 μM (from **step 1**).
- 10 μL Common Adapter at 10 μM (from **step 2**).
- 10 μL 10× AB.
- 60 μL H_2O.

(c) Mix well and spin down.

4. Normalize DNA and prepare sample plates:
   (a) Quantify sample genomic DNA (*see* **Note 1**).
   (b) Prepare sample plates so each well contains 10 μL of DNA at a 20 ng/μL concentration (i.e., 200 ng total). These plates will be used directly for further steps so ensure they are compatible with available thermocyclers.
5. Restriction Digest.

   This protocol uses a double-digest with *Pst*I and a second enzyme *Msp*I. Barcoded adapters will be ligated to the *Pst*I overhang while the common adapter will be ligated to the *Msp*I overhang.

   (a) To each well of the sample plates prepared in **step 4** add (*see* **Note 5**):
   - 3 μL CutSmart buffer (supplied with restriction enzymes).
   - 5 Units *Pst*I HiFi.
   - 5 Units *Msp*I.
   - Complete to 30 μL with H_2O.

   (b) Mix well and spin down.
   (c) Incubate in a thermocycler at 37 °C for 2 h, then hold at 8 °C (*see* **Note 6**).
   (d) Proceed immediately with adapter ligation.
6. Ligate adapters to cut genomic DNA:

   The ligation is carried out directly in the same reaction plate without the need for reaction cleanup.

   (a) To each well of the restriction digest plates prepared in **step 5** add (*see* **Note 7**):
   - 5 μL of 10× T4 DNA ligase reaction buffer (supplied with T4 DNA ligase).
   - 400 Units of T4 DNA ligase.
   - 5 μL from the corresponding well of the working adapter plate prepared in **step 3** (*see* **Note 8**).
   - Complete to 50 μL with H_2O.

(b) Mix well, spin down, and incubate at 22 °C for 2 h, then 65 °C for 20 min and hold at 8 °C when completed (*see* **Note 9**).

7. Pool and clean samples:
   (a) Pool 5 μL from 48 reaction wells into a 1.7 mL tube (columns 1 to 6).
   (b) Repeat **step 1** for the other 48 reaction wells (columns 7 to 12).
   (c) Add 1.2 mL of Qiagen PB buffer to each 1.7 mL tube.
   (d) Mix well using a vortex and spin down.
   (e) Load 750 μL on a Qiaquick column.
   (f) Spin for 15 s.
   (g) Discard flow-through.
   (h) Repeat **steps** 7(**e–g**) until the complete volume from the two tubes has been loaded to the column.
   (i) Wash column with 750 μL of PE, spin 1 min, discard flow-through.
   (j) Rotate column and spin 1 min to remove all traces of PE.
   (k) Transfer column to a new 1.7 mL tube.
   (l) Add 30 μL of EB to the center of the column, let stand for 1 min, then spin 1 min to elute the pooled library.
8. Size the library using a BluePippin:
   (a) Add 10 μL of BluePippin buffer (supplied with 2% gel cassette) to the eluted library from **step** 7.
   (b) Follow BluePippin instructions for loading on a 2% gel cassette.
   (c) We set elute times from 46 to 60 min.
   (d) You should retrieve about 50–60 μL per library that would be sufficient for multiple PCR reactions.
9. PCR amplification and Enrichment.

   Appropriate primers complementary to the ligated adapters are added and PCR is performed to amplify the pool of restriction fragments (*see* **Note 10**).

   (a) For each library prepare the amplification mix:
   - 22.9 μL of H_2O.
   - 10 μL of 5× Q5 buffer (Supplied with Q5 Polymerase).
   - 10 μL of Q5 enhancer solution (Supplied with Q5 Polymerase).
   - 1 μL of 10 mM dNTP.

- 0.3 μL of 10 μM FWD IonExpress Primer.
- 0.3 μL of 10 μM REV IonExpress Primer.
- 5 μL of DNA from **step 8**.
- 0.5 μL of Q5 polymerase.

(b) Mix well and spin down.

(c) Run the following PCR Program:

- 75 °C for 5 min.
- 5 cycles of:
  - 98 °C 10 s.
  - 55 °C 30 s.
  - 72 °C 30 s.
- 7 cycles of:
  - 98 °C 10 s.
  - 65 °C 30 s.
  - 72 °C 30 s.
- 72 °C 5 min.
- Hold at 4 °C.

(d) Add 50 μL of Axygen PCR clean up kit and mix well, transfer to a 1.5 mL tube.

(e) Let stand for 5 min at room temperature.

(f) Put on magnet for 2 min.

(g) Remove the liquid without disturbing the magnetic beads.

(h) While keeping the tube on the magnet, wash the pellet twice with 1 mL of freshly prepared 80% ethanol.

(i) Remove all traces of ethanol and let dry for 10–15 min.

(j) Remove from magnet.

(k) Resuspend dried beads in 30 μL of EB, let stand for 2 min.

(l) Put on magnet and wait for 5 min for beads to pellet.

(m) Transfer your eluted library to a new tube. Be careful not to carry over beads.

10. Quality Control:

(a) We perform a Nanodrop quantification right after purification. Expect between 5 and 20 ng/μL.

(b) The most important quality control is a Bioanalyzer trace (or equivalent). High-quality libraries will look like "Bart Simpson's hairdo," meaning relatively sharp edges with spikes on top. There should be no primer dimers located

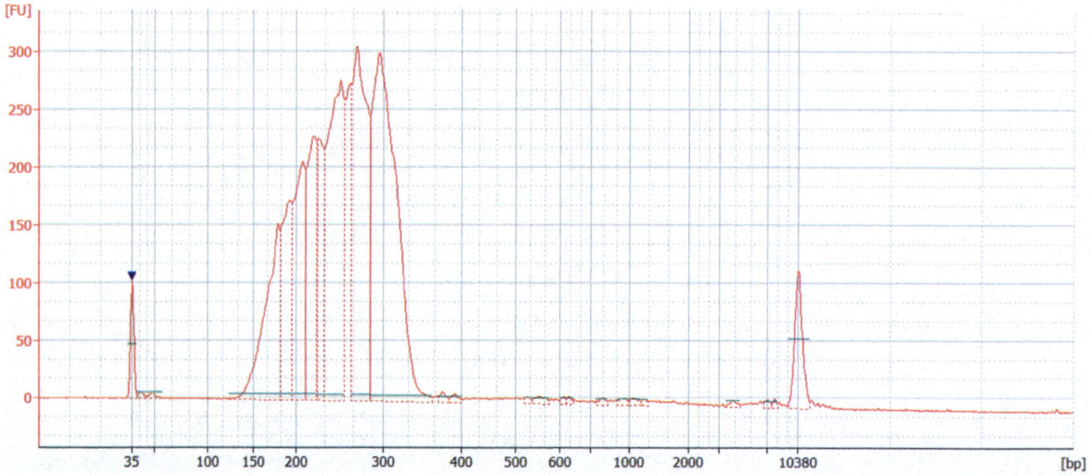

Fig. 1 An example of high-quality library

around 100–110 nt. Background after 400 bases should be flat. A large camelback hump from 500 to 2000 bases is indicative of PCR over cycling (Fig. 1).

(c) Quantify the library with Picogreen or equivalent (*see* **Note 11**). Convert concentration from ng/μL to nM (*see* **Note 12**). Dilute library to 200 pM.

3.3 Sequencing

Typically, each 96-plex library is sequenced on a single Ion PI chip yielding >70 M reads with a median length of 140–160 bp. If deeper coverage is required, the same library can be loaded on additional chips to provide a larger amount of reads per sample.

1. Load Ion CHEF and perform Ion Proton Sequencing:
   The sequencing reaction will proceed from the barcoded adapter. Follow the manufacturer's instructions to load the Ion CHEF and Ion Proton Sequencer.
   (a) Load 25 μL of a 200 pM GBS library. Our experience has shown that it generates good sequencing runs (Fig. 2).
   (b) Run the FastqCreator plugin when the sequencing run is completed to generate the fastq file.
   (c) Compress the fastq file using gzip to move the data from the Ion Server to the data analysis server.

3.4 GBS Data Analysis

This step is carried out by various bioinformatics tools (SABRE, BWA, PLATYPUS, Vcftools, etc.) included in the Fast-GBS pipeline [14]. A comparison of two pipelines UNEAK and Fast-GBS

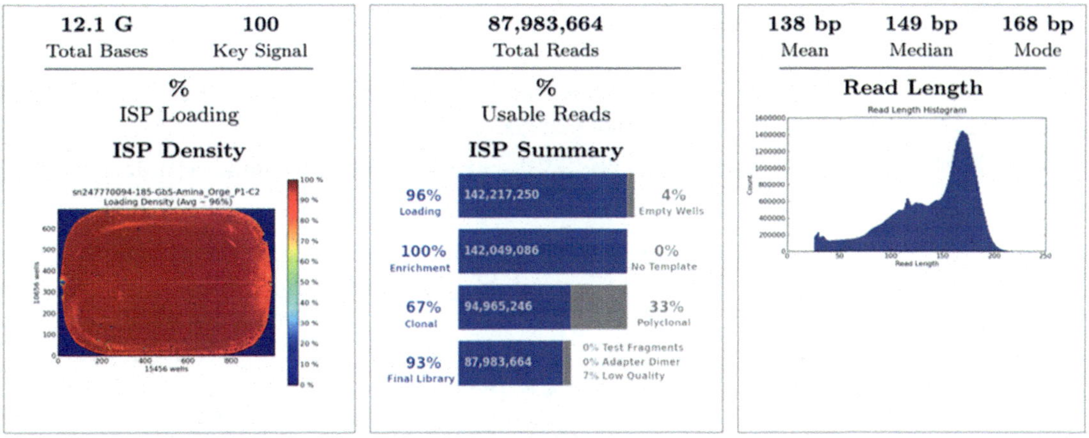

Fig. 2 A run summary for one chip Ion Torrent (Proton)

representing the two major tools (*see* **Note 13**) has shown that Fast-GBS pipeline produces more SNP data.

1. Prior to feeding the fastq files into the pipeline, check the quality (*see* **Note 14**) of the raw sequences with FastQC [15] or Galaxy [16].
2. As explained in the Fast-GBS page [14], we need first to create four directories: **refgenome** (*see* **Note 15**), **data**, **barcodes**, and **results** and put the appropriate files in the first three: (i) the reference genome with the companion index file, (ii) the raw Fastq sequences in compressed format (.gz), and (iii) the barcode sequences with the corresponding sample name.
3. Using the appropriate parameter file, the pipeline Fast-GBS is run. Default options can be used; however, one can change them depending on the nature of the data. Also, some basic filtrations are included in the pipeline by default:
   (a) Minimum read length to keep: 50 nucleotides.
   (b) Minimum size of bam file (per sample): 3000 kb.
   (c) Sequencing depth: minimum number of reads supporting a variant: 2 reads.
   (d) The maximum amount of missing data tolerated per locus: 80%.
4. The Fast-GBS variant file (.vcf file) is stored in the **results** directory already created.

3.5 Further SNP Filtering

1. Using VCFtools [17], quality filters are applied to Fast-GBS variants in order to select SNPs of superior quality. The main filters used in barley are:

(a) Variants detected only on one of the seven chromosomes (*see* **Note 16**).

(b) Variants with a quality filter "PASS" in the VCF file (Variant Call Format).

(c) Preserve SNP markers only, i.e., eliminate indels.

(d) Preserve biallelic SNPs only.

(e) The maximum amount of missing data tolerated per locus. Instead of performing this filtration as a default option in the Fast-GBS pipeline, we can also apply it at this step. The choice of the missing data threshold will vary (50–90%) depending on the population you have and how many markers you need. In a population derived from a biparental cross, linkage disequilibrium (LD) is high and we can tolerate more missing data in our variant file. This may not be the case with an association panel where LD is typically much shorter.

(f) The minor allele frequency (MAF) is the frequency of the less common allele in a population. The choice of the MAF threshold will depend on the nature of your population and eventual use of the data. To describe population structure and kinship, we may be interested in keeping even rare alleles as these may refine the relationships between lines. In such a case, we often use MAF as low as 1%. For an association panel, we more typically use minimal MAF values between 5 and 10%. In the progeny of a biparental cross, we expect alleles to segregate in a 1:1 ratio. By using a MAF in the 10–30% range, we can eliminate most "false" SNPs (e.g., sequencing errors) and markers located in regions with a severe segregation distortion.

2. To obtain a more complete SNP catalog, missing data can be imputed with BEAGLE software [18]. A second filter for the MAF can be performed following imputation as this can result in alterations of allele frequencies.

3.6 Example

To illustrate the type of results obtained with such a protocol, we performed GBS on a collection of six-row barley lines from our breeding program:

1. A collection of 250 lines was used. The cultivar used to build the reference genome cv. Morex was included in this collection to allow us to assess the accuracy of the resulting SNP calls.
2. The 250 lines were subdivided into three sets corresponding to three plates of about 80 lines. The DNA was extracted using the CTAB-based protocol from 5 mg of dried young leaves. We obtained an average DNA concentration of 113 ng/μL.

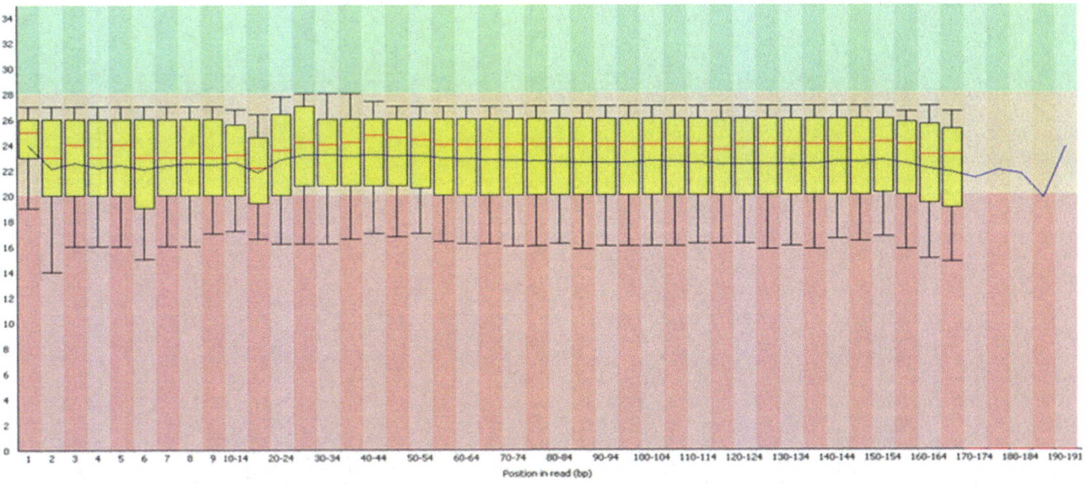

Fig. 3 Read quality for one Ion torrent chip computed with FastQC

3. Three 96-plex GBS libraries were prepared for this population (one for each plate). In order to have a good depth of coverage, each library was sequenced on two PI chips.
4. The six chips were sequenced on an Ion Torrent Proton machine by single-end sequencing. The sequencing generated about 170 million raw reads per library. The FastQC program showed that the reads were of good quality (Fig. 3) and the average sequence length per chip was between 100 and 130 bp.
5. These fastq reads were fed in the Fast-GBS pipeline using the default options.
6. About 1,180,000 “raw” variants (SNPs and indels) were detected by PLATYPUS.
7. Using VCFtools, additional filters were applied to retain only high-quality SNP markers. The thresholds were:
   (a) Less than 80% missing data per locus.
   (b) A minor allele frequency (MAF) > 1%.
8. We added another filtration step for heterozygotes. We only kept loci with less than 10% of heterozygotes. The material used in this study is advanced lines typical in a breeding program.
9. After imputation, a second filter for the MAF was performed with a threshold of 1%.
10. Fast-GBS yielded about 30,000 polymorphic and high-quality SNPs (Fig. 4). The results of GBS and SNP calling with Fast-GBS are summarized in Table 3.

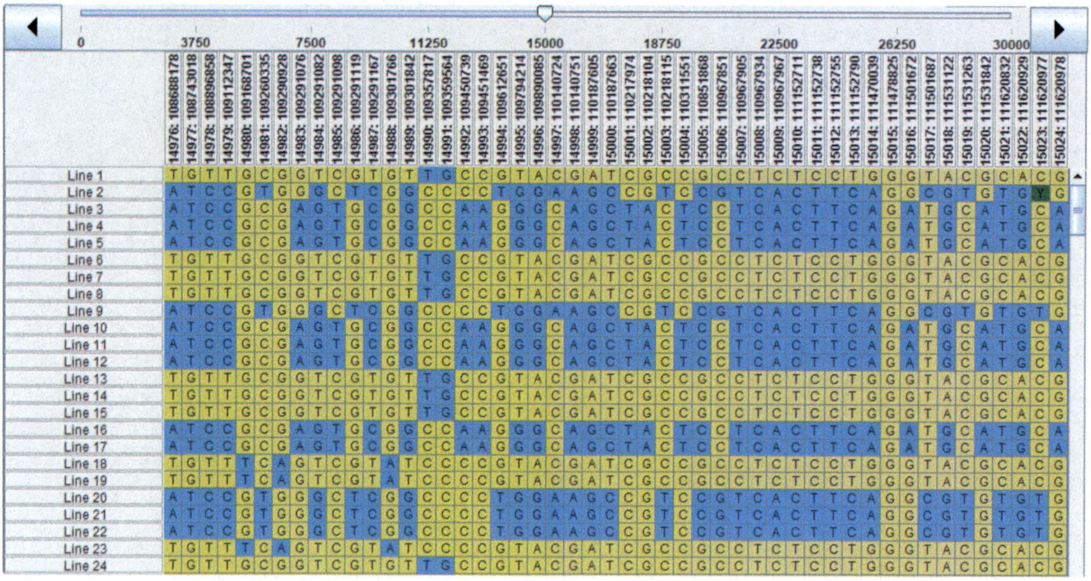

Fig. 4 A sample of 30,000 SNP catalog obtained after filtration and imputation (Tassel software), with ordered SNP loci in columns and lines in rows

Table 3
Summary of genotyping-by-sequencing and SNP calling results

Number of individual	250
Number of libraries	3
Read filtered/individual	~2 M
Fast-GBS variants	1,180,000
Filtered SNPs	30,000
Accuracy of SNP calling before imputation	97%
Accuracy of SNPs called using imputation	95%

11. Using a custom Perl script, we compared Morex genotypes from the reference genome with those called by Fast-GBS for the 30,000 SNPs before and after imputation, we found a high level of accuracy of about 97% and 80% respectively (*see* **Note 17**).
12. Using the PHENOGRAM software [19], we investigated the distribution of the 30,000 SNPs across the physical map of barley (Fig. 5).

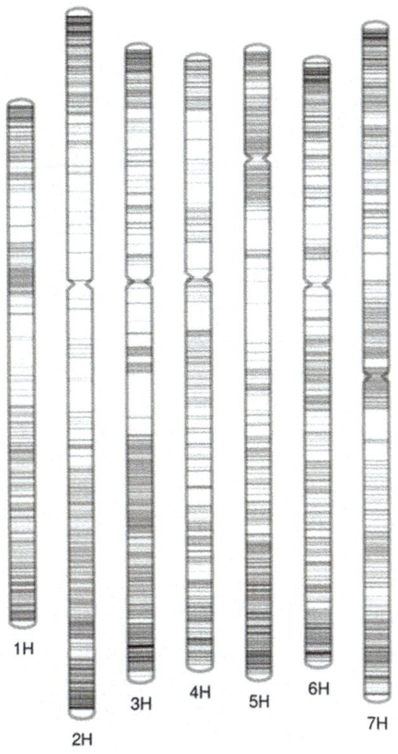

Fig. 5 Distribution of 30,000 SNP loci across the IBSC physical map. The gene-rich distal portions of the chromosomes showed the highest marker density, whereas the gene-poor pericentromeric regions were more sparsely populated

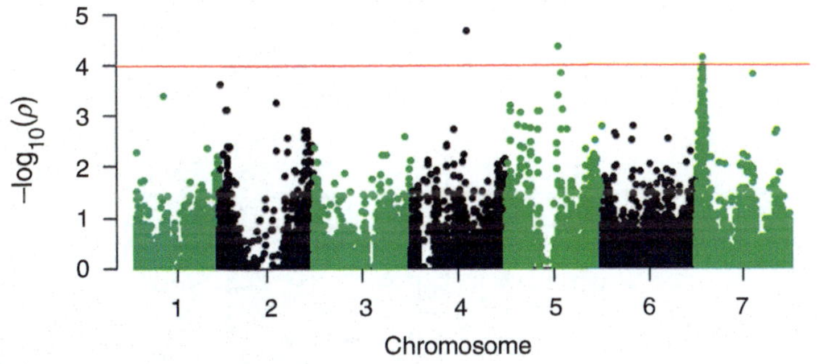

Fig. 6 Manhattan plot for GWAS of heading time. The significance threshold is $-\log10(P) = 4$. The most significant SNPs showed an association with EPS7S and VRN-H3, a major genes for this trait in barley

13. Information provided by GBS-SNP file can be used for several analysis and applications in barley such as genome-wide association analysis (Fig. 6), genetic diversity and structure (Fig. 7), and genomic selection (Fig. 8).

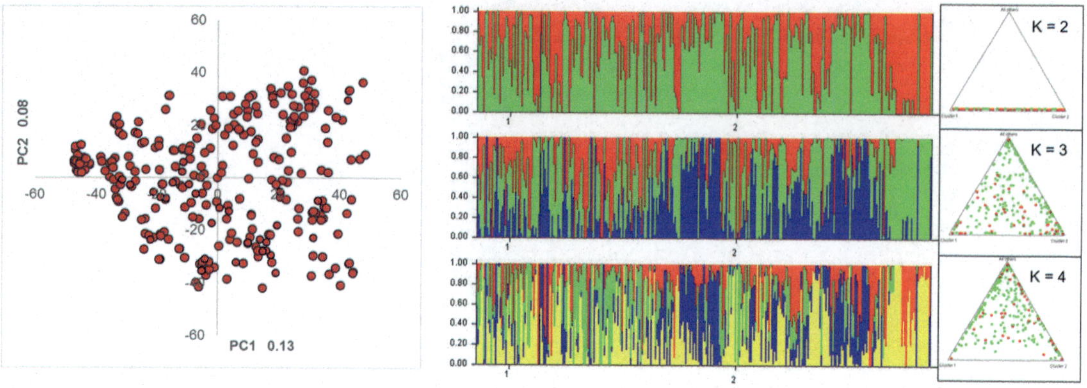

Fig. 7 Population structure analysis using two methods: (i) Principal component analysis and (ii) clustering using STRUCTURE

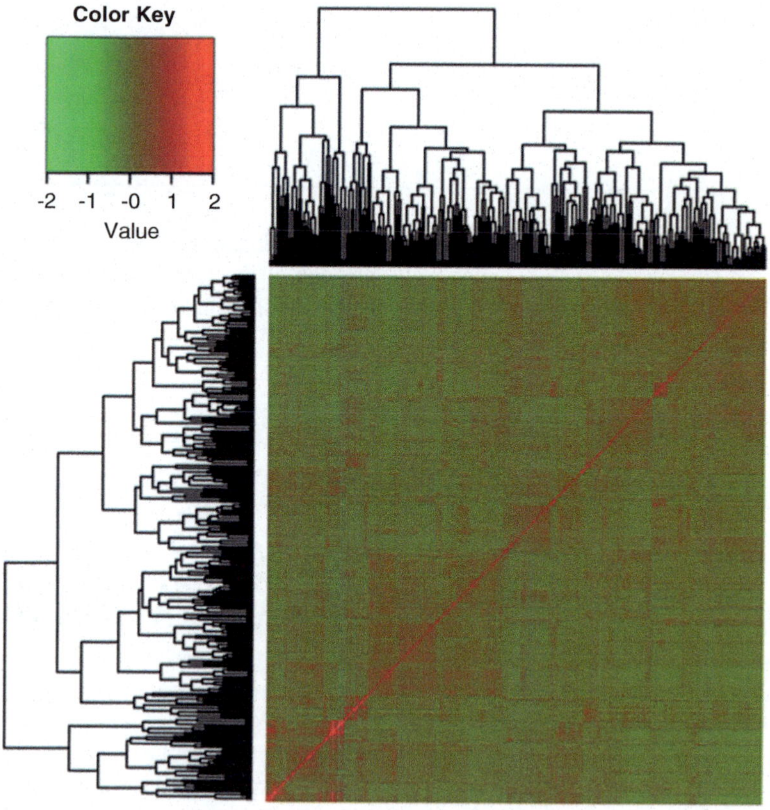

Fig. 8 Heat map representing the realized genomic relationship (G matrix) among lines. A crucial information used in statistical prediction models used in GS

4 Notes

1. DNA concentration and quality is critical to produce a stable number of sequence tags from each sample. It is recommended that DNA be quantified using a florescence-based quantification method such as PicoGreen and Qubit. DNA quality can be assessed using a spectrophotometer and the observation of the 260/230 and 260/280 ratios should be above 1.7. Do not necessarily throw away DNA that does not meet the highest standards, ensure that they are well quantified and they might just work.
2. Different levels of multiplexing can be used: 48-plex, 96-plex, 192-plex, and 384-plex. The choice will depend on the depth of coverage you want to achieve (decreases with increased multiplexing) and the budget you have (cost per sample decreases with increasing multiplexing).
3. Annealed oligonucleotides at a 10 M concentration are very stable and can be stored at −20 °C indefinitely.
4. The original procedure recommended that adapters should be quantified after annealing to ensure that the double strand DNA formation was complete and they are at the correct concentration. Uniform concentration of adapters was believed critical to producing uniform numbers of reads between samples when sequencing the multiplexed library. We have not observed significant differences between wells and no longer perform quantification at this stage. Uniformity could be linked to the choice of the oligonucleotide provider.
5. It is easier to prepare a master mix (buffer, enzymes, and H_2O), then add 20 μL of it to each well of the samples plates. Prepare at least an extra 10% of the master mix. We use an Eppendorf stream repeater with 1 mL combitips advanced to distribute the mastermix.
6. The original procedure had a 20 min at 80 °C step to heat inactivate the restriction enzymes. *Pst*I HiFi and *Msp*I cannot be heat inactivated, so this step is not required. Also note that adapters, by design, do not contain sites for restriction enzymes used and once ligated to a matching end, they are designed not to be recleaved.
7. It is easier to prepare a master mix (buffer, enzymes, and H_2O), then add 15 μL of it to each well of the samples plates. Prepare at least an extra 10% of the master mix. We use an Eppendorf stream repeater with 1 mL combitips advanced to distribute the mastermix. Remember that adapters must be added separately.

8. The original procedure called for adjusting adapter concentration depending on the species. We have used the specified concentrations of adapters with over 100 species covering a large portion of the life kingdom that includes fungi, insects, plants, and animals without a single adjustment. However, restriction enzyme combinations might not be optimal for all species and this becomes particularly true when the restriction enzymes hit highly repeated elements, in this case, changing the restriction enzyme combinations is a better choice than trying to adjust the concentration of adapters. Also note that it is essential that the common adapter is added to at least 20-fold excess compared to the barcoded adapter to cover the difference in cut frequency between *Pst*I and *Msp*I.
9. Completed ligation can be safely stored at −20 °C.
10. Only fragments that have ligated adapters to both a *Pst*I cut-site and an *Msp*I cut-site will amplify. Keep the number of PCR cycles low to avoid undetectable PCR duplication events. It is better to perform multiple PCR reactions to increase yield rather than increasing the number of PCR cycles. We routinely perform 3 or 4 PCR reactions per GBS library. It is highly recommended to physically isolate pre-PCR and post-PCR operations to prevent contamination.
11. It is important to quantify libraries using a standardized methodology as this measurement will be used to load the precise amount of molecules on the sequencing instrument. Therefore, ensure that the methodology is sensitive and falls well within the linear quantification range.
12. To convert ng/μL DNA concentration to nM: [nM DNA] = DNA concentration (ng/μL) $\times$ 10^6 (μL/L)/(Sample fragment size in bp $\times$ 656.4 (g/mole).
13. We previously compared two computational approaches to call SNPs from barley GBS data: (i) de novo (UNEAK) and (ii) genome reference (Fast-GBS) approaches, using a population of 80 six- and two-row barley lines. Fast-GBS yielded twofold more filtered SNPs and a defined position in the genome. Since the barley reference is based on genetic portion of the genome, one can combine the SNP specific to each pipeline (UNEAK and Fast-GBS) through a genotype imputation to get a larger SNP characterization.
14. The scoring system of each base is known as the Phred score. This score ranges from 0 to 64.
15. As SAMtools ceases to index sequences beyond 512 Mb, it is necessary to split chromosomes in two to avoid any loss of coverage at the end of chromosomes (beyond 512 Mb). Thus, the reference genome is loaded as a set of 14 half

chromosomes, each of which is sufficiently small to be fully indexed prior to running the Fast-GBS pipeline.

16. The version of the assembled reference genome composed of only the seven barley chromosomes [2] was used for alignment and SNP calling. The whole reference contains also contigs without a known position or with only partial information of their position.

17. We performed another validation strategy for the SNP called by the reference-based pipeline (Fast-GBS). Using a biparental population (76 doubled haploid lines), we constructed a genetic map with SNP called by Fast-GBS. The Pearson correlation coefficient of the genetic and physical positions order of the SNP was 0.98.

References

1. International Barley Genome Sequencing Consortium (IBGSC) (2012) A physical, genetic and functional sequence assembly of the barley genome. Nature 491:711–716. https://doi.org/10.1038/nature11543
2. International Barley Genome Sequencing Consortium (IBGSC) Barley genome, version 32 ftp://ftp.ensemblgenomes.org/pub/plants/release-32/fasta/hordeum_vulgare/dna/
3. Mascher M, Wu S, Amand PS et al (2013) Application of genotyping-by-sequencing on semiconductor sequencing platforms: a comparison of genetic and reference-based marker ordering in barley. PLoS One 8(10):e76925. https://doi.org/10.1371/journal.pone.0076925
4. Schulte D, Close TJ, Graner A et al (2009) The international barley sequencing consortium—at the threshold of efficient access to the barley genome. Plant Physiol 149:142–147
5. Elshire RJ, Glaubitz JC, Sun Q et al (2011) A robust, simple genotyping-by-sequencing (GBS) approach for high diversity species. PLoS One 6(5):e19379. https://doi.org/10.1371/journal.pone.0019379
6. Poland JA, Rife TW (2012) Genotyping-by-sequencing for plant breeding and genetics. Plant Genome 5(3):92–102. https://doi.org/10.3835/plantgenome2012.05.0005
7. Poland JA, Brown PJ, Sorrells ME et al (2012) Development of high-density genetic maps for barley and wheat using a novel two-enzyme genotyping-by sequencing approach. PLoS One 7(2):e32253. https://doi.org/10.1371/journal.pone.0032253
8. Lu F, Lipka AE, Glaubitz J et al (2013) Switchgrass genomic diversity, ploidy, and evolution: novel insights from a network-based SNP discovery. PLoS Genet 9(1): e1003215. https://doi.org/10.1371/journal.pgen.1003215
9. Catchen J, Hohenlohe PA, Bassham S et al (2013) Stacks: an analysis tool set for population genomics. Mol Ecol 22(11):3124–3140. https://doi.org/10.1111/mec.12354
10. Glaubitz JC, Casstevens TM, Lu F et al (2014) TASSEL-GBS: A high capacity genotyping by sequencing analysis pipeline. PLoS One 9(2): e90346. https://doi.org/10.1371/journal.pone.0090346
11. Sonah H, Bastien M, Iquira E et al (2013) An improved genotyping by sequencing (GBS) approach offering increased versatility and efficiency of SNP discovery and genotyping. PLoS One 8(1):e54603. https://doi.org/10.1371/journal.pone.0054603
12. Torkamaneh D, Laroche J, Bastien M et al (2017) Fast-GBS: a new pipeline for the efficient and highly accurate calling of SNPs from genotyping-by-sequencing data. BMC Bioinformatics 18:5. https://doi.org/10.1186/s12859-016-1431-9
13. Torkamaneh D, Laroche J, Belzile F (2016) Genome-wide SNP calling from genotyping by sequencing (GBS) data: a comparison of seven pipelines and two sequencing technologies. PLoS One 11(8):e0161333. https://doi.org/10.1371/journal.pone.0161333
14. Laroche, J. Fast-GBS main page. https://bitbucket.org/jerlar73/fastgbs

15. Andrews, S. FastQC a quality control tool for high throughput sequence data. http://www.bioinformatics.babraham.ac.uk/projects/fastqc/
16. Afgan E, Baker D, Beek M et al (2016) The Galaxy platform for accessible, reproducible and collaborative biomedical analyses. Nucleic Acids Res 44:W3–W10. https://doi.org/10.1093/nar/gkw343
17. Danecek P, Auton A, Abecasis G (2011) The variant call format and VCFtools. Bioinformatics 27(15):2156–2158. https://doi.org/10.1093/bioinformatics/btr330
18. Browning S, Browning B (2007) Rapid and accurate haplotype phasing and missing-data inference for whole genome association studies by use of localized haplotype clustering. Am J Hum Genet 81(5):1084–1097
19. Wolfe D, Dudek S, Ritchie SD et al (2013) Visualizing genomic information across chromosomes with PhenoGram. BioData Min 6:18

Chapter 16

DNA Methylation Analysis in Barley and Other Species with Large Genomes

Karolina Chwialkowska, Urszula Korotko, and Miroslaw Kwasniewski

Abstract

Detailed DNA methylation analyses in plant species with large and highly repetitive genomes can be challenging as well as costly. Here, we describe a complete protocol for a high-throughput DNA methylation changes analysis using Methylation-Sensitive Amplification Polymorphism Sequencing (MSAP-Seq; Chwialkowska et al., Front Plant Sci. 8: 2056 (2017)). This method allows detailed information about DNA methylation changes in large and complex genomes to be obtained at a relatively low cost. MSAP-Seq is based on conventional MSAP marker analysis and employs all its basic steps such as restriction cleavage with methylation-sensitive restriction enzyme, ligation of universal adapters, and PCR amplification. However, the traditional gel-based amplicon separation is replaced by direct, global sequencing with next-generation sequencing (NGS) methods. Consequently, MSAP-Seq allows for parallel analysis of hundreds of thousands of different CCGG sites and evaluation of their DNA methylation state. This technique especially targets to genic regions, so it is well suited for large genomes with low gene density, such as barley and other plants with large genomes.

Key words DNA methylation, MSAP, Next-generation sequencing, Large genomes, Barley

1 Introduction

DNA methylation is a chemical modification of DNA that results from the covalent addition of the methyl substituent to the fifth carbon of the cytosine aromatic ring. DNA methylation and demethylation pathways are important components of epigenetic machinery regulating gene expression, mobility of transposable elements, and maintaining genome integrity. They are also remarkably involved in the processes of chromatin condensation and decondensation [1]. DNA methylation in plants can occur within three sequence contexts: CG, CHG, and CHH (where H is A, C, or T); however, CG and CHG sequences are much more frequently methylated than CHH [2]. The level of DNA methylation and its pattern may vary considerably between different genomic and genic regions, as well as different plant tissues and organs [3–5]. Global

Wendy A. Harwood (ed.), *Barley: Methods and Protocols*, Methods in Molecular Biology, vol. 1900, https://doi.org/10.1007/978-1-4939-8944-7_16,

DNA methylation levels were found to be positively correlated with genome size, mostly because of transposable elements content [6, 7]. Moreover, a wide range of studies reported extensive DNA methylation changes in response to internal as well as external stimuli [8–11].

Up till now, several different techniques have been applied in the analyses of DNA methylation level, pattern, and their changes. However, most of them, such as detailed large-scale bisulfite sequencing (methylC-seq) methods, are suitable only for model species with small and simple genomes and low repetitive element content. For species with large and complex genomes, typical for many crops, the Methylation-Sensitive Amplification Polymorphism (MSAP) technique, due to its simplicity, reliability, and cost-effectiveness, is the most popular method of choice, aimed at the determination of DNA methylation changes. The MSAP technique is a modification of the Amplified Fragment Length Polymorphism method (AFLP; [12]) and involves restriction cleavage with methylation-sensitive enzymes *Hpa*II and *Msp*I, followed by adapter ligation, PCR amplification, and final polyacrylamide gel-based visualization [13, 14]. Cleavage capability of the *Hpa*II and *Msp*I enzymes is highly affected by the methylation state of the recognized 5′-CCGG-3′ sequence, which allows for a determination of the specific methylation profile for each of the bands analyzed on gel. MSAP is a universal method for DNA methylation studies involving a wide variety of species, not only model ones, regardless of their genome size, complexity, and, importantly, reference genome availability [8, 15–20]. However, MSAP can provide only general information about DNA methylation and its changes, without specific information about analyzed genomic sequences. Attempts to get specific information about the underlying DNA sequences involve laborious steps of band excisions, re-amplifications, and subcloning. These, in turn, are usually performed only on a small scale, and thus the specific genomic information gathered is highly limited.

Taking into consideration the simplicity and usefulness of MSAP, we designed and introduced the Methylation-Sensitive Amplification Polymorphism Sequencing (MSAP-Seq) technique, that is a high-throughput but low-cost method for direct identification of specific genomic sequences, which are undergoing DNA methylation changes in large genomes. This technique is based on the typical MSAP analysis, nonetheless it was significantly upgraded by replacement of the traditional gel-based MSAP amplicon separation with direct, high-throughput sequencing using Next-Generation Sequencing (NGS) methods and automated data analysis, allowing for parallel identification of a large set of sequences that are undergoing DNA methylation changes at relatively low cost (Table 1).

Table 1
MSAP-Seq methodology: advantages and disadvantages

Advantages	Disadvantages
Straightforward and reliable—developed on the basis of the generally accepted MSAP method, adapted for NGS technologies	Restriction enzyme-based—only changes within recognized CCGG sequences can be identified, therefore not precisely targeted to single nucleotide scale
Universal—protocol can be directly applied to any species	Most informative when applied to species with known genome sequences
High-throughput—parallel analyses of hundreds of thousands of different sites	
Targeting gene-rich regions	
Allowing quantitative evaluation of methylation changes	
Supplemented by user-friendly pipeline for bioinformatic analyses	
Suitable even for large and complex genomes with low gene content	
Low cost	

Similar to conventional MSAP, the first step in MSAP-Seq workflow is restriction cleavage with two restriction enzymes: methylation-sensitive *Hpa*II, recognizing CCGG sequence, and methylation-insensitive *Eco*RI, recognizing GAATTC sites (Fig. 1). Then, restriction site-specific adaptors are ligated and the obtained fragments are amplified with partially selective primers, allowing for the adjustment of the number of sequences that are generated. Amplicons are subsequently purified and fragmented by sonication to create short tags that can be easily sequenced using high-throughput sequencing technologies. Purified sequence tags are used for library preparation. As the MSAP-Seq method was optimized for popular Illumina NGS platforms, MSAP-Seq libraries are generated including typical steps: end repair of DNA fragments, adenylation, ligation of barcoded adapter followed by purification, PCR amplification and a final purification. At the end, the MSAP-Seq libraries are evaluated regarding their size distribution and concentration. Sequencing-ready MSAP-Seq libraries are finally subjected to cluster generation and sequencing using Illumina platform of choice. Finally, sequencing results are subjected to analysis with dedicated MSEQER software, developed for automated MSAP-Seq data analysis (Korotko et al. 2017—unpublished). During the data analysis, preprocessed reads are mapped to the appropriate reference genome, allowing for the identification of specific genomic sequences present in the samples. Deep

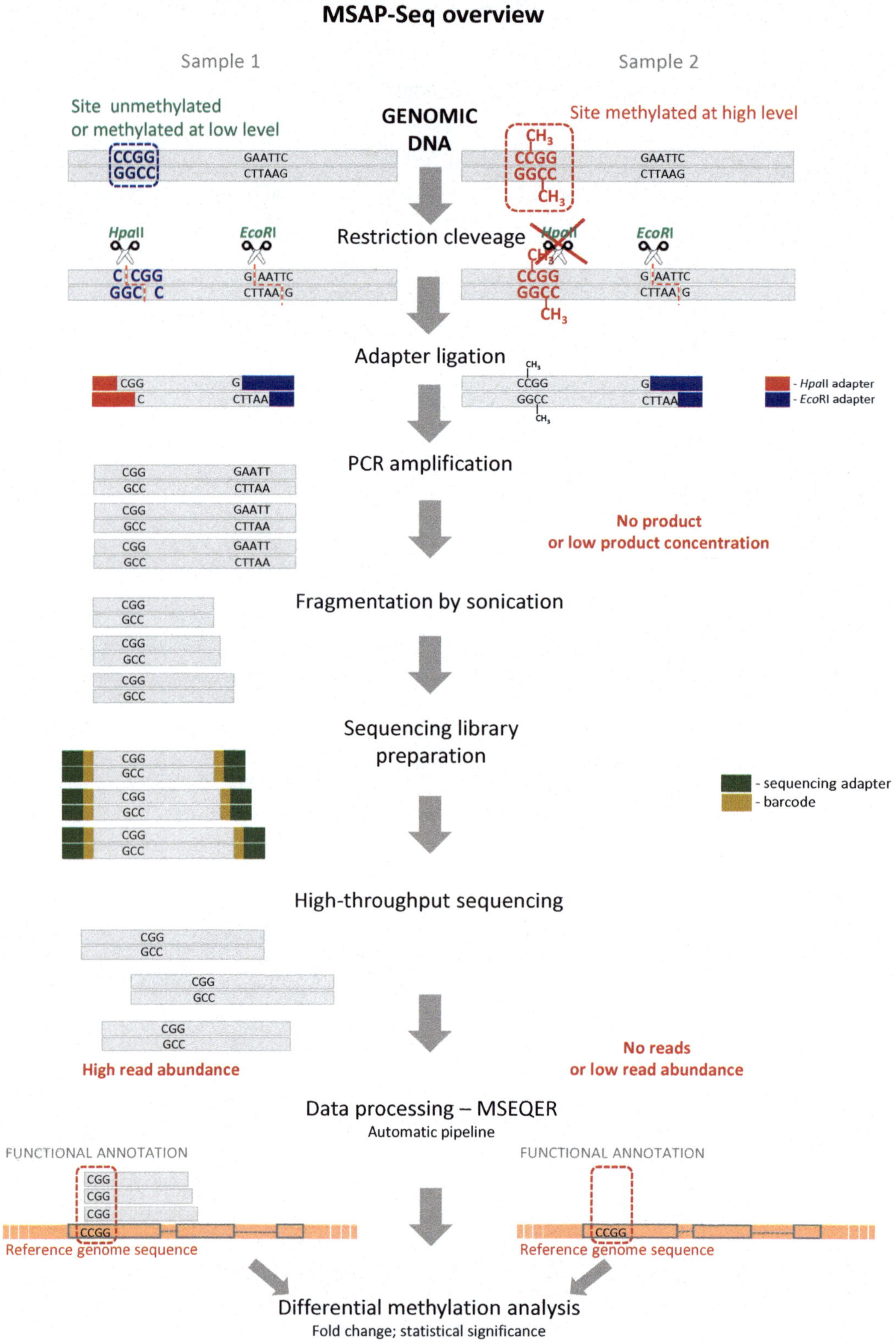

Fig. 1 An overview of the MSAP-Seq assay (Adapted from Chwialkowska et al., 2017 under Frontiers copyright rules)

sequencing of MSAP-Seq amplicons allows for quantitative characterization of observed DNA methylation changes by the evaluation of fold change values of the abundance of the normalized reads in analyzed samples.

We have successfully applied MSAP-Seq methodology in numerous analyses of DNA methylation pattern in *Hordeum vulgare*, a crop with large and complex genome of 5.3 Gbp. These studies involved the comparative analysis of DNA methylation changes in barley leaf and root in response to water-deficiency stress [4, 21] as well as the comprehensive evaluation of DNA methylation dynamics under drought stress conditions. Following the presented protocol, 14–28 million reads were obtained on average per one sequenced sample [4, 21]. As much as 85% of processed and filtered reads were mapped to the barley reference genome. This resulted in about 250,000 different CCGG sites in the barley genome which were included in the analyses of differential methylation. Importantly, 75% of all barley genes were represented in the analysis by at least one site. Moreover, we have found that more than one-fifth of all sequences represent genes. Therefore, as only around 2% of barley genome contains genic sequences, it proves that the MSAP-Seq method is especially enriched for gene-containing regions. Statistical analyses revealed that several thousands of different CCGG sites undergo DNA methylation changes under water-deficiency conditions in barley. All in all, the MSAP-Seq method is extremely high-throughput in comparison to the low capacity of traditional MSAP and, additionally, cheaper than other methods involving methylome analyses coupled with NGS. MSAP-Seq is a very reasonable choice for DNA methylation studies in large genome species with low gene content. MSAP-Seq methodology can be easily applied to a wide variety of methylome-based studies, such as the popular evaluation of the influence of different environmental factors on DNA methylation changes, or methylome diversity analyses among different genotypes or natural populations.

2 Materials

2.1 Components for MSAP Amplicons Generation

2.1.1 Reagents

1. Sterile ddH_2O.
2. *Hpa*II enzyme (10 U/μL).
3. *Eco*RI enzyme (20 U/μL).
4. 10× buffer dedicated for both restriction enzymes (*see* **Note 1**).
5. Mineral oil, molecular biology grade.
6. T4 DNA ligase (1 U/μL) with dedicated 10x buffer (*see* **Note 1**).
7. 5 mM dNTPs (*see* **Note 2**).
8. Taq DNA Polymerase (2 U/μL) with dedicated 10× buffer (*see* **Note 1**).

9. Agarose, molecular biology grade.
10. 0.5× TBE buffer (44.5 mM Tris, 44.5 mM boric acid, 1 mM EDTA, pH 8.3).
11. 6× loading buffer (40% sucrose, 0.25% bromophenol blue and xylene cyanol).
12. GeneRuler 100 bp DNA Ladder.
13. Ethidium bromide solution for agarose gel staining (10 mg/mL) (*see* **Note 3**).
14. Agencourt® AMPure® XP magnetic beads (Beckman Coulter, catalog number A63880) (*see* **Note 4**).
15. 80% ethanol (*see* **Note 5**).
16. Sterile 1× TE buffer (10 mM Tris–HCl, 1 mM EDTA, pH 8.0).
17. 5 pmol/μL of double-stranded *Eco*RI adapter combined from two oligos: 5′- CTCGTAGACTGCGTACC-3′ and 5'-AATT GGTACGCAGTCTAC-3′ (*see* **Note 6**).
18. 50 pmoL/μL of double-stranded *Hpa*II adapter combined from two oligos: 5'-GACGATGAGTCTAGAA-3′ and 5'-CG TTCTAGACTCATC-3′ (*see* **Note 6**).
19. 50 ng/μL primer *Eco*RI-AC: 5′- GACTGCGTACCAATTCAC -3′ (*see* **Note 7**).
20. 50 ng/μL primer *Hpa*II-TG: 5′- GATGAGTCTAGAA CGGTG-3′ (*see* **Note 7**).

2.1.2 Equipment

1. Microcentrifuge.
2. Vortex.
3. Thermomixer.
4. Thermocycler.
5. Agarose gel electrophoresis system.
6. Magnetic rack.
7. Sonication device.
8. Microvolume spectrophotometer.

2.2 Components for MSAP-Seq Library Preparation

2.2.1 Reagents

1. NEXTflex™ Rapid DNA-Seq Kit Bundle with DNA Barcodes 1–24 (BIOO Scientific, catalog number 5144-03).
2. High-Sensitivity DNA kit (Agilent, catalog number 5067-4626) or High-Sensitivity D1000 ScreenTape (Agilent, catalog number 5067-5584).
3. Agencourt® AMPure® XP magnetic beads (Beckman Coulter, catalog number A63881) (*see* **Note 3**).
4. 80% ethanol (*see* **Note 4**).

2.2.2 Equipment

1. Microcentrifuge.
2. Vortex.
3. Thermocycler.
4. Magnetic rack.
5. Agilent 2100 Bioanalyzer or Agilent TapeStation system.

3 Methods

Adapted from Chwialkowska et al., 2017 under Frontiers copyright rules.

3.1 MSAP Amplicons Generation

Restriction Enzymes Digestion

1. Dilute each of the genomic DNA samples to a concentration of 50 ng/μL and use 10 μL of this dilution for restriction digestion (*see* **Note 8**).
2. Prepare double digestion reactions in a PCR tube combining the following reagents on ice (volumes per one sample).

7.625 μL	ddH_2O
10 μL	50 ng/μL DNA sample
2 μL	10× buffer dedicated for restriction enzymes
0.25 μL	*Hpa*II enzyme (10 U/μL)
0.125 μL	*Eco*RI enzyme (20 U/μL)
20 μL	TOTAL

3. Incubate at 37 °C for 6 h in a thermocycler or water bath; remember to add a drop of mineral oil if you do not use a thermocycler with heated lid.
4. Heat inactivate enzymes by incubation at 80 °C for 20 min in a thermocycler with heated lid.

Adapter Ligation

5. Combine in a separate tube the following reagents on ice to prepare ligation mix (volumes per one sample).

7.2 μL	ddH_2O
1.2 μL	50 pmol/μL adapter *Hpa*II
1.2 μL	5 pmol/μL adapter *Eco*RI
1.2 μL	10× buffer for T4 DNA ligase
1.2 μL	T4 DNA ligase (1 U/μL)
12 μL	TOTAL

6. Add 12 μL of the ligation mix directly to the sample after restriction, mix, and quick-spin (if you applied mineral oil remember to add it to the solution under the oil).
7. Incubate at 37 °C for 16 h in a thermocycler or water bath.

PCR Amplification

8. Prepare PCR mix on ice in a separate tube (volumes per one sample).

32.5 μL	ddH_2O
5 μL	Sample after the adapter ligation step
1.5 μL	50 ng/μL primer *Eco*RI-AC
1.5 μL	50 ng/μL primer *Hpa*II-TG
2 μL	5 mM dNTPs
5 μL	10× buffer for Taq DNA Polymerase
2.5 μL	Taq DNA Polymerase (2 U/μL)
50 μL	TOTAL

9. Perform PCR under the following program:

94 °C – 30 s 56 °C – 40 s 72 °C – 50 s	×30 cycles
4 °C – pause	

10. Optionally: examine 10 μL of amplification products on a 1.5% agarose gel (expect smear of bands in a range of 200–800 bp).

Post-PCR Purification with Magnetic Beads

11. Resuspend Agencourt® AMPure® XP magnetic beads (Beckman Coulter) and aliquot 72 μL per sample in new 1.5 mL tubes.
12. Add 40 μL of amplification products to the 72 μL of magnetic beads.
13. Mix thoroughly by pipetting for 10 times and incubate for 5 min at room temperature.
14. Place onto a magnetic stand for 2 min or until the beads settle and supernatant is clear.
15. Aspirate the cleared solution from the tube and discard. Take care not to aspirate any beads.
16. With a tube still placed on a magnetic stand add 200 μL of 80% ethanol to each sample and incubate for 30 s at room

temperature (do not disturb the beads). Carefully remove ethanol by pipette and discard.

17. Repeat previous step for a total of two washes. Ensure all ethanol has been removed.
18. Leave the tube open and let dry at room temperature for 5 min or until bead pellet is visibly dry (it will become matte). Do not overdry the beads (they cannot crack).
19. Remove the tube from the magnetic stand and resuspend dried beads with 53 μL of 1× TE buffer.
20. Mix thoroughly by pipetting for 10 times and incubate for 5 min at room temperature.
21. Place the sample onto a magnetic stand for 2 min or until the beads settle and supernatant is clear.
22. Transfer 50 μL of sample into a new tube.
23. Cool samples in a fridge and keep on ice until fragmentation step. You can also freeze them at −20 °C and thaw on ice before sonication.

Fragmentation by Sonication

24. Perform fragmentation by sonication using 50 μL of MSAP sample under the following conditions if using Bioruptor® Plus (Diagenode):

LOW power		×10 cycles
30 s	ON	
30 s	OFF	

Post-fragmentation Purification with Magnetic Beads

25. Resuspend Agencourt® AMPure® XP magnetic beads (Beckman Coulter) and aliquot 90 μL per sample in new 1.5 mL tubes.
26. Add 50 μL of sample after amplification to the 90 μL of magnetic beads.
27. Mix thoroughly by pipetting for 10 times and incubate for 5 min at room temperature.
28. Place the mixture onto a magnetic stand for 2 min or until the beads settle and supernatant is clear.
29. Aspirate the cleared solution from the tube and discard.
30. With a tube on a magnetic stand add 200 μL of 80% ethanol to each sample and incubate for 30 s at room temperature (do not disturb the beads). Carefully remove ethanol by pipette and discard.
31. Repeat previous step for a total of two washes. Ensure all ethanol has been removed.

32. Leave the tube open and let dry at room temperature for 5 min or until bead pellet is visibly dry (it will become matte). Do not overdry the beads (they cannot crack).
33. Remove the tube from the magnetic stand and resuspend dried beads with 37 μL of ddH_2O.
34. Mix thoroughly by pipetting for 10 times and incubate for 5 min at room temperature.
35. Place the sample onto a magnetic stand for 2 min or until the beads settle and supernatant is clear.
36. Transfer 35 μL of sample into a new tube.
37. Evaluate the amplicon concentration using, for example, NanoDrop (Thermo Fischer Scientific). Ideally you should have around 1 μg or more of DNA fragments in 32 μL volume (maximum volume to be processed during library preparation).

3.2 MSAP-Seq Library Preparation

Protocol prepared with the application of the NEXTflex™ Rapid DNA-Seq Kit (BIOO Scientific) compatible with Illumina sequencing systems.

End Repair and Adenylation

1. Add the following components on ice in a PCR tube (volumes per one sample).

32 μL	Purified MSAP sample suspended in H_2O
15 μL	NEXTflex™ End-Repair and Adenylation Buffer Mix
3 μL	NEXTflex™ End-Repair and Adenylation Enzyme Mix
50 μL	TOTAL

2. Incubate on a thermocycler using the following program:

   22 °C – 20 min

   72 °C – 20 min

   4 °C – pause

Adapter Ligation

3. Thaw NEXTflex™ Ligase Enzyme Mix to room temperature, then vortex for 5–10 s. Do not spin down the tube, as this may cause components of the mix to separate and affect performance.
4. Prepare ligation mix on ice by adding the following reagents (volumes per one sample).

   At this step you can barcode each of the samples separately; up to 24 samples can be run on one Illumina sequencing lane, so you can use 24 different barcodes.

50 μL	DNA sample after end repair and adenylation
47.5 μL	NEXTflex™ Ligase Enzyme Mix
2.5 μL	appropriate NEXTflex™ Barcode (different for each of the samples)
100 μL	TOTAL

5. Thoroughly mix the reaction by pipetting up and down 15 times and visually inspect tubes to ensure proper homogenization as the ligase enzyme mix is very viscous.
6. Incubate at 22 °C for 15 min in a thermocycler.
7. Resuspend Agencourt® AMPure® XP magnetic beads (Beckman Coulter) and aliquot 60 μL per sample in new 1.5 mL tubes.
8. 100 μL of sample after adapter ligation add to the magnetic beads.
9. Mix thoroughly by pipetting for 10 times and incubate for 5 min at room temperature.
10. Place the tube on the magnetic stand at room temperature for 5 min or until the beads settle and supernatant is clear.
11. Aspirate the cleared solution from the tube and discard.
12. With a tube on a magnetic stand add 200 μL of 80% ethanol to each sample and incubate for 30 s at room temperature (do not disturb the beads). Carefully remove ethanol by pipette and discard.
13. Repeat previous step for a total of two washes. Ensure all ethanol has been removed.
14. Leave the tube open and let dry at room temperature for 5 min or until bead pellet is visibly dry. Do not overdry the beads (they cannot crack).
15. Remove the tube from the magnetic stand and resuspend dried beads with 52 μL of Resuspension Buffer. Mix thoroughly by pipetting until homogenized.
16. Incubate for 5 min at room temperature.
17. Place the sample onto a magnetic stand for 5 min or until the beads settle and supernatant is clear.
18. Do not discard the sample in this step. Transfer 50 μL of clear sample to a new tube.
19. Add 40 μL of AMPure XP Beads to each sample. Mix thoroughly until homogenized.
20. Incubate sample at room temperature for 5 min or until the beads settle and supernatant is clear.
21. Aspirate the cleared solution from the tube and discard.

22. With a tube on a magnetic stand add 200 μL of 80% ethanol to each sample and incubate for 30 s at room temperature (do not disturb the beads). Carefully remove ethanol by pipette and discard.
23. Repeat previous step for a total of two washes. Ensure all ethanol has been removed.
24. Leave the tube open and let dry at room temperature for 5 min or until bead pellet is visibly dry. Do not overdry the beads (they cannot crack).
25. Remove the tube from the magnetic stand and resuspend dried beads with 22 μL of Resuspension Buffer. Mix thoroughly until homogenized.
26. Incubate resuspended beads at room temperature for 5 min.
27. Place the tube on the magnetic stand at room temperature for 5 min.
28. Transfer 20 μL of clear sample to a new tube.
29. You can pause at this step with samples stored at −20 °C. To restart thaw frozen samples on ice.

PCR Amplification

30. Combine the following reagents on ice in a PCR tube (volumes per one sample).

5 μL	Purified DNA sample after adapter ligation
31 μL	Nuclease-free water
12 μL	NEXTflex™ PCR Master Mix
2 μL	NEXTflex™ Primer Mix
50 μL	TOTAL

31. Thoroughly mix the reaction by pipetting and place in thermocycler under the following PCR program:

98 °C – 2 min	
98 °C – 30 s 65 °C – 30 s 72 °C – 60 s	×6 cycles
4 °C – pause	

32. Resuspend Agencourt® AMPure® XP magnetic beads (Beckman Coulter) and aliquot 40 μL per sample in new 1.5 mL tubes.
33. Add 50 μL of sample after adapter ligation to the magnetic beads.

34. Mix thoroughly by pipetting for 10 times and incubate for 5 min at room temperature.
35. Place the tube on the magnetic stand at room temperature for 5 min or until the beads settle and supernatant is clear.
36. Aspirate the cleared solution from the tube and discard taking care not to disturb beads. Some liquid may remain in wells.
37. With a tube on a magnetic stand add 200 μL of 80% ethanol to each sample and incubate for 30 s at room temperature (do not disturb the beads). Carefully remove ethanol by pipette and discard.
38. Repeat previous step for a total of two washes. Ensure all ethanol has been removed.
39. Remove the tube from the magnetic stand and let dry at room temperature for 5 min or until bead pellet is visibly dry. Do not overdry the beads (they cannot crack).
40. Resuspend dried beads with 21 μL of Resuspension Buffer. Mix thoroughly until homogenized.
41. Incubate resuspended beads at room temperature for 5 min.
42. Place the tube on the magnetic stand at room temperature for 5 min.
43. Transfer 20 μL of clear sample to a new well.
44. Examine your library distribution by gel electrophoresis or Agilent Bioanalyzer and High-Sensitivity DNA Chips (Fig. 2).
45. Quantify DNA library templates with Qubit (Thermo Fisher Scientific) or using qPCR, which is recommended for optimal cluster density. This can be performed using any qPCR quantification kit with the NEXTflex™ Primer Mix.
46. The library is now ready for cluster generation with the standard Illumina protocol. Proceed to cluster generation or store at −20 °C.

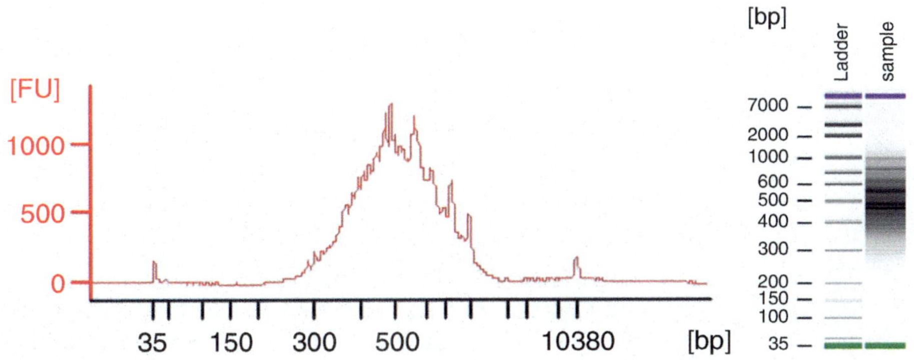

Fig. 2 MSAP-Seq library validation with Agilent Bioanalyzer and the High-Sensitivity DNA kit (Agilent)

47. For cluster generation, the libraries should be diluted to the 15 pM and then sequenced in the Illumina system (e.g., Illumina HiSeq 1500) in a concentration of one sample per 1/24 lane (up to 24 samples at one lane).

Data Analysis

48. After sequencing is complete proceed to data analysis with dedicated MSEQER software available at http://mseqer.us.edu.pl/.
49. Upload your files and continue to MSAP-Seq analysis:
    (a) Select appropriate reference genome.
    (b) If you were using paired-end reads—match pairs of files.
    (c) Mark biological replicates.
    (d) Provide sequences of adapters for trimming. If you are using adapters provided in this protocol, just paste to the left box: GATGAGTCTAGAA (Adapter *Hpa*II) and to the right box: GACTGCGTACCAATTCAC (Adapter *Eco*RI).
    (e) Run the analysis.

4 Notes

1. Ensure all the components are thawed before use. Mix them thoroughly (buffers should be vortexed) and short spin.
2. Deoxyribonucleotides tend to decompose under freezing-thawing cycles. Prepare fresh dNTPs dilutions and thaw a new one just before use.
3. Use ethidium bromide in agarose gel in final concentration of 0.5 μg/mL.
4. Allow Agencourt AMPure XP Beads to come to room temperature before use—30 min out of fridge is enough.
5. 80% ethanol should be freshly prepared just before use as ethanol is hygroscopic and evaporates quickly what lowers its concentration in the solution.
6. To prepare adapters for the first time combine both oligos in one tube, denature them by heating in thermomixer at 65 °C for 15 min, and then allow to renature by slowly cooling—simply by placing the tubes on a lab bench at RT. After cooling the adapters can be stored at −20 °C.
7. The number and composition of selective nucleotides can be modified and optimized with regard to the number of fragments that will be generated. Here there are used two selective

nucleotides: AC for *Eco*RI-related primer and TG for *Hpa*II-related.

8. Use high-quality genomic DNA. Check its integrity by 0.8% agarose gel electrophoresis.

Acknowledgments

This work was supported by the European Union within The 7th Framework Programme under project no. 289300 "EURoot: Enhancing resource Uptake from Roots under stress in cereal crops," the Ministry of Science and Higher Education grant 2486/7.PR/2012/2, and by the Polish National Science Centre (NCN) grant no. 2014/13/N/NZ2/01153. KC and UK were granted a scholarship co-funded by the European Union within the framework of the European Social Fund "DoktoRIS Scholarship Program for Innovative Silesia." KC was a beneficiary of the ETIUDA scholarship funded by the Polish National Science Centre (NCN) under grant no. 2016/20/T/NZ2/00577.

References

1. Richards E, Elgin S (2002) Epigenetic codes for heterochromatin formation and silencing: rounding up the usual suspects. Cell 108:489–500
2. Cokus S, Feng S, Zhang X, Chen Z, Merriman B, Haudenschild C, Pradhan S, Nelson S, Pellegrini M, Jacobsen S (2008) Shotgun bisulphite sequencing of the Arabidopsis genome reveals DNA methylation patterning. Nature 452:215–219
3. Braszewska-Zalewska A, Wolny E, Smialek L, Hasterok R (2013) Tissue-specific epigenetic modifications in root apical meristem cells of *Hordeum vulgare*. PLoS One 8:e69204
4. Chwialkowska K, Nowakowska U, Mroziewicz A, Szarejko I, Kwasniewski M (2016) Water-deficiency conditions differently modulate the methylome of roots and leaves in barley (*Hordeum vulgare* L.). J Exp Bot 67:1109–1121
5. Roessler K, Takuno S, Gaut B (2016) CG methylation covaries with differential gene expression between leaf and floral bud tissues of *Brachypodium distachyon*. PLoS One 11: e0150002
6. Ausin I, Feng S, Yu C, Liu W, Yu Kuo H et al (2016) DNA methylome of the 20-gigabase Norway spruce genome. Proc Natl Acad Sci U S A 113:E8106–E8113
7. Takuno S, Ran J-H, Gaut B (2016) Evolutionary patterns of genic DNA methylation vary across land plants. Nat Plants 2:15222
8. Tan M-P (2010) Analysis of DNA methylation of maize in response to osmotic and salt stress based on methylation-sensitive amplified polymorphism. Plant Physiol Biochem 48:21–26
9. Boyko A, Blevins T, Yao Y, Golubov A, Bilichak A et al (2010) Transgenerational adaptation of arabidopsis to stress requires DNA methylation and the function of dicer-like proteins. PLoS One 5:e9514
10. Wang W, Pan Y, Zhao X, Dwivedi D, Zhu L, Ali J, Fu B, Li Z (2011) Drought-induced site-specific DNA methylation and its association with drought tolerance in rice (*Oryza sativa* L.). J Exp Bot 62:1951–1960
11. Meyer P (2015) Epigenetic variation and environmental change. J Exp Bot 66:3541–3548
12. Vos P, Hogers R, Bleeker M, Reijans M, van de Lee T, Hornes M, Frijters A, Pot J, Peleman J, Kuiper M (1995) AFLP: a new technique for DNA fingerprinting. Nucleic Acids Res 23:4407–4414
13. Reyna-Lopez G, Simpson J, Ruiz-Herrera J (1997) Differences in DNA methylation patterns are detectable during the dimorphic transition of fungi by amplification of restriction polymorphisms. Mol Gen Genet 253:703–710

14. Xiong Z, Xu C, Saghai Maroof M, Zhang Q (1999) Patterns of cytosine methylation in an elite rice hybrid and its parental lines, detected by a methylation-sensitive amplification polymorphism technique. Mol Gen Genet 261:439–446
15. Filek M, Janiak A, Szarejko I, Grabczynska J, Machacková I, Krekule J (2006) Does DNA methylation pattern mark generative development in winter rape? Z Naturforsch C 61:387–396
16. Guzy-Wrobelska J, Kaliciak A, Szarejko I, Machackova I, Krekule J, Barciszewska M (2013) Vernalization and photoperiod-related changes in the DNA methylation state in winter and spring rapeseed. Acta Physiol Plant 35:817–827
17. Marconi G, Pace R, Traini A, Raggi L, Lutts S, Chiusano M, Guiducci M, Falcinelli M, Benincasa P, Albertini E (2013) Use of MSAP markers to analyse the effects of salt stress on DNA methylation in rapeseed (*Brassica napus* var. oleifera). PLoS One 8:e75597
18. Gautam M, Dang Y, Ge X, Shao Y, Li Z (2016) Genetic and epigenetic changes in oilseed rape (*Brassica napus* L.) extracted from intergeneric allopolyploid and additions with Orychophragmus. Front Plant Sci 7:438
19. Gimenez M, Yañez-Santos A, Paz R, Quiroga M, Marfil C, Conci V, García-Lampasona S (2016) Assessment of genetic and epigenetic changes in virus-free garlic (*Allium sativum* L.) plants obtained by meristem culture followed by in vitro propagation. Plant Cell Rep 35:129–141
20. Wang B, Zhang M, Fu R, Qian X, Rong P, Zhang Y et al (2016) Epigenetic mechanisms of salt tolerance and heterosis in Upland cotton (*Gossypium hirsutum* L.) revealed by methylation-sensitive amplified polymorphism analysis. Euphytica 208:477–491
21. Chwialkowska K, Korotko U, Kosinska J, Szarejko I, Kwasniewski M (2017) Methylation sensitive amplification polymorphism sequencing (MSAP-Seq)-a method for high-throughput analysis of differentially methylated CCGG sites in plants with large genomes. Front Plant Sci 8:2056

Chapter 17

High-Resolution RT-PCR Analysis of Alternative Barley Transcripts

Craig G. Simpson, John Fuller, Paulo Rapazote-Flores, Claus-Dieter Mayer, Cristiane P. G. Calixto, Linda Milne, Pete E. Hedley, Clare Booth, Robbie Waugh, and John W. S. Brown

Abstract

Assembly of the barley genome and extensive use of RNA-seq has resulted in an abundance of gene expression data and the recognition of wide-scale production of alternatively spliced transcripts. Here, we describe in detail a high-resolution reverse transcription-PCR based panel (HR RT-PCR) that confirms the accuracy of alternatively spliced transcripts from RNA-seq and allows quantification of changes in the proportion of splice isoforms between different experimental conditions, time points, tissues, genotypes, ecotypes, and treatments. By validating a selection of barley genes, use of the panel gives confidence or otherwise to the genome-wide global changes in alternatively spliced transcripts reported by RNA-seq. This simple assay can readily be applied to perform detailed transcript isoform analysis for any gene in any species.

Key words HR RT-PCR, Alternative splicing, RNA-seq

1 Introduction

Improving barley performance in the field depends on understanding how traits develop and the underlying regulation of expressed genes. The formation of alternative transcripts by the process of alternative splicing (AS) has a strategic role in the multiple layers of transcriptional, mRNA stability and coordinated posttranscriptional networks that function to deliver gene products at the right time and place [1–4]. AS increases the protein coding capacity of eukaryote genomes and leads to the regulation and fine tuning of gene expression [2, 5, 6]. During AS, different splice site choice results in different messenger RNA (mRNA) isoforms: the size of exons can vary due to the use of alternative 5'or 3'splice sites; exons can be alternatively included or excluded, termed exon skipping; and introns can remain in the mRNA, designated intron retention. The selection of alternative splice sites is dictated by *cis*-acting

Wendy A. Harwood (ed.), *Barley: Methods and Protocols*, Methods in Molecular Biology, vol. 1900,
https://doi.org/10.1007/978-1-4939-8944-7_17,

motifs located in the precursor mRNAs (pre-mRNAs) that serve as recognition sites for RNA-binding proteins and accessory factors involved in the splicing process. Variation in AS is often tissue-specific and the pattern of AS can be developmentally and environmentally regulated [4, 5, 7, 8]. AS produces alternative proteins that vary, for example, in their function, their interaction with other proteins, and/or in their subcellular localization. Furthermore, alternative splice isoforms can be recognized as "aberrant" and degraded by nonsense mediated decay (NMD), ultimately changing transcript abundance [2].

Large-scale sequencing of multiple RNA samples by next-generation sequencing (NGS) technologies allows global *de novo* detection of individual gene transcript variants and the potential to quantify changes in transcript variant abundances. Using NGS across a range of eight different barley tissues, 55% of genes had alternative transcripts [9]. In germinating seed of four different barley varieties, AS was found in 14–20% of intron-containing genes, including those involved in cell wall polysaccharide metabolism [10, 11]. To support and validate gene transcript changes identified by NGS, we have used a medium-throughput AS panel based on reverse transcription-PCR and separation of fluorescently labeled amplicons by capillary sequencing. High-resolution RT-PCR has the power to resolve products that differ by a single base pair and detects statistically significant AS changes between different samples [12]. The panel was originally developed for Arabidopsis and has been widely used to demonstrate AS in natural plant variants, different plant organs; in plants grown under different conditions and in genes encoding core components of the circadian clock [13–15]. Moreover, the analysis of mutants defective in candidate splicing regulators such as serine-arginine rich proteins or the subunits of the cap-binding complex, as well as transgenic plants overexpressing hnRNP (heterogeneous nuclear ribonucleoproteins) proteins has demonstrated the global impact of these regulators on plant AS [12, 16–18]. The HR RT-PCR procedure is highly transferable to different genes and plant species. In barley, it has recently been used to show conserved temperature responsive alternatively spliced isoform switching in circadian clock genes between barley and Arabidopsis [19].

We have assembled a panel of primers specific to selected barley genes that produce amplicons across barley AS events. We used these to confirm the accuracy of predicted AS events from the RNA-seq assembly and to determine the proportions of alternative transcripts produced.

2 Materials

2.1 RNA Isolation

1. Qiagen RNeasy Mini Kit (50).
2. Promega RQ1 RNase-Free DNase.
3. Thermo Scientific, NanoDrop 2000 UV-Vis Spectrophotometer.
4. TE: 10 mM Tris and 1 mM EDTA, pH 8 HCl.
5. RNasin.
6. Phenol/Chloroform: Dissolve 250 g phenol with 100 mL Tris HCl pH 8 at 40 °C for 1 h. Add an equal volume of chloroform, mix, and store at 4 °C in a dark bottle.
7. Ethanol/Sodium Acetate pH 4.8 mix: Mix 19 Vol 100% Ethanol with 1 Vol 3 M Sodium acetate pH 4.8.

2.2 cDNA Synthesis

1. Clontech, RNA to cDNA EcoDry double primed premix.

2.3 PCR Reagents

1. Taq DNA polymerase and 10× Buffer.
2. dNTP's—Dilute 100 mM stock of each deoxynucleotide (dATP, dCTP, dGTP, and dTTP) to 20 mM. Mix each nucleotide together with sterile distilled water to produce a 1.25 mM working dNTP stock.
3. Primers to detect AS events—Dilute 100 μM stock to 20 μM.
4. PCR plates (Thermo-Fast 96, Semi-skirted).

2.4 Product Separation

1. 500 LIZ Size standard (ABI) for reproducible sizing of RT-PCR fragments. The 500 marker contains 16 single-stranded labeled fragments of different lengths.
2. Hi Di Formamide (ABI).

2.5 Software

1. Applied Biosystems. Genemapper v3.7 or above.
2. Microsoft Excel.
3. Genstat or R statistical analysis software.

3 Methods

3.1 Gene Selection

Any gene of interest that shows AS may be tested for changes in splicing. The AS panel will increase as more alternatively spliced genes are identified and tested. Initially, we selected 94 AS events from an NGS RNA-seq experiment of eight different organs. Each AS event was supported by at least 100 splice junction reads and represented the major types of AS (exon skipping; alternative 5′ and 3′ splice sites; and intron retention). At present, the barley AS panel consists of 215 genes covering 250 AS events. For our AS analysis, we included the AS event found in barley rubisco activase as an AS control. This was the first plant gene identified to show AS, is highly expressed in most photosynthetic tissues and produces two transcripts through AS that are often found in an approximately 1:1 ratio [20]. Protein phosphatase 2A subunit A2 (MLOC_2967; HORVU5Hr1G051850) was also included in our analyses as a transcriptional control to normalize transcription to steady state levels, if required [19].

3.2 Primer Design

Primers are designed by selection of sequences within exon sequences upstream and downstream of the AS event(s). To study intron retention events, primers are designed across a constitutively spliced intron and the alternatively retained intron (*see* **Note 1**). This avoids false positive intron retention results that may occur due to contaminating DNA. Selected primers are usually 19–25 nt long, about 50% GC and produce spliced PCR products that are predicted to be less than 750 bp in length, which are readily detectable in the sequencing run (ABI3730) using the 500 bp marker. Both 5′ and 3′ primers contain one or two G or C nucleotides at their 5′ and 3′ ends to avoid primer dimerization. Primers are BLAST screened against the barley genome to avoid primers that have a perfect match with other regions in the genome. The 5′ forward primer is labeled at its 5′ end with a 6-FAM (6-Carboxyfluorescein) fluorescent dye (*see* **Note 2**).

3.3 RNA Extraction

1. Plant material (*see* **Note 3**)—Extract total RNA from up to 100 mg of any selected barley tissue using available RNA extraction kits. We use the RNeasy Plant Mini Kit (Qiagen) following the manufacturer's instructions (*see* Qiagen RNeasy manual).
2. Determine RNA concentrations by Nanodrop (Thermo Scientific).
3. Most RNA extraction kits leave very little contaminating DNA, but nevertheless can be picked up by sensitive PCR. Our AS analysis does not usually take into consideration any unspliced RNA products that will produce the same PCR product size as contaminating DNA. If the AS analysis needs to consider unspliced products, RNA preparations are further treated with RQ1 DNase to remove remaining DNA. A maximum of 50 μg of RNA (50 μL) is added to TE, 10 mM $MgCl_2$, 100 U RNasin and 10 U of RQ1 DNase in a final volume of 100 μl. Incubate the reaction at 37 °C for 20 min and terminate the reaction by adding 1/50th volume of 0.5 M EDTA pH 8 and 1/50th volume 10% (w/v) SDS. Extract RNA with an equal volume of Phenol/Chloroform pH 8 mix and precipitate with 2.5 vols of Ethanol/Sodium acetate pH 4.8 held at −20 °C. After pelleting by centrifugation, the RNA pellets are washed with 70% ethanol at −20 °C, air-dried and resuspended at a concentration of 1 μg/μL (*see* **Note 4**).

3.4 First Strand cDNA Synthesis

1. Add 5 μg of total RNA (enough for 100 PCR reactions, i.e., one 96-well plate) to sterile distilled water to a volume of 20 μL.
2. Transfer the sample to 1 microfuge tube supplied which contains the "RNA to cDNA EcoDry double primed" first strand cDNA synthesis bead and leave at room temperature for 1 min (*see* **Note 5**).
3. Gently mix the sample by pipetting, spin briefly in a microfuge to collect the sample and incubate at 37 °C for 1 hour then at 70 °C for 10 min. Add sterile distilled water to give a final volume of 100 μL.

3.5 PCR

1. For each 25 μL PCR reaction the following is prepared (make a mastermix).

	× 1	× 100
10 × buffer	2.5 μL	250 μL
1.25 mM dNTP's (200 μM[a])	4 μL	400 μL
Taq DNA polymerase	0.125 μL	12.5 μL
SDW	16.375 μL	1637.5 μL

[a]200 μM is the final dNTP concentration

2. For a 96-well plate reaction, add the complete first strand reaction mix to the 100× PCR reaction mix. Add 24 μL of the resulting mix to each well of a 96-well plate containing 1 μL of the 96 different forward and reverse gene-specific primer pairs (400 nM per primer per reaction). This gives a total PCR reaction volume of 25 μL in each well.
3. Mix the samples by vortexing, spin briefly to collect the samples on the bottom of the well and place on a PCR machine (Perkin Elmer 9700) using the following cycle program:

1 cycle	94 °C – 2 min
24 cycles (*see* **Note 6**)	94 °C – 15 s 50 °C – 30 s 70 °C – 1 min
1 cycle	70 °C – 10 min

Store at 4 °C until ready to use

3.6 Separation and Analysis of the Spliced Products

1. Mix the labeled RT-PCR products from the RT-PCR reactions with Hi Di Formamide (Applied Biosystems) and the 500 LIZ labeled size marker (Applied Biosystems) (*see* **Note 7**). For the 96 reactions in the 96-well plate, prepare the following mix:

	× 1	× 100
500 LIZ Size standard	0.05 μL	5 μL
Hi Di Formamide	8.95 μL	895 μL

2. Aliquot 9 μL of the mix into each well of a 96-well plate and add 1 μL of each RT-PCR reaction. Store the remaining sample at −20 °C for downstream cloning and sequencing.
3. Inject samples, separate by capillary electrophoresis, and detect on an ABI3730 DNA Analyzer (Life Technologies) or equivalent. Set up the platform for fragment analysis using a 36 cm capillary array, POP7 polymer, and dye filter set G5. Run samples containing the LIZ 500 marker using the manufacturer's "GeneMapper36_POP7" Run Module (run time 1200s, run voltage 15 Kv). Subsequently, the peak (RT-PCR product) sizes and areas are calculated and analyzed with Life Technologies GeneMapper v3.7 (*see* **Note 8**).
4. RT-PCR products are accurately identified with ±1–2 bp resolution. Extract the relative fluorescent peak areas for RT-PCR products with expected sizes for the alternatively spliced products and tabulate in Microsoft Excel (Table 1) (*see* **Note 9**).

3.7 Basic Statistics

1. Calculate the relative proportion of the transcripts in the different AS events by dividing the value for each alternatively spliced product by the sum of the values for all the spliced products of that event.
2. For an accurate statistical measurement of AS proportions, three biological repeats are routinely performed for all experiments. Mean AS proportions with standard deviations and standard errors are calculated for the three separate biological repetitions (*see* **Note 10**).
3. For each alternatively spliced transcript, ANalysis Of VAriance (ANOVA) is used in turn to compare all the organs after an angular transformation of the individual AS proportions (*see* **Note 11**). In this case, analysis was performed on the arcsine scale and *p*-values are not adjusted for multiple comparisons (as an overall ANOVA *p*-value ≤ 0.001 is considered highly significant). The *p*-values are obtained from t-tests that are based on the residual variance estimate from all five samples tested. ANOVA assumes a completely randomized experimental design. AS events with significant variation ($p \leq 0.05$) are routinely selected (*see* **Note 12**).

3.8 Example

A HR RT-PCR analysis is shown for a single primer pair that covers an AS event discovered in a barley clathrin adaptor complex subunit protein gene (MLOC_54446; HORVU5Hr1G027080) with

Table 1
Extracted RT-PCR product length and peak areas for a single primer pair across five barley organs

Sample file name	Length detected (bp)	Peak area (RFU)	Proportion	Mean of 3 repeats (Sample File Reps 1–3)	SD	SE
INF1						
110_Infl_Repl	174.12	21,750	0.61	0.58	0.0235	0.0136
110_Infl_Repl	179.07	13,995	0.39	0.42	0.0235	0.0136
110_Infl_Rep2	174.13	7662	0.56			
110_Infl_Rep2	179.08	5906	0.44			
110_Infl_Rep3	174.17	10,550	0.57			
110_Infl_Rep3	179.14	7909	0.43			
INF2						
110_Inf2_Repl	174.52	9259	0.39	0.45	0.0546	0.0315
110_Inf2_Repl	179.52	14,374	0.61	0.55	0.0546	0.0315
110_Inf2_Rep2	174.57	8823	0.49			
110_Inf2_Rep2	179.67	9055	0.51			
110_Inf2_Rep3	174.71	9757	0.48			
110_Inf2_Rep3	179.81	10,689	0.52			
LEA						
110_Lea_Repl	174.45	5613	0.68	0.66	0.0268	0.0155
110_Lea_Repl	179.48	2611	0.32	0.34	0.0268	0.0155
110_Lea_Rep2	174.42	2702	0.66			
110_Lea_Rep2	179.39	1417	0.34			
110_Lea_Rep3	174.52	2854	0.63			
110_Lea_Rep3	179.52	1684	0.37			
EMB						
110_Emb_Repl	174.05	20,120	0.81	0.79	0.0174	0.0100
110_Emb_Repl	179.17	4675	0.19	0.21	0.0174	0.0100
110_Emb_Rep2	174.08	21,418	0.79			
110_Emb_Rep2	179.07	5795	0.21			
110_Emb_Rep3	174.02	15,452	0.78			
110_Emb_Rep3	179.20	4415	0.22			
NOD						
110_Nod_Repl	174.13	4083	0.63	0.58	0.0470	0.0271
110_Nod_Repl	179.24	2411	0.37	0.42	0.0470	0.0271
110_Nod_Rep2	174.19	4444	0.57			
110_Nod_Rep2	179.31	3287	0.43			
110_Nod_Rep3	174.14	4362	0.54			
110_Nod_Rep3	179.24	3790	0.46			

Sample file name indicates primer number, barley organ, and repeat number

SD = Standard deviation = $\sqrt{\sum_{i=1}^{n} (x_i - \bar{x})^2/(n-1)}$ where n is the number of repeats, x_i is the individual value, and $\bar{x}$ is the mean of the repeats

SE = Standard error of mean. SE = SD/$\sqrt{n}$ where n is the number of repeats similarity to the Arabidopsis gene At2g20790 (Fig. 1a). This is one example from many hundreds of genes that can be examined simultaneously using the HR RT-PCR panel. It highlights the resolution obtained by this method and quantification of the proportions of the alternative transcripts.

RNA-seq analysis identified the alternative 5′ splice site in this gene at intron 3, leading to alternative transcripts that differ by only 5 nt (Fig. 1b). The shorter transcript using the distal 5′ splice site produces a transcript that can be translated into a protein that matches the clathrin adaptor complex subunit protein. The longer transcript leads to a frame shift, introduces a premature termination codon, and will result in a severely truncated protein.

HR RT-PCR validated the two HR RT-PCR products as 174–179 bp in length (Fig. 1c). RNA was extracted from five barley organs which included: whole developing inflorescence tissue 30 (INF1) and 50 (INF2) days after planting; leaf tissue, 17 days after planting (LEA); mesocotyl and seminal root tissue, 4 days after germination (EMB); and stem at the third internode, 42 days after planting (NOD) [9]. This was followed by HR RT-PCR and capillary electrophoresis to identify the different transcripts. Data collected on RT-PCR product length and peak area was extracted and tabulated for the three repeats and the proportions determined (Table 1). Mean values of the proportions were determined over the three repeats, followed by standard deviations and standard errors (Table 1). A graph of the data shows the proportional changes in alternative 5′ splice site selection in the five different barley organs (Fig. 1d). Pairwise ANOVA between the five barley organs identified the tissues that showed significant changes in alternative 5′ splice site selection in this gene (Table 2). In this example, INF1 and NOD tissues have identical AS proportions and, therefore, show no significant differences. Comparison of LEA with INF1 and NOD showed a significant change in AS to $p \leq 0.05$. The remaining comparisons were all significant to $p \leq 0.01$ highlighting the variation in AS that occurs in different barley organs. The overall conclusion from the data is that the shorter transcript, which codes for the functional protein, is more prevalent in leaf and embryo organs compared to the internode and inflorescence organs.

There are many different alogorithms to assemble and quantify individual transcripts based on read numbers from RNA-seq data that can be used to determine proportions of AS [22]. Computational systematic analysis of alternative splicing in large data sets can further be determined by calculating inclusion levels of alternative splicing events using algorithms as described for SUPPA [23]. These AS proportions can be directly compared with the alternative splice proportions derived from HR RT-PCR. In this simple example, AS proportions between the five different barley organs were highly comparable to RNA-seq data between the assays

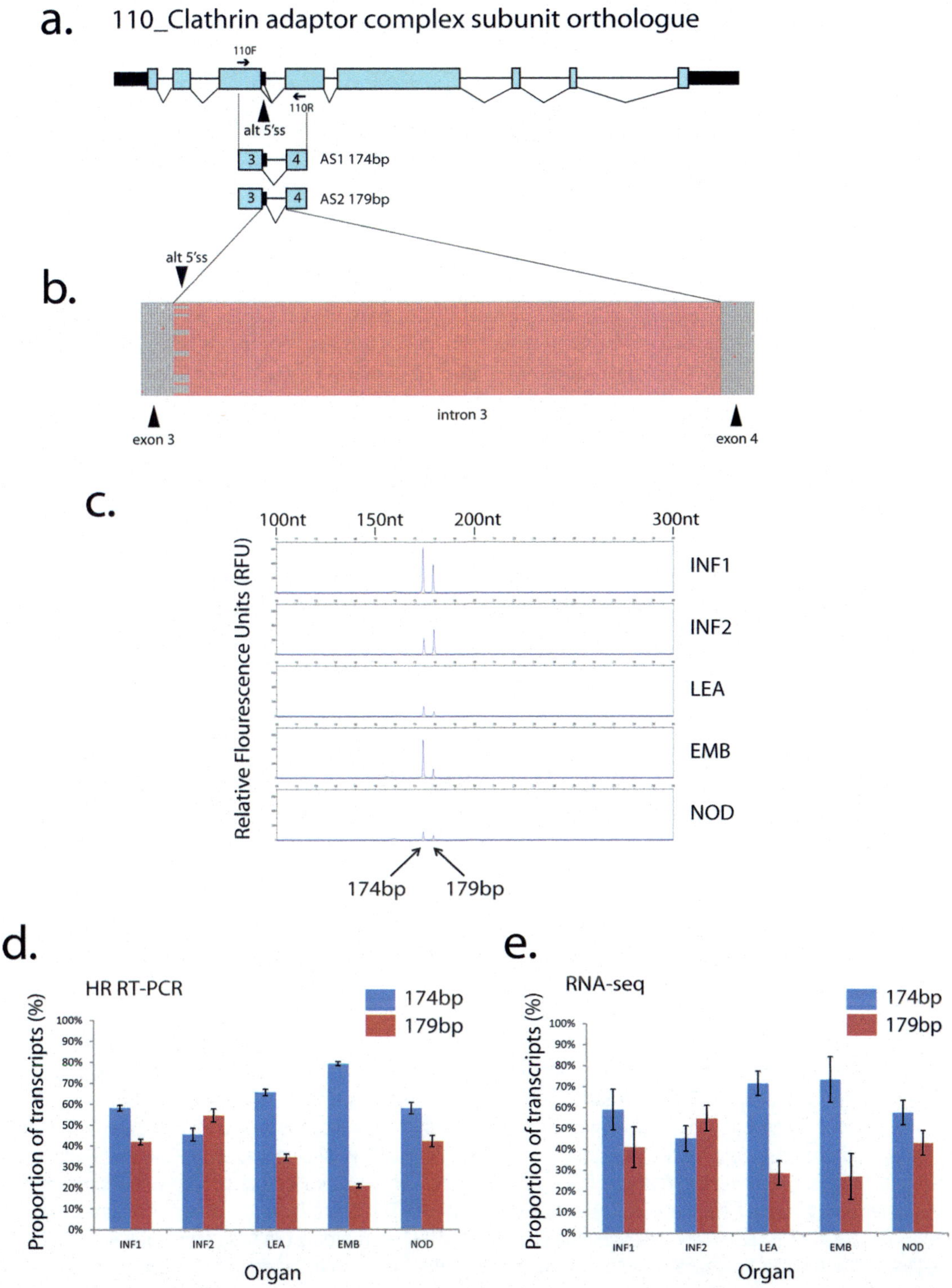

Fig. 1 High-resolution RT-PCR analysis of a single alternative splicing event. (**a**) Schematic representation of a barley clathrin adapter complex subunit orthologue protein (MLOC_54446; HORVU5Hr1G027080). Coding regions are shown as an open blue box, untranslated regions are shown as a black box, and the introns are

Table 2
Analysis of variance between barley organs

	INF1	INF2	LEA	EMB
INF2	0.001959			
LEA	0.027415	0.000051		
EMB	0.000018	0.000000	0.000515	
NOD	0.995754	0.001942	0.027673	0.000018

Values given as *p* values = the significance value from the t-test analysis of variance

with a Pearson correlation value of 0.971 (Fig. 1e). This is not always the case and there are many examples where correlation is poor between the two methods. Different RNA-seq algorithms mis-map reads, mis-assemble transcripts, and generate redundancy which affects accurate quantification. This has recently been addressed in Arabidopsis by quantifying alternative transcripts against a reference transcript database that contains only well-supported transcripts and a similar approach is underway for barley [24, 25]. The HR RT-PCR method described here will be important in the development of the reference transcript database for barley and to validate quantification of individual transcripts produced by different RNA-seq analysis methods.

4 Notes

1. PCR is very sensitive and despite careful removal of contaminating DNA, enough DNA may remain to produce products that are indistinguishable from unspliced pre-mRNA. To overcome this, retained intron event primers are designed across a constitutively spliced intron upstream or downstream of the retained intron. Retained intron spliced products are therefore easily distinguished from the DNA-derived products, which would contain both introns.

Fig. 1 (continued) shown as a line. The alternative 5′ splice site event is indicated and the products of HR RT-PCR with primer pair 110F and R (arrowed) are shown with expected RT-PCR product sizes. (**b**) Tablet visualization [21] showing individual sequence reads covering intron 3 between exon 3 and exon 4. The exon regions are shown in gray while the intron sequence removed is shown in red. Individual reads show alternative selection of an alternative 5′ splice site (arrow). (**c**) Electropherograms showing representative examples of the output from the ABI 3730 sequencer. The HR RT-PCR products are identified as peaks 174 bp and 179 bp (*see* Fig. 1a). The X-axis indicates length of HR RT-PCR product in base pairs (bp) and the Y-axis indicates the relative fluorescence units. Results are shown for the splicing analysis of five different barley organs: inflorescence (INF1 and INF2); leaf (LEA); embryo (EMB); and internode (NOD) tissues. See text for more detailed information. (**d**, **e**) Graphs indicates the proportion of spliced products (174 bp—blue bar and 179 bp—red bar) expressed as a percentage (%) across the barley organs from an HR RT-PCR (**d**) and an RNA-seq experiment (**e**). Error bars represent standard errors between three biological repeats

2. Different fluorescent standard dye sets are available for genotyping applications. We have successfully used 6-FAM labeled primers in association with size markers labeled with ROX (6-Carboxyl-X-Rhodamine) or the proprietary LIZ label. As these are used as size markers, we avoid labeling primers with these fluorescent labels. We have further attempted to multiplex AS analysis using HEX (6-carboxy-1,4-dichloro-2′,4′,5′,7-′-tetra-chlorofluorescein) labeled primers. Depending on the amounts of transcripts made, we found overlap in the emission spectra between 6-FAM and HEX labeled primers, which led to the presence of unexpected peaks in the other labeled RT-PCR products and confused downstream analysis. We have returned to single fluorescent dye label analysis.
3. AS is affected at different developmental stages and tissues. It is also affected by environmental abiotic and biotic stresses, and circadian times. High-resolution RT-PCR AS analysis is highly sensitive to these changes. It is therefore important that biological repeat plants including control plants are grown at the same time and in the same conditions with the exception of the condition tested.
4. In some cases, DNA may still remain after a single DNase treatment. A second DNase treatment will remove any remaining DNA.
5. The RNA to cDNA EcoDry Double primed cDNA synthesis beads contain both oligo dT and random hexamers. We have found an improvement in RT-PCR peaks using this double primed system.
6. A total of 24 PCR cycles were selected on the basis that PCR is in the logarithmic phase of amplification and allows us to screen both highly and more poorly transcribed genes. We previously tested the variation in AS results in technical replicates and the majority showed less than 1% variability with a small number reaching up to 3%. We set a value slightly higher at 5% variation as a cutoff value for statistically significant changes in AS (*see* **Note 12**). Samples taken at >24 cycles showed that the more highly expressed genes were moving out of the exponential phase of PCR amplification [26].
7. We routinely use the 500 LIZ Size standard to calculate RT-PCR product sizes. For RT-PCR products greater than 750 bp the 1200 LIZ Size standard can be used. 1 μL of each reaction is mixed with 0.5 μL of the standard and 8.5 μL of Hi Di Formamide. Samples containing the LIZ 1200 marker are run using the "3730_36cm_POP7_GS1200Lizv2_1" Run Module (run time 6000 s, run voltage 6.1 Kv).

8. Genmapper software is routinely updated. Check Life Technologies website for updates (https://www.lifetechnologies.com/uk/en/home.html).
9. The nature of AS means that you may obtain expected peaks (RT-PCR products) that vary substantially in their peak areas. The Genemapper software may be set to select a minimum peak size and reduce the number of small background peak areas that may be extracted and tabulated. However, we set the peak level low (200 relative fluorescent units) to capture as many of the alternatively spliced products as possible. Background peaks are therefore removed from the analysis.
10. The HR RT-PCR AS analysis measures changes in the proportions of the different alternative spliced transcripts found. Although, between the repeats, there may be apparent changes in the overall amounts of transcript produced, the proportion of the different alternatively spliced transcripts is maintained.
11. An angular transformation is often used with proportions to bring the data closer to a normal distribution. It is performed on the raw proportions before any statistical analysis is done. If p is the proportion, then the transformation is $x = \arcsin(\sqrt{p})$.
12. In a number of cases where there is very little variation between the repeats, the standard errors may be very low and even small changes in the proportion of splicing may be identified as statistically significant with $p \leq 0.05$. We therefore select examples that are statistically significant and show >5% splicing change (*see* **Note 6**).

Acknowledgments

Work is supported by grants from the Biotechnology and Biological Sciences Research Council (BB/I00663X/1: to RW), and the Scottish Government Rural and Environment Science and Analytical Services division.

References

1. Zhou HL, Luo G, Wise JA et al (2014) Regulation of alternative splicing by local histone modifications: potential roles for RNA-guided mechanisms. Nucleic Acids Res 42:701–713
2. Reddy ASN, Marquez Y, Kalyna M et al (2013) Complexity of the alternative splicing landscape in plants. Plant Cell 25:3657–3683
3. Naftelberg S, Schor IE, Ast G et al (2015) Regulation of alternative splicing through coupling with transcription and chromatin structure. Annu Rev Biochem 84:165–198
4. Filichkin S, Priest HD, Megraw M et al (2015) Alternative splicing in plants: directing traffic at the crossroads of adaptation and environmental stress. Curr Opin Plant Biol 24:125–135
5. Staiger D, Brown JWS (2013) Alternative splicing at the intersection of biological timing, development, and stress responses. Plant Cell 25:3640–3656
6. Lee Y, Rio DC (2015) Mechanisms and regulation of alternative Pre-mRNA splicing. Annu Rev Biochem 84:291–323

7. Mastrangelo AM, Marone D, Laidò G et al (2012) Alternative splicing: enhancing ability to cope with stress via transcriptome plasticity. Plant Sci 185–186:40–49
8. Capovilla G, Pajoro A, Immink RG et al (2015) Role of alternative pre-mRNA splicing in temperature signaling. Curr Opin Plant Biol 27:97–103
9. International Barley Sequencing Consortium (2012) A physical, genetic and functional sequence assembly of the barley genome. Nature 491:711–716
10. Zhang Q, Zhang X, Pettolino F et al (2016) Changes in cell wall polysaccharide composition, gene transcription and alternative splicing in germinating barley embryos. J Plant Physiol 191:127–139
11. Zhang Q, Zhang X, Wang S et al (2016) Involvement of alternative splicing in barley seed germination. PLoS One 11:e0152824
12. Simpson CG, Fuller J, Maronova M et al (2008) Monitoring changes in alternative precursor messenger RNA splicing in multiple gene transcripts. Plant J 53:1035–1048
13. Marquez Y, Brown JWS, Simpson CG et al (2012) Transcriptome survey reveals increased complexity of the alternative splicing landscape in Arabidopsis. Genome Res 22:1184–1195
14. James AB, Syed NH, Bordage S et al (2012) Alternative splicing mediates responses of the arabidopsis circadian clock to temperature changes. Plant Cell 24:961–981
15. James A, Syed N, Brown J et al (2012) Thermoplasticity in the plant circadian clock: how plants tell the time-perature. Plant Signal Behav 7:1219–1223
16. Raczynska KD, Simpson CG, Ciesiolka A et al (2010) Involvement of the nuclear cap-binding protein complex in alternative splicing in Arabidopsis thaliana. Nucleic Acids Res 38:265–278
17. Streitner C, Köster T, Simpson CG et al (2012) An hnRNP-like RNA-binding protein affects alternative splicing by *in vivo* interaction with target transcripts in Arabidopsis thaliana. Nucleic Acids Res 40:11240–11255
18. Simpson CG, Lewandowska D, Liney M et al (2014) Arabidopsis PTB1 and PTB2 proteins negatively regulate splicing of a mini-exon splicing reporter and affect alternative splicing of endogenous genes differentially. New Phytol 203:424–436
19. Calixto CP, Simpson CG, Waugh R et al (2016) Alternative splicing of barley clock genes in response to low temperature. PLoS One 11:e0168028
20. Rundle SJ, Zielinski RE (1991) Organization and expression of two tandemly oriented genes encoding ribulosebisphosphate carboxylase/oxygenase activase in barley. J Biol Chem 266:4677–4685
21. Milne I, Stephen G, Bayer M et al (2013) Using Tablet for visual exploration of second-generation sequencing data. Brief Bioinform 14:193–202
22. Steijger T, Abril JF, Engström PG et al (2013) Assessment of transcript reconstruction methods for RNA-seq. Nat Methods 10:1177–1184
23. Alamancos GP, Pagès A, Trincado JL et al (2015) Leveraging transcript quantification for fast computation of alternative splicing profiles. RNA 21:1521–1531
24. Zhang R, Calixto CPG, Marquez Y et al (2016) AtRTD2: A Reference Transcript Dataset for accurate quantification of alternative splicing and expression changes in Arabidopsis thaliana RNA-seq data. bioRxiv. https://doi.org/10.1101/051938
25. Brown JWS, Calixto CP, Zhang R (2017) High-quality reference transcript datasets hold the key to transcript-specific RNA-sequencing analysis in plants. New Phytol 213:525–530
26. Kim SH, Koroleva OA, Lewandowska D et al (2009) Aberrant mRNA transcripts and the nonsense-mediated decay proteins UPF2 and UPF3 are enriched in the nucleolus. Plant Cell 21:2045–2057

Chapter 18

Exome Capture for Variant Discovery and Analysis in Barley

Micha Bayer, Jenny A. Morris, Clare Booth, Allan Booth, Niki Uzrek, Joanne R. Russell, Robbie Waugh, and Pete E. Hedley

Abstract

Exome capture is a reduced representation approach that selectively captures sequence from only the gene-bearing regions of a genome. It is based on probes targeted at these regions and, compared with whole genome shotgun sequencing, leads to a significant reduction in cost and data processing effort while still providing insights into the most relevant part of a genome. An exome capture array for barley was released in 2013 and this has opened the door to numerous studies that have put this technology to good use. In this chapter we detail the laboratory protocols required for enrichment and sequencing, and provide detailed step-by-step instructions for the bioinformatics analysis of the resulting data.

Key words Barley exome capture, Library preparation, Enrichment, Data analysis, Bioinformatics

1 Introduction

Next-generation sequencing (NGS) has enabled unprecedented access to sequence variants in genomes, which can subsequently form the basis for identifying causal gene differences responsible for important traits of interest. Affordable access to NGS in recent years has ensured that even large-genome species such as barley, which has a 5.1 Gbp genome [1], can be mined efficiently for single nucleotide polymorphisms (SNPs), indels, and deletions. Whole exome capture (EC) is a means of targeted re-sequencing only of gene regions, which account for less than 2% of the entire genome in barley [1]. Data generated can be used directly for population studies, QTL mapping and diversity analysis, or variants can be extracted to use on other lower-plexity platforms, including SNP chips, custom amplicon sequencing, and KASP assays (LGC Genomics). The primary advantage of EC over other variant assays, which tend to be "closed" platforms dependent upon previously characterized SNPs, is that it is not limited to known variants and is therefore an excellent means of novel unbiased SNP discovery.

Wendy A. Harwood (ed.), *Barley: Methods and Protocols*, Methods in Molecular Biology, vol. 1900,
https://doi.org/10.1007/978-1-4939-8944-7_18,

A custom EC assay (SeqCap EZ Library; Roche Diagnostics) was established in barley through a consortium funded effort and covers 60 Mbp of exome, utilizing long oligonucleotide probes designed to c. 40,000 genes annotated in the cultivar Morex 2012 reference sequence [1, 2]. The EC barley assay has been utilized in several projects [3–7] capturing a range of material, from cultivated to wild barleys, demonstrating the utility and robustness of the design. Recently, using a comprehensive set of EC data we have designed an Illumina SNP chip, representing 44,000 loci, for which there was an excellent assay translation rate.

Here we describe laboratory-based processing of barley DNA samples for EC and also provide a detailed bioinformatics workflow for data quality control and filtering to ensure that only the most robust and reliable variants are identified. The barley assay allows DNAs to be multiplexed prior to capture enrichment and sequencing, thereby increasing the efficiency and reducing costs of the processing.

2 Materials

2.1 Genomic DNA Extraction

1. DNeasy Plant Mini Kit (Qiagen), or similar.
2. Minigel electrophoresis system.
3. SYBR®Safe DNA gel stain, or similar.
4. 1 kb λ-DNA molecular ladder, or similar.
5. UV transilluminator.
6. General laboratory solutions: TBE, bromophenol blue.

2.2 Illumina Library Construction

1. Kapa Library Preparation Kit (Part No: 07137923001; Roche).
2. DNA vacuum concentrator.
3. Ultra Sonicator (Covaris M220, or similar).
4. MicroTUBE AFA fiber screw-cap (6 mm × 16 mm, 50 μL, Part No: 500096; Covaris).
5. Ethyl alcohol.
6. DynaMag Magnet, or similar.
7. General laboratory equipment: heat block; water baths; microcentrifuge; spectrophotometer; vortex.
8. Bioanalyzer, or similar fragment analyzer.
9. AMPure XP beads (Part No: A63880; Beckman Coulter).
10. General laboratory solutions: absolute ethanol; TE buffer (pH 8.0); elution buffer 10 mM Tris–HCl (pH 8.0).

Table 1
List of software and versions

Tool	Version	URL
GATK	3.4.0	https://software.broadinstitute.org/gatk/
bamtools	2.2.3	https://github.com/pezmaster31/bamtools
bwa mem	0.7.10	http://bio-bwa.sourceforge.net/
samtools	1.3.1	http://samtools.sourceforge.net/
vcflib	20140627	https://github.com/vcflib/vcflib
Picard toolkit	1.138	http://broadinstitute.github.io/picard/

The list of software shown specifies versions used in the pipeline implemented by the authors. Scientific software is subject to ongoing development and later or earlier versions of the software used here may not behave as expected, or may even lack some of the parameters used here. While most software developers attempt to preserve backward compatibility during the development process, there is no guarantee that a different version will provide the same outcome as the versions used here. With the exception of GATK, which is cross-platform, all the tools used here are designed to run on the Linux operating system. All tools are executed through command line statements

2.3 Exome Capture

1. SeqCap EZ Developer Library, Barley Exome Design (Part No 120426_Barley_BEC_D04; Roche (*see* **Note 1**).
2. SeqCap Adapter Kit (Part No: 07141530001; Roche).
3. SeqCap Hybridization and Wash Kit (Part No: 05634261001; Roche).
4. SeqCap EZ Accessory Kit (Part No: 07145594001; Roche).
5. SeqCap HE-Oligo Kit A (Part No: 06777287001; Roche).
6. SeqCap HE-Oligo Kit B (Part No: 06777317001; Roche).
7. SeqCap Pure Capture Bead Kit (Part No: 06977952001; Roche).

2.4 Data Analysis

1. Linux server or compute cluster (CentOS 6.8 used here but most Linux distributions should work), with access through command line interface. We recommend one or more multi-processor machines with at least 64 GB of RAM.
2. List of software as shown in Table 1. All software is available freely for academic users.

3 Methods

3.1 Genomic DNA Extraction

1. Use young seedling leaf tissue (approximately 3–5 cm) for gDNA extractions (*see* **Note 2**).
2. Isolate DNA using DNeasy plant mini-preparation kits, or similar, following the manufacturer's instructions.
3. Visually assess the DNA by gel electrophoresis. Briefly a 1.5% agarose-gel is prepared with 1× Tris/Borate/EDTA (TBE)

(pH 8.0) containing 10 μL SYBR®Safe DNA gel stain, placed in the electrophoresis chamber and covered with about 5 mm 1× TBE once it was cooled and solidified. For each sample, 5 μL of DNA is loaded in the wells together with 2.5 μL 1× bromophenol blue buffer. A 1 kb λ-DNA molecular ladder is used as standard for determining the DNA concentration by visual comparison. Electrophoresis is conducted at a constant voltage of 100 V for 45 min and the results visualized and recorded using a UV transilluminator (*see* **Note 3**).

4. DNA is quantified using Pico-green according to the manufacturer's instruction.

3.2 Illumina Library Construction

Illumina-compatible whole genome shotgun libraries are made which include barcoding (indexing) to allow multiplexing prior to exome capture and downstream read identification (*see* **Note 4**). Here, the Kapa library construction is described as follows.

3.2.1 Fragmentation of Genomic DNA

1. Dilute gDNA to a final concentration of 2 ng/μL in TE buffer (10 mM Tris–HCl (pH 8.0), 0.1 mM EDTA).
2. Transfer 53 μL of the input gDNA (~100 ng) to a microTUBE AFA fiber screw-cap.
3. Set Covaris M220 instrument to generate an average size range of 180–220 bp. The following settings have been successfully used: Peak Incident Power, 50 W; Duty Factor, 20%; Cycles per Burst, 200; Temperature, 20 °C; Duration, 280 s (*see* **Note 5**).
4. Transfer the fragmented gDNA to a 0.2 mL PCR tube.
5. Run 1 μL on a Bioanalyzer 2100 DNA High-Sensitivity chip. Successfully fragmented DNA is shown in Fig. 1a, with an average fragment size between 180 and 250 bp.

3.2.2 End Repair of Fragments

1. Prepare the End Repair Master Mix as follows: water, 8 μL; 10× KAPA End Repair Buffer, 7 μL; KAPA End Repair Enzyme, 5 μL.
2. Assemble each End Repair reaction as follows in well(s) of a 96 well plate: Fragmented gDNA, 50 μL; End Repair Master Mix, 20 μL.
3. Mix by pipetting, spin briefly and incubate at 20 °C for 30 min in a thermocycler. Allow the AMPure XP beads to warm to room temperature for at least 30 min, prior to next step.
4. Prepare fresh 80% ethanol using ethyl alcohol in a screw-capped tube.
5. Proceed immediately to the next step once the End Repair reaction time is finished.
6. To each 70 μL End Repair reaction, add 120 μL Agencourt AMPure XP beads. Mix thoroughly by pipetting.

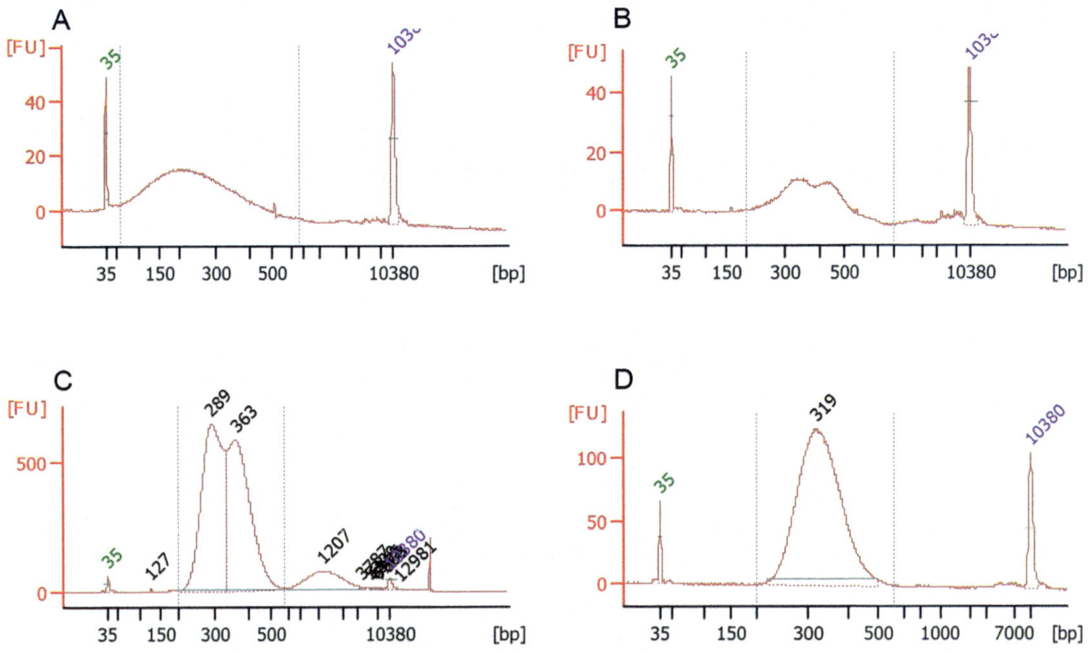

Fig. 1 Bioanalyzer traces showing examples of successful sample preparation through the exome capture process: (**a**) Post-fragmentation gDNA; (**b**) Post size-selection library; (**c**) Pre-capture library; (**d**) Final captured library

7. Incubate the plate at room temperature for 10 min to allow the DNA to bind to the beads.
8. Put the plate on a magnet to capture the beads. Incubate until the liquid is clear.
9. Carefully remove and discard the supernatant.
10. Keeping the plate on the magnet, add 200 μL of 80% ethanol. Incubate the plate on the magnet at RT for 30 s.
11. Carefully remove and discard the ethanol.
12. Repeat ethanol wash.
13. Seal the plate and spin briefly (up to 2000 rpm), then place again on magnet. Remove all residual ethanol without disturbing the beads.
14. Allow the beads to dry at room temperature (~3 min, *see* **Note 6**). Remove the plate from the magnet and proceed immediately.

3.2.3 A-Tailing of Fragments

1. Prepare the A-Tailing Master Mix as follows for each library preparation: water, 42 μL; 10× Kapa A-Tailing Buffer, 5 μL; Kapa A-Tailing Enzyme, 3 μL.
2. Thoroughly resuspend the beads by pipetting, then seal plate.

3. Incubate at 30 °C for 30 min in a thermocycler.
4. Prepare the Indexed Adapter required in the Adapter Ligation step as follows (*see* **Note 7**). Briefly spin required adaptors (from the SeqCap Adapter Kit A and/or B) to pellet contents. Add 50 μL cold, PCR-grade water (included in the SeqCap Adapter kit). Briefly vortex and spin down the resuspended Index Adapter tubes, then keep on ice. Following use, store at −20 °C for future use.
5. Equilibrate the PEG/NaCl SPRI solution (Kapa kit) to room temperature, protected from the light.
6. After the 30 min at 30 °C, add 90 μL PEG/NaCl SPRI solution to each 50 μL A-tailing reaction with beads (*see* **Note 8**).
7. Mix thoroughly by pipetting and incubate at room temperature for 10 min to allow the DNA to bind to the beads.
8. Place the plate on a magnet to capture the beads. Incubate until the liquid is clear.
9. Carefully remove and discard the supernatant.
10. Keeping the plate on the magnet, add 200 μL 80% (v/v) ethanol.
11. Incubate at room temperature for 30 s.
12. Carefully remove and discard the ethanol.
13. Repeat ethanol wash.
14. Briefly spin the plate, place it on the magnet, and remove all residual ethanol without disturbing the beads.
15. Allow the beads to dry at room temperature (*see* **Note 9**). Once dried, remove the plate from the magnet and proceed immediately.

3.2.4 Adapter Ligation

1. Prepare the Ligation Master Mix as follows for each library preparation: Water, 32 μL; 5× Kapa Ligation Buffer, 10 μL; Kapa T4 DNA Ligase, 5 μL; Indexed Adapter, 3 μL.
2. Thoroughly resuspend the beads by pipetting.
3. Seal the plate and incubate at 20 °C for 15 min in a thermocycler, then proceed immediately to the next step.
4. Add 50 μl PEG/NaCl SPRI solution to each 50 μL ligation reaction/beads.
5. Mix thoroughly by pipetting and incubate at room temperature for 10 min to allow the DNA to bind to the beads.
6. Place the plate on a magnet to capture the beads. Incubate until the liquid is clear.
7. Carefully remove and discard the supernatant.

8. Keeping the plate on the magnet, add 200 μL 80% (v/v) ethanol.
9. Incubate the plate at room temperature for 30 s.
10. Carefully remove and discard the ethanol.
11. Repeat ethanol wash.
12. Briefly spin the plate, place on the magnet, and remove all residual ethanol without disturbing the beads.
13. Allow the beads to dry at room temperature.
14. Resuspend the beads in 100 μL of elution buffer (EB) and incubate for 2 min at room temperature to allow the DNA to elute off the beads (*see* **Note 10**).
15. For a safe stopping point, store resuspended beads at 4 °C for up to 24 h.

3.2.5 Size Selection

Libraries are size selected using beads to ensure an optimal size range (250 bp–450 bp) is achieved for exome capture and sequencing.

1. Add 60 μL PEG/NaCl SPRI solution to the resuspended beads.
2. Pipette to mix and incubate at room temperature for 10 min to allow library fragments >450 bp to bind to the beads.
3. Place the plate on a magnet to capture the beads. Incubate until liquid is clear.
4. Carefully transfer 155 μL of the supernatant(s) containing library fragments <450 bp to a new plate.
5. Discard the old plate with the beads carrying library fragments >450 bp.
6. Vortex the AMPure XP beads (previously equilibrated at room temperature) and add 20 μL to the plate containing 155 μL of the previous supernatant.
7. Thoroughly resuspend the beads by pipetting and incubate at room temperature for 10 min to allow library fragments >250 bp to bind to the beads.
8. Place the plate on a magnet to capture the beads. Incubate until the liquid is clear.
9. Carefully remove and discard the supernatant.
10. Keeping the plate on the magnet, add 200 μL of 80% (v/v) ethanol. No need to mix.
11. Incubate the plate at RT for 30 s.
12. Carefully remove and discard the ethanol.
13. Repeat ethanol wash.

14. Briefly spin the plate, place it on the magnet, and remove all residual ethanol without disturbing the beads.
15. Allow the beads to dry at room temperature. Remove the plate from the magnet, then proceed immediately to the next step.
16. Thoroughly resuspend the beads in 25 μL of elution buffer (EB) buffer and incubate at room temperature for 2 min to allow the DNA to elute off the beads.
17. Place the plate on a magnet to capture the beads. Incubate until the liquid is clear.
18. Transfer 20 μL of the clear supernatant to a new plate (*see* **Note 11**). Keep aside a further 2 μL aliquot in a labeled 0.2 mL tube on ice. This will be run on Bioanalyzer 2100 (*see* Fig. 1b) to check size selection.

3.2.6 Pre-capture PCR Amplification

1. Prepare lyophilized Pre-LM-PCR Oligos for use the first time: briefly spin and add 550 μL PCR-grade water (from the kit) to the tube labeled "Pre-LM-PCR Oligo 1 & 2 (LP1)." Vortex briefly and spin.
2. Prepare the LM-PCR Master Mix on ice according to the following: Kapa HiFi HotStart Ready Mix, 25 μL; Pre LM-PCR Oligos 1 & 2 (5 μM), 5 μL.
3. Pipette 30 μL of the LM-PCR Master Mix into each 20 μL sample(s) on the plate (including a negative water control, *see* **Note 12**).
4. Mix well by pipetting five times. Do not vortex.
5. Amplify samples in thermocycler using the following Pre-capture LM-PCR program: Step 1: 98 °C, 45 s; Step 2: 98 °C, 15 s; Step 3: 60 °C, 30 s; Step 4: 72 °C, 30 s; Step 5: repeat Steps 2–4 8 times (9 cycles in total); Step 6: 72 °C, 60 s; Step 7: hold at 4 °C.

3.2.7 Pre-capture Purification

1. Allow an aliquot of AMPure XP Beads to equilibrate to room temperature. Vortex the beads for 10 s.
2. Add 90 μL AMPure XP Beads to 50 μL amplified library and the negative water control.
3. Pipette to mix and incubate at room temperature for 10 min to allow the DNA to bind the beads.
4. Place the plate containing the bead-bound DNA on the magnet and allow the solution to clear.
5. Remove and discard the supernatant being careful not to disturb the beads.
6. Keeping on the magnet, add 200 μL of 80% (v/v) ethanol to each well and incubate at room temperature for 30 s.
7. Remove and discard the 80% ethanol.

8. Repeat ethanol wash.
9. Briefly spin the plate, place it on the magnet, and remove all residual ethanol without disturbing the beads.
10. Allow the beads to dry at room temperature for 1 min.
11. Remove the plate from the magnet and add 52 μL PCR-grade water. Pipette up and down to mix to ensure that all of the beads are resuspended.
12. Incubate at RT for 2 min.
13. Place the plate back on the magnetic and allow the solution to clear.
14. Remove 50 μL supernatant that now contains the amplified sample library and transfer into a fresh plate. Seal the plate with lids and keep on ice. Transfer the water control sample to a labeled tube.
15. Measure A260/A280 on a NanoDrop (DNA setting) to determine the concentration. The sample library yield should be >1 μg, with a A260/A280 of 1.7–2.0. The minimum required is 20 ng/μL.
16. Run 1 μL pre-capture LM-PCR product on a Bioanalyzer 2100 DNA High-Sensitivity chip (Fig. 1c, *see* **Note 13**). Also run the negative water control.
17. This is a safe stopping point. The library can be stored at −20 °C for up to 1 year.

3.2.8 Hybridizing Sample and SeqCap EZ Probes

1. In case of multiplexing, mix together in a 1.5 mL microfuge tube, equal amounts (by mass) of each amplified DNA sample library to obtain a single pool with a combined mass of at least 1.25 μg ("Multiplex DNA Sample Library Pool").
2. Spin the lyophilized SeqCap HE Universal and required SeqCap HE Index oligo tubes briefly.
3. Add 120 μL PCR-grade water to the SeqCap HE Universal Oligo tube (1 mM final concentration). Vortex 5 s and spin.
4. Add 10 μL PCR-grade water to each required SeqCap HE Index Oligo tube (1 mM final concentration). Vortex 5 s and spin (*see* **Note 14**).
5. Mix together the HE oligos so that the resulting Multiplexing Hybridization Enhancing Oligo Pool contains, by mass, 50% SeqCap HE Universal Oligo 1 and 50% of a mixture of the appropriate SeqCap HE Index oligos (*see* **Note 15**).
6. Add 10 μL Developer Reagent to a new 1.5 mL tube.
7. Add 1 μg Multiplex DNA Sample Library to the 1.5 mL tube containing Developer Reagent.

8. Add 2 μL (2000 pmoL) of the specific Multiplex Hybridization Enhancing Oligo pool.
9. Close the tube and make seven holes in the tube's cap with a syringe needle.
10. Dry the Multiplex DNA Sample Library Pool/Developer reagent/Multiplex Hybridization Enhancing Oligo Pool in a DNA vacuum concentrator at 60 °C for a minimum of 20 min.
11. Once the sample is dry, cover the holes with a sticker or piece of tape.
12. Add the following to each sample: 2× SC Hybridization Buffer, 7.5 μL; SC Hybridization Component A, 3 μL.
13. Vortex the sample for 20 s and centrifuge at maximum speed for 10 s.
14. Place each sample at 95 °C heat block for 10 min.
15. During the denaturation step, equilibrate the appropriate number of 4.5 μL SeqCap EZ capture probe pool aliquots (one per library or pooled library) to ice temperature.
16. After denaturation, centrifuge samples at maximum speed for 10 s.
17. Important: work quickly through the following steps.
18. Quickly transfer the library sample to the aliquot of SeqCap EZ probe pool (in a 0.2 mL tube).
19. Carefully pipette up and down 3–4× to mix. Avoid introducing bubbles.
20. Incubate in a thermocycler at 47 °C for 16–20 h (*see* **Note 16**).

3.2.9 Washing and Recovering Captured DNA Sample

1. Remove the SeqCap capture beads from the Hyb + wash kit and an aliquot of Ampure XP beads, and equilibrate to room temperature.
2. Dilute 10× SC Wash Buffers (I, II, and III), 10× Stringent Wash Buffer, and 2.5× Bead Wash Buffer with PCR-grade water to create 1× working solutions.
3. Place the working solutions at the appropriate temperatures (Stringent Wash Buffer & Wash Buffer I at 47 °C; Wash Buffer I and all others at room temperature).
4. Vortex the SeqCap beads for 15 s.
5. Aliquot 100 μL SeqCap beads for each capture into a single 1.5 mL microfuge tube (*see* **Note 17**).
6. Place the tube on a 1.5 mL tube magnet. When the liquid becomes clear (<5 min), remove the supernatant, being careful to leave all of the beads in the tube.
7. Add 200 μL 1× Bead Wash Buffer.

8. Remove the tube from the magnet and vortex for 10 s.
9. Place the tube back into the magnet and, once clear, remove the supernatant.
10. Repeat wash steps.
11. Resuspend the beads in 100 μL 1× Bead Wash Buffer.
12. Aliquot 100 μL resuspended beads into a new 0.2 mL tube.
13. Hold the tube against a plate magnet to bind the beads, remove, and discard the liquid.
14. Proceed to next step quickly as possible. Do not allow the Capture Beads to dry out.
15. Place the Capture Beads at 47 °C on a PCR block, then quickly transfer the hybridization samples to the Capture Beads.
16. Mix by carefully pipetting 10×. Avoid introducing bubbles and leaving liquid droplets up the side of the tube.
17. Leave the tube(s) containing the beads and the hybridized sample(s) at 47 °C for a minimum of 45 min (up to 3 h).
18. Vortex for 3 s at 15 min intervals, to keep the beads in suspension. Ensure samples are kept at 47 °C (*see* **Note 18**).
19. After 45 min maintain tubes on the 47 °C block, and add 100 μL of Wash Buffer I pre-heated to 47 °C.
20. Mix by vortexing 10 s. Hold the tube against the plate magnet and remove Wash Buffer I.
21. Return the tube to the 47 °C block and add 200 μL of 47 °C Stringent Wash Buffer to the beads.
22. Keep the tube on the block and pipette 10× to mix. The temperature should not drop much below 47 °C.
23. Incubate at 47 °C for 5 min.
24. Hold the tube against the plate magnet and remove the Stringent Wash Buffer.
25. Return the tube to the 47 °C block and add 200 μL of 47 °C Stringent Wash Buffer to the beads.
26. Keep the tube on the block and pipette up and down 10× to mix. The temperature should not drop much below 47 °C.
27. Incubate at 47 °C for 5 min, then transfer sample to a 1.5 mL tube.
28. Place the tube in the 1.5 mL magnet to bind the beads and remove the Stringent Wash Buffer.
29. Remove the tube from the magnet and, at room temperature, add 200 μL Wash Buffer I.

30. Mix by vortexing for 2 min.
31. Place back on the magnet to bind the beads and remove Wash Buffer I.
32. Remove from the magnet and add 200 μL of Wash Buffer II.
33. Mix by vortexing for 1 min.
34. Place back on the magnet to bind the beads and remove Wash Buffer II.
35. Remove from the magnet and add 200 μL of Wash Buffer III.
36. Mix by vortexing for 30 s.
37. Place back on the magnet to bind the beads and remove Wash Buffer III.
38. Remove from the magnet and add 50 μL PCR-grade water to the bead-bound captured sample. Vortex briefly to mix, then sit on ice.

3.2.10 Post-capture PCR

1. Prepare lyophilized Post-LM-PCR Oligos 1&2 (LP2) for use the first time: add 480 μL PCR-grade water to the tube.
2. Vortex briefly to resuspend the oligos and spin down. They should be stored at −20 °C after use.
3. Prepare the Post-capture LM-PCR Master Mix in a 1.5 mL tube as follows: Kapa HiFi HotStart Ready Mix (KH), 50 μL; Post-LM-PCR Oligos 1&2 (LP2, 5 μM) 10 μL.
4. Vortex the bead-bound captured DNA from the previous stage to ensure homogenous mixture of beads.
5. Aliquot the following mix into two wells of a PCR plate (per capture). Mix well by pipetting up and down (do not vortex). Bead-bound Captured gDNA, 20 μL; LM-PCR Master Mix, 30 μL.
6. Amplify samples in thermocycler using the following Post-capture LM-PCR program: Step 1: 98 °C, 45 s; Step 2: 98 °C, 15 s; Step 3: 60 °C, 30 s; Step 4: 72 °C, 30 s; Step 5: repeat Steps 2–4 13 times (for a total of 14 cycles); Step 6: 72 °C, 60 s; Step 7: hold at 4 °C.
7. Allow your aliquot of AMPure XP Beads (or the equivalent ones from the SeqCap Pure Capture Bead Kit) to warm to RT while the PCR is running.

3.2.11 Post-capture Purification

1. Equilibrate AMPure XP Beads to room temperature.
2. Pool the corresponding amplified captured Multiplex DNA Sample Library into a 1.5 mL microcentrifuge tube (*see* **Note 19**).

3. Vortex the AmPure XP Beads for 10 s before use to ensure a homogenous mixture.
4. Add 180 μL AMPure XP Beads to the 100 μL pooled amplified captured Multiplex DNA Sample Library/beads.
5. Vortex briefly and incubate at room temperature for 10 min to allow the DNA to bind the beads.
6. Place the tube on a 1.5 mL magnet and allow the solution to clear.
7. Once clear, remove and discard the supernatant being careful not to disturb the beads.
8. Add 200 μL of 80% (v/v) ethanol, leaving the tube in the magnet. No mixing required.
9. Incubate at room temperature for 30 s.
10. Remove and discard the ethanol.
11. Repeat the ethanol wash.
12. Briefly spin the tube, place it on the magnet, and remove all residual ethanol without disturbing the beads.
13. Allow the beads to dry on the magnet at room temperature with the tube lid open (*see* **Note 20**).
14. Remove the tube from the magnet and resuspend the DNA using 52 μL EB buffer. Pipette up and down 10× to mix.
15. Incubate at room temperature for 2 min.
16. Place the tube back on the magnetic and allow the solution to clear.
17. Transfer 50 μL supernatant (containing the amplified sample library) to a new 1.5 mL microfuge tube and place on ice.
18. Measure A260/A280 on a NanoDrop spectrophotometer to determine the concentration and quality (*see* **Note 21**). Use EB as a blank.
19. Run 1 μl post-capture LM-PCR product on a Bioanalyzer 2100 High-Sensitivity DNA chip (*see* **Note 22**).
20. A successfully constructed library should look like Fig. 1d, with an average fragment size 150–500 bp. Store the library at −20 °C, ready for quantification (*see* **Note 23**) and sequencing.

3.3 Illumina sequencing

Standard paired-end Illumina sequencing is performed on the EC-enriched library pool. For barley EC, we recommend 2× 75 bp read length sequencing with a minimum predicted on-target coverage of 50× (*see* **Note 24**).

4 Data Analysis

4.1 Conventions

See **Notes 25** and **26** for general comments regarding hardware and software choices.

Command line statements are enclosed in boxes, as shown in this example:

```
myCommand parameter1 parameter2
```

Longer commands use the newline escape character ("\") to indicate that text on subsequent lines is meant to be on the same line but has been wrapped for readability:

```
myCommand \
parameter1 \
parameter2 \
parameter3 \
parameter4 \
parameter5 \
```

4.2 Quality Control

A commonly used tool for the purpose of quality control of raw Illumina data is FASTQC (http://www.bioinformatics.babraham.ac.uk/projects/fastqc/). Command line invocation is by means of the wrapper script provided and involves just a single parameter, the name of the FASTQ formatted read file (uncompressed or GZIP compressed):

```
fastqc myData.fastq
```

(*See* **Note 27** for more detail).

4.3 Read Mapping

The exome capture reads are mapped to the reference sequence using a short read mapper. We recommend the BWA-MEM algorithm for this (http://bio-bwa.sourceforge.net/) (*see* **Note 28** for more detail).

First, the reference sequence has to be indexed using the "bwa index" command:

```
bwa index myReferenceSequence.fasta
```

This builds a number of index files that allow the mapping algorithm to quickly and efficiently match each read to the appropriate part of the genome.

The read mapping algorithm itself is invoked as follows:

```
bwa mem myReferenceSequence.fasta readsR1.fastq readsR2.fastq \
-R "@RG\tID:mySampleName\tSM:mySampleName" > alignment.sam
```

The standard output stream of BWA is raw SAM format and in this example is redirected to a file called alignment.sam using the ">" character. Use of the –R flag and its argument will result in read group information to be added to the SAM output (*see* http://www.htslib.org/ and http://samtools.github.io/hts-specs/SAMtags.pdf for more information on the SAM format and the RG tag). This allows downstream software to associate reads with sample name and other information (*see* **Note 29**).

The initial SAM output is then converted to compressed, indexed BAM format using the samtools view command [8] while simultaneously removing unmapped reads to save disk space (-F 4 flag). BAM is the *de facto* standard for storing short read alignments and most downstream analysis software requires its input data to be in this format.

An additional filtering step uses the bamtools toolkit (https://github.com/pezmaster31/bamtools) to remove reads that contain excessive numbers of mismatches, based on their alignment score (AS) flag in the SAM/BAM output. Mismatch cutoffs in read mapping are essential for the accuracy of downstream analysis as read mismapping caused by overly relaxed mismatch parameters can lead to dramatically increased false positive rates in variant calling [9]. The minimum alignment score cutoff is calculated as (read length in bp)—(maximum allowed number of mismatches * default mismatch penalty). The default mismatch penalty for BWA-MEM is 5, and the cutoff of 80 in the example below is based on a read length of 100 and a maximum of 4 mismatches per read (4% mismatch rate).

The initial BAM output is then piped into the "samtools sort" utility which sorts reads by both contig and start position, a requirement for further downstream analysis:

```
samtools view -F 4 -b -h alignment.sam | \
bamtools-2.2.3 filter -tag "AS:>=80" | \
samtools sort -o alignment.sorted.bam -
```

In this example, the resulting BAM file would be called "alignment.sorted.bam." The trailing dash ("–") indicates that the input for the samtools sort command is its standard input stream, i.e., the output of bamtools. Please refer to the bamtools manual page for details on the other command line options used above (http://www.htslib.org/doc/samtools.html).

4.4 BAM File Preprocessing and GVCF File Production

The GATK Best Practices workflow [10] involves several further preprocessing steps before the alignment data can be used for the final variant calling stage.

The first of these is the removal of duplicate reads. These represent nonindependent observations which may skew the downstream analysis and should therefore either be removed or flagged up as duplicates. In order to reduce disk storage requirements, removal is preferable, and we use the "samtools rmdup" command for this purpose [8]:

```
samtools rmdup alignment.sorted.bam alignment.rmduped.bam
```

The output file from this command is alignment.rmduped.bam.

The second BAM file preprocessing step consists of the local realignment of reads around indels. This adjusts the placement of reads that have been aligned suboptimally around indels, and thereby removes base mismatches that could be misinterpreted as variants in the downstream analysis. Both of the steps above are designed to keep the false positive SNP rate to a minimum [11]. This stage consists of two separate steps, both of which involve GATK tools. In the first of these, the BAM file is scanned and a list of target sites is identified for realignment:

```
java -jar GenomeAnalysisTK.jar \
-T RealignerTargetCreator \
-R myReferenceSequence.fasta \
-I alignment.rmduped.bam \
-o target_intervals.list
```

In the second step, the actual realignment itself is carried out and a new BAM file is produced:

```
java -jar GenomeAnalysisTK.jar \
-T IndelRealigner \
-R myReferenceSequence.fasta \
-I alignment.rmduped.bam \
-targetIntervals target_intervals.list \
-o alignment.realigned.bam
```

We then need to index the newly created realigned BAM file:

```
samtools index alignment.realigned.bam
```

The realigned BAM file is then run through the actual variant caller component of the GATK, the HaplotypeCaller, to produce an initial VCF output file:

```
java -jar GenomeAnalysisTK.jar \
-T HaplotypeCaller \
-R myReferenceSequence.fasta \
-I alignment.realigned.bam \
-o initialVariants.vcf \
-dontUseSoftClippedBases
```

This has to be filtered using the vcffilter tool (https://github.com/vcflib/vcflib#vcffilter) to produce a second VCF file containing only high quality variants with a variant quality score of ≥20. This score is phred-based, and a value of 20 equates to a likelihood of 1% of a variant having been called in error:

```
vcffilter -f "QUAL > 20" initialVariants.vcf \
> initialVariants.filteredQ20.vcf
```

This filtered VCF file is a requirement for the next step in the pipeline—the base quality score recalibration (BQSR) (*see* **Note 30**).

The first step of the BQSR procedure consists of computing a recalibration table which is then used as input for the second stage:

```
java -jar GenomeAnalysisTK.jar \
-T BaseRecalibrator \
-R myReferenceSequence.fasta \
-I alignment.realigned.bam \
-knownSites initialVariants.filteredQ20.vcf \
-o recalibrationTable.txt
```

In the second step, the recalibration table is applied to the input BAM file and a new, recalibrated BAM file is produced:

```
java -jar GenomeAnalysisTK.jar \
-T PrintReads \
-R myReferenceSequence.fasta \
-I alignment.realigned.bam \
-BQSR recalibrationTable.txt \
-o alignment.recalibrated.bam
```

During this final run of the HaplotypeCaller a GVCF file is produced, which is the endpoint of the single sample processing stage:

```
java -jar GenomeAnalysisTK.jar \
-T HaplotypeCaller \
-R myReferenceSequence.fasta \
-I alignment.recalibrated.bam \
-o finalVariants.g.vcf \
-ERC GVCF \
--variant_index_type LINEAR \
--variant_index_parameter 128000 \
-dontUseSoftClippedBases
```

The output from this step is a GVCF format file (*see* **Note 31**).

4.5 Variant Calling

GATK provides a cohort protocol for the purpose of comparing multiple samples to one another during the final variant and genotype calling stage. This is implemented as a tool that combines multiple GVCF files into one or more cohort GVCF files which are then processed by the joint genotyper tool. Below is a hypothetical example where three samples are combined into one cohort GVCF file:

```
java -Xmx50g -jar GenomeAnalysisTK.jar \
-T CombineGVCFs \
-R myReferenceSequence.fasta \
-o cohort.1.g.vcf \
--disable_auto_index_creation_and_locking_when_reading_rods \
--variant sample1.final.variants.g.vcf \
--variant sample2.final.variants.g.vcf \
--variant sample3.final.variants.g.vcf
```

(*see* **Note 32**).

In the final stage, we use the joint genotyper (GATK's "GenotypeGVCFs" command) to call variants and genotypes for all the samples in our cohort files (here, two cohorts are shown as an illustration):

```
java -Xmx100g -jar GenomeAnalysisTK.jar \
-T GenotypeGVCFs \
-R myReferenceSequence.fasta\
-o jointGenotyperSNPs.vcf \
-nt 32 \
-V cohort.1.g.vcf \
-V cohort.2.g.vcf
```

The final output from this step in our example would be a file named "jointGenotyperSNPs.vcf" (*see* **Note 33**).

4.6 Filtering

Filtering of raw variant calls is recommended in order to remove potential false positives (*see* **Note 34**). Types of filters that should be routinely applied include variant likelihood ("QUAL" in VCF files, *see* the VCF file format specification at http://samtools.github.io/hts-specs/VCFv4.2.pdf) and a depth filter that removes sites where mismapping of reads to a secondary location leads to excess coverage and false positive variants [12].

Using the vcffilter library as above, we can combine multiple filters into a single statement:

```
vcffilter \
-f "QUAL > 30" \
-f "DP > 100"\
jointGenotyperSNPs.vcf \
> jointGenotyperSNPs_filtered.vcf
```

We recommend a QUAL filter of >30 as shown (*see* **Note 35**).

Depth filter cutoffs need to be appropriate for the read depth expected. Our recommendation would be to remove sites with coverage greater than 1.5× mean read depth. The latter can be calculated as follows:

(# reads mapped * read length)/size of exome capture space in bp.

Example:

(40,000,000 reads mapped * 100 bp)/60,000,000 bp exome = 66.7 × mean coverage.

In this case an appropriate read depth cutoff would be 66.7 * 1.5 = 100. If this is a per-sample value, then the cutoff for the final joint genotyper SNPs (which will be based on multiple samples) needs to be multiplied by the number of samples to provide the correct joint coverage (i.e., if the final variant calls were from 40 samples our coverage cutoff should be 40 × 100 = 4000).

4.7 Visualization

Data visualization is of paramount importance in quality control of NGS data processing. It can be used to spot excessive numbers of read errors, abnormal variant distribution patterns or general problems with the underlying read mapping such as excess numbers of read mismatches or lack of coverage.

We use the Tablet assembly viewer for visualization of BAM files, variants, and exonic regions [13, 14] (*see* **Note 36**). The partial Tablet screenshot in Fig. 2 shows a typical example of mapped exome capture reads from a single sample, in a region with multiple exons. Forward and reverse reads in read pairs are shown in green and blue, respectively. A BED file with exon coordinates has been imported as a feature track, and these are shown above the main canvas as light green bars. Likewise, a VCF file with initial variant positions has been imported too, and these are shown below the exon annotation as individual dark blue markers. On the main canvas itself, variants are visible as thin white vertical lines.

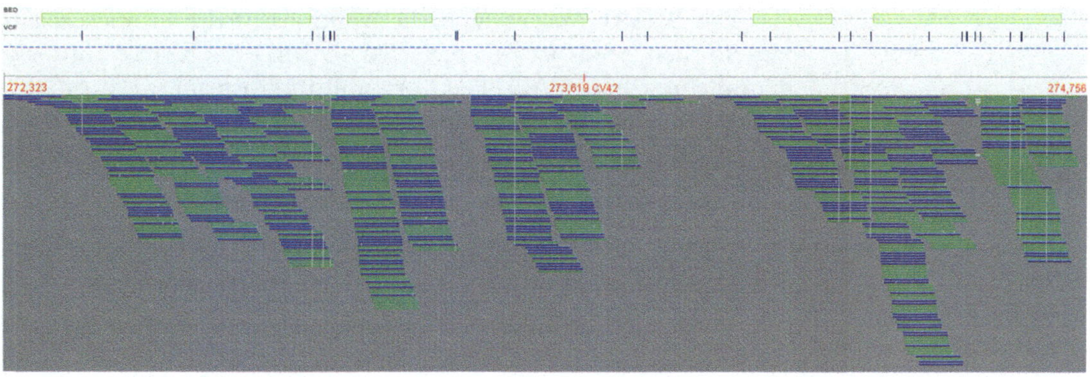

Fig. 2 Partial screenshot of the Tablet assembly viewer software, showing on the main canvas mapped exome capture reads from a single sample (green = forward, blue = reverse) and SNPs (thin white vertical lines). The annotation tracks above the main canvas show exon annotation (top track, light green horizontal bars) and SNP information imported from the VCF file (second track from top, thin blue vertical marks)

4.8 On-Target Rate Computation

Exome capture is an enrichment approach, which means variable outcomes may be achieved in terms of the degree of enrichment. It is therefore of interest to quantify the success of the capture process itself by computing the proportion of bases that map to the enrichment target regions (here exons). A popular tool for this is CalculateHsMetrics from the Picard suite of utilities (http://broadinstitute.github.io/picard/). We can use this to compute for each sample's BAM file the percentage of bases mapped in the exonic target regions.

This tool requires a targets file which contains start and end locations for any targets used in the capture itself. The current barley exome capture array [2] was designed from the Morex v.3 assembly from 2012 [1] and although the design file itself is readily available from the manufacturer, Nimblegen (https://sftp.rch.cm/diagnostics/sequencing/nimblegen_annotations/ez_barley_exome/barley_exome.zip), it cannot be used for on-target computation with other assemblies/reference sequences. To derive the locations of the exon targets on other reference sequences, the target sequences have to first be mapped onto the new reference using the BLASTN command line tool run [15, 16]:

```
blastn \
-query barley_mapping_sequence.fa \
-db myReferenceSequence.fasta \
-max_target_seqs 1 \
-max_hsps_per_subject 1 \
-evalue 1e-10 \
-perc_identity 90 \
-out BLAST_output.txt
```

The BLAST output with the positions of the exome capture targets then needs to be converted to BED file format (https://genome.ucsc.edu/FAQ/FAQformat.html#format1). We can use Linux's built-in awk command line tool for this:

```
awk '{ print $2 "\t" $9 "\t" $10 "\t" $1}' BLAST_output.txt \
> exon_coords.bed
```

A sequence dictionary for the reference sequence then has to be created using Picard's CreateSequenceDictionary tool:

```
java -jar picard.jar CreateSequenceDictionary \
R=myReferenceSequence.fasta \
O=myReferenceSequence.dict
```

The BED file with the exon coordinates then needs to be converted to the intervals file format expected as input by the Picard CalculateHsMetrics tool (http://broadinstitute.github.io/picard/):

```
java -jar picard.jar BedToIntervalList \
INPUT=exon_coords.bed \
SEQUENCE_DICTIONARY=myReferenceSequence.dict \
OUTPUT=exon_coords.intervals
```

Once these steps are complete, we can run the actual CalculateHsMetrics tool:

```
java -jar picard.jar CalculateHsMetrics \
BAIT_INTERVALS=exon_coords.bed \
TARGET_INTERVALS=exon_coords.bed \
INPUT=alignment.recalibrated.bam \
OUTPUT=picard_Hs_metrics.txt \
VALIDATION_STRINGENCY=LENIENT
```

The output file picard_Hs_metrics.txt contains the value we are interested in—the proportion of bases among all reads that are on target. This is labeled "PCT_USABLE_BASES_ON_TARGET." See http://broadinstitute.github.io/picard/picard-metric-definitions.html#HsMetrics for a full explanation of all the metrics contained in the output file.

5 Notes

1. Nimblegen offer a custom design service for exome capture and smaller targeted genome regions. The Barley Exome Design [2] uses a custom SeqCap EZ Developer Library which was established as an international collaborative effort to reduce design costs. Other cheaper options are available, including those produced by Agilent Technologies (SureSelect) and Mycroarray (MYbaits).
2. Use of young (<2 weeks old) barley seedling material for gDNA extraction is important. This produces the best possible DNA quality in combination with Qiagen DNeasy Extraction Kits. Older material can suffer from lower yields and polysaccharide contamination, resulting in DNA less amenable to downstream processing.
3. Visualizing DNA on a gel should not be used to quantify the DNA for library preparation but is a good "first step" to determining whether the DNA is intact and of good quality. If a discrete band is not observed, this generally means that the DNA has been degraded and unsuitable for downstream analysis.
4. Any Illumina-compatible whole genome shotgun library protocols can be used so long as standard Illumina adapters and indices are utilized. This is to ensure efficient blocking of adapter sequences is achieved downstream during the exome capture process. The numbers of samples which can be multiplexed prior to EC is dependent upon the final Illumina sequencing platform and read length selected, which ultimately determines the predicted coverage. We recommend pooling no greater than 12 barley lines prior to EC per single NextSeq 500 or HiSeq 2500 lane.
5. The exact time for fragmentation may need optimizing for different species, we have increased it to 280 s for barley. During the sonication process, briefly spin down the tube every 30 s, as well as at the end.
6. Overdrying the beads may result in dramatic yield loss. You should notice a "crack" in the beads once dried and they will no longer be shiny.
7. Use Illumina's Experiment Manager software to check compatibility of index adaptors.
8. Pipette PEG/NaCl SPRI solution carefully as it is quite viscous.
9. Overdrying the beads may result in dramatic yield loss.

10. This is a safe stopping point. Store the resuspended beads at 4 °C for up to 24 h.
11. Do not transfer any beads with the supernatant.
12. It is important to set up a water control here and process in the same way as the samples including bead stages. Therefore, pipette 20 μL PCR- grade water into an empty well alongside your samples.
13. A successfully constructed library should have an average fragment size between 250 and 500 bp. Sometimes dual peaks are observed, this can be due to slight overloading of the Bioanalyzer (as shown in Fig. 1c), or daisy-chaining due to excess PCR cycling.
14. Once resuspended the Hybridization Enhancing (HE) oligos can be aliquoted into smaller volumes to minimize the number of freeze/thaw cycles. Store them at −20 °C.
15. The total combined mass of the Multiplex Hybridization Enhancing Oligo Pool should be 2000 pmoL, which is the amount required for a single Sequence Capture experiment. Example: If a multiplex DNA Sample Library Pool contains four DNA sample libraries prepared with SeqCap Adapters Indexes 2, 4, 6, and 8, respectively, then the Multiplex Hybridization Enhancing Oligo Pool would contain the following: SeqCap HE Universal Oligo, 1000 pmoL (1 μL of 1000 μM); SeqCap HE Index 2 Oligo, 250 pmoL (0.25 μL of 1000 μM); SeqCap HE Index 4 Oligo, 250 pmoL (0.25 μL of 1000 μM); SeqCap HE Index 6 Oligo, 250 pmoL (0.25 μL of 1000 μM); SeqCap HE Index 8 Oligo, 250 pmoL (0.25 μL of 1000 μM); Total 2000 pmoL (2 μL of 1000 μM).
16. The heated lid of the thermocycler should be turned on and set to maintain 57 °C. Samples can be left for up to 72 h if needed.
17. Low-binding microfuge tubes are recommended. Enough beads for six captures can be prepared in a single tube.
18. Work quickly keeping samples at 47 °C to ensure only "on target" fragments get through the process.
19. It is recommended to prepare larger pool volumes due to the difficulty of accurately pipetting small volumes. These pools can be dispensed into individual single-use aliquots that can be stored at −20 °C until needed.
20. Some beads might stick a little to the sides of wells; however, just take as much as possible. Turn the tube to spread wet beads around and help drying. High volume of beads for this step, so will take 5 min or longer. Over drying of the beads can result in yield loss.

21. The sample library yield should be >500 ng, with an A260/A280 between 1.7 and 2.0.
22. It is probable that a small aliquot of the samples will require diluting 1:10 to be within range of the Bioanalyzer chip.
23. It is best to quantify captured libraries using qPCR (Kapa Library Quantification Kit) due to presence of "daisy chain" concatamers, which do not affect sequencing but can lead to unreliable Qubit quantification.
24. We recommend sequencing on NextSeq or HiSeq equipment. Please calculate read coverage based upon individual service provider's specifications for output.
25. Parallel processing of samples. The workflow described here is highly compute-intensive and hence time-consuming, but it was the most accurate option available at the time of writing. Run times for a single barley exome capture sample vary between 1 and 2 CPU days with our setup, depending on coverage. This means that large datasets with multiple samples have to be processed in parallel if the data analysis is to complete in a reasonable time frame. The tools described above are a mixture of multi- and single-threaded programs, but single-threaded programs are rate-limiting as they produce computational bottlenecks. Therefore, running a single sample with multiple threads is a poor use of resources, and a better approach is to process each sample as a single thread, with many of these jobs running in parallel. There are a number of options available to achieve this, including individual multiprocessor servers, compute clusters and cloud computing. Our setup for processing datasets with hundreds of exome capture samples is a small in-house compute cluster with 144 compute cores and up to 256 GB RAM per node. This has generally been adequate for dealing with data on this scale, albeit with careful management of resources and occasional bottlenecks.
26. Most projects involving exome capture are aimed at small variant discovery. However, a more recent trend is to also use exome capture data for the discovery of structural variants (SV) such as larger (kbp-scale) deletions or insertions. A potential caveat here is that SV discovery with exome capture sequencing is more challenging than with whole genome data, due to the uneven and sparse read coverage. A number of tools have been published in recent years that have been specifically designed for this purpose: ExomeDepth [17], Split-read [18], CoNIFER [19], EXCAVATOR [20], and Scalpel [21, 22]. As is often the case, different tools can generate very different result sets [21], and care must be taken in the interpretation of results. Two recent reviews have also summarized the tools available for this purpose [23, 24].

27. FASTQC comes with a both command line interface and a graphical user interface, with the latter implemented as cross-platform, standalone desktop client software. It provides easy-to-interpret summary plots of metrics that quantify important quality traits such as base qualities, duplication levels, contaminant sequences, and others. This allows analysists to gauge whether sequence quality is adequate before proceeding to the analysis stage. Further details on the use of this tool and the interpretation of its output are provided on the FASTQC manual page at http://www.bioinformatics.babraham.ac.uk/projects/fastqc/.
28. Throughout our protocol, we are following the recommendations of the Genome Analysis Toolkit (GATK) Best Practices [10], which for the purpose of read mapping recommends the use of raw, untrimmed data, rather than carrying out quality-trimming and adapter removal first. This enables accurate removal of read duplicates, which relies on matching start and end coordinates of untrimmed reads. Parts of reads that are mismatched with the reference sequence are soft-clipped in BWA-MEM, i.e., the read is included in the BAM file in its entirety, but the CIGAR string entry [8] specifies which part of the read has actually been aligned and is thus suitable for downstream analysis.
29. It is computationally preferable to write the SAM output to disk, although in theory pipes could be used to stream the output directly into the downstream tools. However, the *bwa mem* command can be parallelized with the –t option which means that reads are getting mapped in parallel threads. The downstream tools are single-threaded and thus represent a computational bottleneck if everything is done as a single pipe operation, and even with the overhead of disk access, writing the SAM file to disk is still the quickest option.
30. The base quality score recalibration (BQSR) step's purpose is to remove the bias found in raw base quality scores, which tends to be associated with a base's position in the read and also the identity of neighboring bases [11]. It produces a final BAM file where base qualities have been adjusted as appropriate, and this is the file that is used as input for the second (and here final) run of the variant caller module, the HaplotypeCaller. This procedure requires a truth set of known variants which can be used to guide the recalibration. In the absence of a publicly available benchmark dataset (e.g., those available for human genomics), users are encouraged to produce their own calibration datasets by means of a bootstrapping approach that produces a high quality call set. The recommendation is to aim for convergence between expected and observed base qualities,

which can require several iterations of the BQSR, but we found that a single iteration provides an acceptable degree of convergence.

31. This is a proprietary GATK file format which represents a variant of the standard VCF format which contains additional information designed to allow easy comparison of multiple samples in cohorts (http://gatkforums.broadinstitute.org/gatk/discussion/4017/what-is-a-gvcf-and-how-is-it-different-from-a-regular-vcf). Note the `.g.vcf` file extension, which is required by the GATK components.
32. In practice, small numbers of samples (below ~ 50, based on our experience) do not require this additional step, and instead their GVCF files can be fed into the joint genotyper directly. We also found that for larger numbers of samples, a cohort size of around 20 represents a good compromise between memory consumption and speed.
33. The GenotypeGVCFs component is compute-intensive and can be multi-threaded using the –nt option as shown. It is also memory-intensive, and in our example we have increased the default amount of allocated memory to 100 GB using the –Xmx option. The exact amount of memory required varies with the number of cohorts and their size, but as a guide we recommend between 0.5 and 1 GB of RAM per sample for a cohort size of 20.
34. The type of filters and their parameters depend to a large extent on the intended purpose of the dataset. A trade-off exists between completeness of the callset and reliability of the constituent variants. Aggressive filtering generally means fewer false positives but more false negatives, whereas little filtering leads to an inverse outcome. If the purpose of the callset is SNP discovery for, e.g., a genotyping panel, where reliability of SNPs is of paramount importance, then an aggressive filtering approach should be taken. Conversely, if the callset is intended for the identification of a single SNP linked to a given phenotype, then a conservative filtering approach should be taken that removes very few SNPs, in order not to remove the SNP of interest along with false positives.
35. The variant quality is a phred-like quality score (https://en.wikipedia.org/wiki/Phred_quality_score) that quantifies the likelihood of a variant being genuine rather than artifactual. A score of 30 signifies a 0.1% chance of the variant having been called in error, and we deem this a sufficiently low probability.
36. Tablet is standalone desktop software that can be installed easily on Windows, Mac, and Linux platforms using ready-made installers bundled with their own version of Java (*see* https://ics.hutton.ac.uk/tablet/download-tablet/).

References

1. Mayer KFX, Waugh R, Langridge P et al (2012) A physical, genetic and functional sequence assembly of the barley genome. Nature 491(7426):711
2. Mascher M, Richmond TA, Gerhardt DJ et al (2013) Barley whole exome capture: a tool for genomic research in the genus Hordeum and beyond. Plant J 76(3):494–505
3. Pankin A, Campoli C, Dong X et al (2014) Mapping-by-sequencing identifies HvPHYTOCHROME C as a candidate gene for the early maturity 5 locus modulating the circadian clock and photoperiodic flowering in barley. Genetics 198(1):383–396
4. Wendler N, Mascher M, Nöh C et al (2014) Unlocking the secondary gene-pool of barley with next-generation sequencing. Plant Biotechnol J 12(8):1122–1131
5. Nice LM, Steffenson BJ, Brown-Guedira GL et al (2016) Development and genetic characterization of an advanced backcross-nested association mapping (AB-NAM) population of wild × cultivated barley. Genetics 203(3):1453–1467
6. Hisano H, Sakamoto K, Takagi H et al (2017) Exome QTL-seq maps monogenic locus and QTLs in barley. BMC Genomics 18(1):125
7. Russell J, Mascher M, Dawson IK et al (2016) Exome sequencing of geographically diverse barley landraces and wild relatives gives insights into environmental adaptation. Nat Genet 48 (9):1024–1030
8. Li H, Handsaker B, Wysoker A et al (2009) The sequence alignment/map format and SAMtools. Bioinformatics 25(16):2078–2079
9. Ribeiro A, Golicz A, Hackett C et al (2015) An investigation of causes of false positive single nucleotide polymorphisms using simulated reads from a small eukaryote genome. BMC Bioinformatics 16(1):382
10. Van der Auwera GA, Carneiro MO, Hartl C et al (2013) From FastQ data to high confidence variant calls: the genome analysis toolkit best practices pipeline. Curr Protoc Bioinformatics 43:11.10.1–11.1033
11. DePristo MA, Banks E, Poplin R et al (2011) A framework for variation discovery and genotyping using next-generation DNA sequencing data. Nat Genet 43(5):491–49+
12. Ribeiro A, Golicz A, Hackett CA et al (2015) An investigation of causes of false positive single nucleotide polymorphisms using simulated reads from a small eukaryote genome. BMC Bioinformatics 16 (1):1–16
13. Milne I, Stephen G, Bayer M et al (2013) Using tablet for visual exploration of second-generation sequencing data. Brief Bioinform 14(2):193–202
14. Milne I, Bayer M, Cardle L et al (2010) Tablet-next generation sequence assembly visualization. Bioinformatics 26(3):401–402
15. Altschul SF, Gish W, Miller W et al (1990) Basic local alignment search tool. J Mol Biol 215(3):403–410
16. Camacho C, Coulouris G, Avagyan V et al (2009) BLAST+: architecture and applications. BMC Bioinformatics 10:421
17. Plagnol V, Curtis J, Epstein M et al (2012) A robust model for read count data in exome sequencing experiments and implications for copy number variant calling. Bioinformatics 28(21):2747–2754
18. Karakoc E, Alkan C, O'Roak BJ et al (2012) Detection of structural variants and indels within exome data. Nat Methods 9 (2):176–178
19. Krumm N, Sudmant PH, Ko A et al (2012) Copy number variation detection and genotyping from exome sequence data. Genome Res 22(8):1525–1532
20. Magi A, Tattini L, Cifola I et al (2013) EXCAVATOR: detecting copy number variants from whole-exome sequencing data. Genome Biol 14(10):R120
21. Narzisi G, O'Rawe JA, Iossifov I et al (2014) Accurate de novo and transmitted indel detection in exome-capture data using microassembly. Nat Methods 11(10):1033–1036
22. Fang H, Bergmann EA, Arora K et al (2016) Indel variant analysis of short-read sequencing data with Scalpel. Nat Protoc 11 (12):2529–2548
23. Tattini L, D'Aurizio R, Magi A (2015) Detection of genomic structural variants from next-generation sequencing data. Front Bioeng Biotechnol 3:92
24. Guan P, Sung W-K (2016) Structural variation detection using next-generation sequencing data: a comparative technical review. Methods 102:36–49

Index

A

B

C

D

E

Wendy A. Harwood (ed.), *Barley: Methods and Protocols*, Methods in Molecular Biology, vol. 1900,
https://doi.org/10.1007/978-1-4939-8944-7, © Springer Science+Business Media, LLC, part of Springer Nature 2019

Printed by Printforce, the Netherlands